Forthcoming titles

Environmental Biology of Fishes

Malcolm Jobling

The Norwegian College of Fisheries Science
University of Tromsø
Norway

CHAPMAN & HALL

London · Glasgow · Weinheim · New York · Tokyo · Melbourne · Madras

Published by Chapman & Hall, 2–6 Boundary Row, London SE1 8HN

Chapman & Hall, 2–6 Boundary Row, London SE1 8HN, UK

Blackie Academic & Professional, Wester Cleddens Road, Bishopbriggs, Glasgow G64 2NZ, UK

Chapman & Hall GmbH, Pappelallee 3, 69469 Weinheim, Germany

Chapman & Hall USA, One Penn Plaza, 41st Floor, New York NY10119, USA

Chapman & Hall Japan, ITP-Japan, Kyowa Building, 3F, 2-2-1 Hirakawacho, Chiyoda-ku, Tokyo 102, Japan

Chapman & Hall Australia, Thomas Nelson Australia, 102 Dodds Street, South Melbourne, Victoria 3205, Australia

Chapman & Hall India, R. Seshadri, 32 Second Main Road, CIT East, Madras 600 035, India

First edition 1995

© 1995 Malcolm Jobling

Typeset in 10/12 Photina by Acorn Bookwork, Salisbury, Wilts

Printed in Great Britain by T.J. Press (Padstow) Ltd., Padstow, Cornwall

ISBN 0 412 58080 2

A catalogue record for this book is available from the British Library

Library of Congress Catalog Card Number: 94-72450

∞ Printed on permanent acid-free text paper, manufactured in accordance with ANSI/NISO Z39.48-1992 and ANSI/NISO Z39.48-1984

Contents

Preface

To most, the word fish will be evocative, conjuring up a variety of thoughts and memories. To the gamefisherman, the word will probably mean salmon and a favourite riverside beat; to the coarsefisher the word may bring back memories of pleasant summer days angling for roach in idle backwaters, or dreams of the record pike; to ancients in their dotage the word fish may bring reminiscences of schoolboy days – short pants, jam-jars, bent pins, bright gay minnows and pugnacious sticklebacks; and few who have taken college zoology courses will ever forget the fish dissection class, with the sad grey corpses of the spurdog and the sharp, cloying aroma of formalin pervading the atmosphere. Salmon, roach, pike, minnows, sticklebacks and spurdogs are all easily recognized as being fish, and few would classify them in any other animal group.

The fish are, however, a very heterogeneous assemblage and the distinctions between fish and members of other animal groups are not always so clearcut. Until the 16th century many natural historians classified a wide variety of aquatic animals as 'fish' – not only were aquatic mammals such as the whales, dolphins, seals and hippopotamuses considered as fish, but also crocodiles, snails, crabs and sea urchins were placed into this group. As late as the middle of the 19th century some zoologists placed fish, amphibians and reptiles together in a single animal class, and even up to the present day the fishmonger offers for sale not only fish, but also a range of crustaceans, molluscs and other aquatic organisms.

At present there are several hundreds, if not thousands, of publications relating to fish biology in the catalogues, and sufficient books to cover several metres of library shelving have been published within the last decade or so. Despite this there are relatively few books that have been written with the aim of meeting the requirements of college and university students. Some introductory texts do exist, but whilst they may be admirable in their way, each is inevitably deficient in some areas.

The majority of authors of introductory texts tend to adopt a rigid 'systems approach' in which anatomy and morphology are divorced from considerations of ecology and behaviour. In addition, the majority of authors of introductory texts tend to give short shrift to descriptions of physiological principles and behaviour, and both are usually either ignored or given only superficial coverage.

Thus, the authors of texts in fish biology usually give little consideration to 'what fish do and how they do it'. Books in which a more integrated presentation has been attempted have usually been written for the layman, rather than the college student. Such books tend to have the disadvantage of being anecdotal, and seldom, if ever, have the in-depth coverage required for courses at college and university level.

Consequently, there appears to be a niche to be filled by a book offering an integrated approach to fish biology from the physiological and ecological viewpoint, but it would clearly be impossible to provide such a treatment without some coverage of anatomy and morphology. In presenting fish biology to students, I have tried several approaches during the course of the last decade-and-a-half, and the current volume has a basis in what I have found to be reasonably successful. In preparing this book I have been ably assisted by numerous colleagues and students, who have commented upon earlier versions of the manuscript. Especial thanks go to Hilka Falseth, Liss Olsen and Frøydis Strand who prepared the figures, and the efforts of Chuck Hollingworth and Martin Tribe should also not go unmentioned.

Whilst there is no pretence that this book is a panacea, it is hoped that college students and teachers will find much of interest. The book has been written primarily to meet the needs of the undergraduate student taking courses in fish biology as part of a degree programme in zoology, biology, or aquatic biosciences, but it is to be hoped that students of aquaculture, fisheries and resource management, and comparative physiology will also find much to whet their appetites.

Malcolm Jobling
Tromsø, Norway

Series foreword

Among the fishes, a remarkably wide range of biological adaptations to diverse habitats has evolved. As well as living in the conventional habitats of lakes, ponds, rivers, rock pools and the open sea, fish have solved the problems of life in deserts, in the deep sea, in the cold antarctic, and in warm waters of high alkalinity or of low oxygen. Along with these adaptations we find the most impressive specializations of morphology, physiology and behaviour. For example, we can marvel at the high-speed swimming of the marlins, sailfish and warm-blooded tunas, air-breathing in catfish and lungfish, parental care in the mouth-brooding cichlids and viviparity in many sharks and toothcarps.

Moreover, fish are of considerable importance to the survival of the human species in the form of nutritious, delicious and diverse food. Rational exploitation and management of our global stocks of fishes must rely upon a detailed and precise insight of their biology.

The *Chapman & Hall Fish and Fisheries Series* aims to present timely volumes reviewing important aspects of fish biology. Most volumes will be of interest to research workers in biology, zoology, ecology and physiology but an additional aim is for the books to be accessible to a wide spectrum of non-specialist readers ranging from undergraduates and postgraduates to those with an interest in industrial and commercial aspects of fish and fisheries.

Environmental Biology of Fishes by Malcolm Jobling comprises volume 16 in the *Chapman & Hall Fish and Fisheries Series*. In this book we find a wide-ranging synoptic review of the fundamentals of fish design and physiological adaptation to environment that contributes to their remarkable success. Malcolm Jobling's new book will find its niche as a logically organized textbook that provides students of fish biology with a stimulating, scholarly and imaginative perspective of current knowledge. I hope it will soon become widespread among the habitat of fish biologist's bookshelves.

Professor Tony J. Pritcher
Editor, Chapman & Hall Fish and Fisheries Series
Director, Fisheries Centre, University of British Columbia
Vancouver, Canada

Chapter one

Fish: an introduction

1.1 FISH ORIGINS AND DIVERSITY

The possession of paired limbs and jaws, thought to be derived from modified gill arches, distinguishes the fish and tetrapods from the jawless vertebrates (Agnatha) (Fig. 1.1). The gnathostomes, or jawed vertebrates, are further distinguished from the agnathans by the presence of three semicircular canals in the ear (the agnathans possess either one or two vertical semi-circular canals), and by the fact that the vertebrae of adult fish and tetrapods possess centra (only the notochord is present in agnathans).

Fossil remains of the agnathans occur in the rocks of the late Cambrian and middle Ordovician (approximately 500 million years ago), but this group of vertebrates appears to have been at its most diverse during the Silurian and Devonian periods (420–350 million years ago). At present the agnathan group is represented by fewer than 80 extant species of hagfish (Myxini) and lampreys (Cephalaspidomorphi) (Fig. 1.1).

The extant representatives of the agnathan group are elongate, worm-like animals that lack external signs of paired fins and have a skeleton without bone. There is a single median nasal, or olfactory, opening and the outer body surface is naked, i.e. scaleless. Both the hagfish and the lampreys possess a circular mouth, a complex toothed tongue and pouch-like gills which open to the exterior through circular openings. Whilst these features may suggest that the extant agnathans are quite closely related, in reality the hagfish and lampreys have no more than a few characters in common.

Thus, the hagfish and lampreys almost certainly represent divergent lines of development, and they have long separate histories, stretching back almost 250 million years. Consequently, hagfish and lampreys are probably only very distantly related to each other. In reality, the lampreys may be more closely related to the gnathostomes than either is to the hagfish. This latter theory has arisen due to the fact that the lampreys and gnathostomes have several features in common, features which are absent from the hagfish. These features include neural and haemal arches, which can be

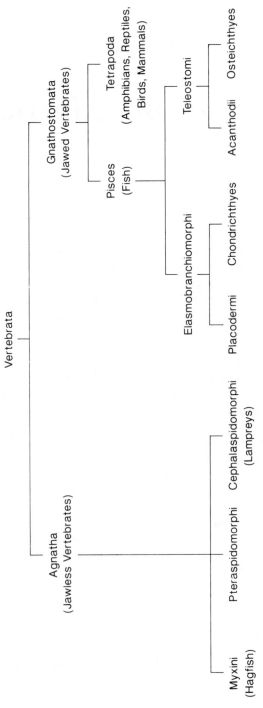

Fig. 1.1 General scheme of classification of the major groups of vertebrates.

considered as indicating the development of a true backbone, radial muscles allowing fin flexion, neuromast organs in the lateral line, extrinsic eye muscles and nervous regulation of the heart.

Some of the extinct agnathans were unlike their 'modern' counterparts in that their bodies were covered by a dermal skeleton. The body covering consisted of either minute scales or larger plates. Collectively these extinct agnathans are often called ostracoderms, which means 'bony-skinned'. It is thought that one of the ostracoderm groups gave rise to the gnathostomes.

There are in the region of 45–50 000 extant species of jawed vertebrates (Gnathostomata). The four tetrapod classes (Amphibia, Reptilia, Aves and Mammalia) comprise a total of approximately 21 500 different species, and the remaining species of jawed vertebrates are all fish. The jawed vertebrates first begin to appear in the fossil record during the Silurian period, some 400–450 million years ago. The most ancient of the jawed vertebrates appear to have been fish of a now-extinct line, the acanthodians (Acanthodii). These fish existed from Silurian up until Permian times (from about 420 until 240 million years ago).

Aristotle (384–322 BC) appears to have been the first to have made the distinction between the cartilaginous (Elasmobranchiomorphi) and bony (Teleostomi) fish, and these two phylogenetic lines are generally recognized as being valid today. Both the Teleostomi and the Elasmobranchiomorphi comprise two classes, one extant and one represented only by fossil forms. The two classes of the Teleostomi are the extinct acanthodians (Acanthodii) and the modern bony fish (Osteichthyes) (Fig. 1.2), whereas the Elasmobranchiomorphi comprise the modern-day cartilaginous fish (Chondrichthyes) and the extinct placoderms (Placodermi) (Fig. 1.3). Representatives of both the extant cartilaginous (Chondrichthyes) and bony (Osteichthyes) classes of fish are represented in the fossil record from the Devonian period (350 million years ago), so the two phylogenetic lines appear to have had parallel development.

The class Osteichthyes may be divided into four subclasses: Dipneusti, Crossopterygii, Branchiopterygii and Actinopterygii (Fig. 1.2). The degree of divergence between the dipnoans (lungfish) and crossopterygians may not be as wide as has been suspected by some workers, and these fish are usually considered to represent a common major line in the development of the osteichthyans – the 'lobe-finned' fish. Similarly, there is uncertainty as to whether or not the last two subclasses (Branchiopterygii and Actinopterygii) are sufficiently distinct to merit separation, and many workers opt to combine the two within the actinopterygian subclass. Irrespective of whether the two subclasses are combined or not, these fish can be considered to represent the other major line of the osteichthyan development – the 'ray-finned' fish. Representatives of both the developmental lines of the osteichthyans are found in the fossil record as early as the Devonian period

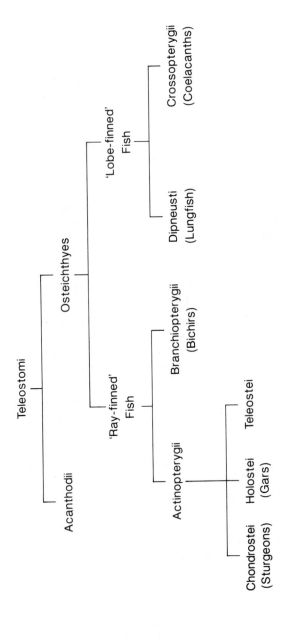

Fig. 1.2 General scheme of classification of the Teleostomi. The acanthodians represent an extinct lineage, whereas the osteichthyan lineage is represented by over 20 000 extant species of fish.

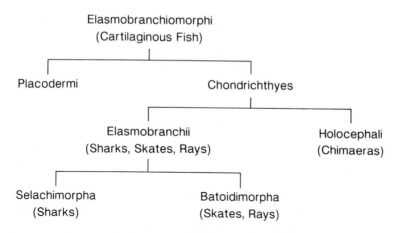

Fig. 1.3 General scheme of classification of the cartilaginous fish – the Elasmobranchiomorphi. The placoderms are an extinct lineage, and the modern-day cartilaginous fish – the chondrichthyans – are represented by about 1000 species of sharks, skates, rays and holocephalans.

(about 350 million years ago), so it is clear that the ray-finned and lobe-finned fish became distinct from each other at a relatively early stage in the history of the bony fish.

In the broadest sense, fish may be considered as being jawed vertebrates that are adapted to living in water, and they possess gills throughout their lives. The gills, which have a respiratory function, open to the exterior via some form of slits or opercular opening. The paired limbs, typical of the jawed vertebrates, are present in the form of fins. Such is the diversity of the fish group that no other characters can be said to be universally possessed by all members of this group of animals.

Other features are possessed by so many fish species that they can be considered as being typically piscine, but because not all members of the group display these features they are not definitive characters. For example, the skin is usually covered with scales, but some fish have dermal teeth or bony plates, and others are entirely lacking in scales. Similarly, the fish heart is typically two-chambered and circulates only venous blood. Cardiac muscle cells (myocytes) are found in both the atrium and the ventricle of the heart, and these are the two 'true' chambers. Myocytes may, however, also infiltrate into the sinus venosus and the conus, or bulbus, arteriosus of some species. Thus, it is possible to consider these chambers as constituting part of the heart. In addition, the circulatory systems of a small number of air-breathing fish species are modified so that the heart pumps 'mixed' blood (i.e.

both venous blood returning to the heart from the systemic circulation and blood that has been returned to the heart after having passed through some air-breathing respiratory organ).

1.2   FISH DISTRIBUTIONS

Different species of fish inhabit markedly different habitats and, as a group, the fish represent a very heterogeneous assemblage. Thus, fish occur in lakes, streams, estuaries and oceans throughout the world, and they can be found in almost every conceivable aquatic habitat ranging from hot soda springs where temperatures can exceed 40°C to the waters beneath the Antarctic ice-sheet where the temperature is below 0°C. They live in high mountain lakes and in deep ocean troughs, in stagnant and fast-flowing waters, and even in the depths of caves, where there is total darkness. Some fish are exclusively marine, others are found only in fresh water, and a small number of species undertake migrations between the two radically different types of environment.

Freshwater systems cover about 1% of the world's surface but, because of the shallowness of most streams, pools and lakes, such systems contain no more than 0.01% of the total water resources. Nevertheless, about 40% (8500–9000 species) of the world's total number of fish species are normally found living in freshwater rivers and lakes. The largest numbers of species are found in tropical regions, with freshwater species being most numerous in the river drainages of southeastern Asia and South America. The Amazonian system, for example, is thought to have a fish fauna comprising at least 1500 species.

There are just over 100 species (about 0.5% of the total) of fish that are distinctly diadromous, that is, they show migrations between marine and freshwater habitats for the purposes of feeding and breeding. Amongst the diadromous species most are anadromous, spawning in fresh water but having feeding grounds in the marine environment. A few are catadromous, spawning in the oceans but passing much of the life cycle in freshwater habitats.

The anadromous species appear to be more frequent in higher latitudes, whereas the proportion of catadromous species tends to increase in temperate and subtropical areas. These patterns of geographic distribution have been suggested to be related to the levels of productivity in marine and freshwater environments at different latitudes. In subtropical areas the productivity of freshwater habitats may be high, whereas that of the open ocean may be relatively low. Thus, there may be a larger available food base in freshwater than in open-ocean habitats, and any diadromous fish species might, therefore, be expected to be catadromous. At high latitudes, many

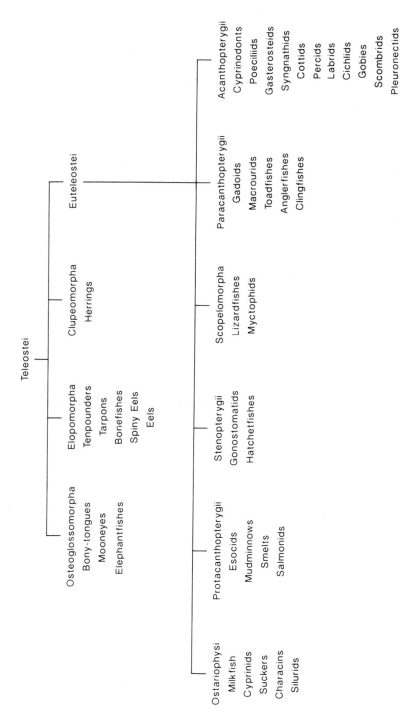

Fig. 1.4 General scheme of classification of the teleosts, giving examples of typical representatives of each lineage.

fresh waters are relatively oligotrophic and levels of productivity are comparatively low compared with those in marine systems. Thus, the food base may be greater in the sea than in fresh water, and this might be expected to lead to there being a larger proportion of anadromous than catadromous species amongst the diadromous fish species found at high latitudes.

The oceans cover approximately 70% of the earth's surface and contain about 97% of its water. Approximately 13 500 species of fish (just under 60% of the total number of species) inhabit marine environments, with the vast majority of these being found in shallow tropical and subtropical coastal waters (about 8500 species). The coastal and shelf waters of temperate and polar regions also provide habitats for comparatively large numbers of species (1000–2000 species), but there are relatively few that occur in the epipelagic zone (surface to 200 m depth) of the open oceans (about 300–350 species).

The remaining fish species are found in the deeper waters of the seas and oceans. Thus, about 10–12% of the total number of fish species inhabit the mesopelagic (from 200 m down to 1000 m), bathypelagic (1000 m to about 3000 m) or the abyssopelagic (deepest ocean troughs) regions of the oceans.

Thus, the vast majority of fish species are to be found in shallow marine and freshwater habitats of tropical and subtropical regions. Furthermore, most of the species will be seen to be representatives of the osteichthyan, rather than the chondrichthyan, lineage. In other words, of the two extant classes of fish – the Chondrichthyes (cartilaginous fish) and the Osteichthyes (bony fish) – the cartilaginous fish (holocephalans, sharks, skates and rays) (Fig. 1.3) are represented by relatively few species (fewer than 1000 species).

Amongst the bony fish, the dipnoans, crossopterygians and brachiopterygians number no more than 15 species, leaving the actinopterygians as by far the largest group of fish in terms of species numbers (Fig. 1.2). There are relatively few species of sturgeons and paddlefish (25 or so) and even fewer holosteans (about 10 extant species). Thus the teleosts, with around 22 000 species, are almost as numerous, in terms of species, as all the other members of the jawed vertebrate lineages put together. Consequently, any discussion of fish biology will largely be concerned with descriptions of teleost diversity (Fig. 1.4) and distributions, and discussions about their anatomy, physiology and behaviour.

FURTHER READING

Books

Grzimek, B. (ed.) (1973) *Grzimek's Animal Encyclopedia*, 4, 5, Van Nostrand Reinhold, New York.

Halstead, L.B. (1969) *The Pattern of Vertebrate Evolution*, Oliver & Boyd, Edinburgh.

Love, M.S. and Cailliet, G.M. (eds) (1979) *Readings in Ichthyology*, Goodyear, Santa Monica.

Nelson, J.S. (1984) *Fishes of the World*, 2nd edn, Wiley, New York.

Thomson, K.S. (1991) *Living Fossil: The Story of the Coelacanth*, W.W. Norton, New York.

Wheeler, A. (1985) *The World Encyclopedia of Fishes*, Macdonald, London.

Review papers and articles

Forey, P.L. (1990) The coelacanth fish: progress and prospects. *Sci. Prog.* **74**, 53–67.

Forey, P. and Janvier, P. (1993) Agnathans and the origin of jawed vertebrates. *Nature, Lond.*, **361**, 129–34.

Gorr, T. and Kleinschmidt, T. (1993) Evolutionary relationships of the coelacanth. *Amer. Scient.* **81**, 72–82.

Hardisty, M.W. (1983) The significance of lampreys for biological research. *Endeavour*, **7**, 110–15.

Rosen, D.E. (1982) Teleostean interrelationships, morphological function and evolutionary inference. *Amer. Zool.*, **22**, 261–73.

Chapter two

Environmental links – sensory systems

2.1 INTRODUCTION

Fish and other animals are in continuous receipt of, and respond to, stimuli of various kinds from the environment. The stimuli may elicit immediate responses or may provide information that leads to changes in the longer term. For example, seasonal changes in photoperiod are the most reliable environmental signal for time of year. Thus, most animals that show seasonal changes in behaviour, such as in migration or reproduction, use photoperiodic information for proper seasonal phasing. In order to make use of this information it is clear that the animal must have appropriate detection and receptor systems (sense organs), must integrate and coordinate the information, and finally must have effector systems capable of acting upon the information received.

The environmental stimuli impinging upon the organism will include light, mechanical disturbances, electrical stimuli and a myriad of chemicals. Fish have sensory systems capable of photoreception, mechanoreception, chemoreception and, in addition, some fish species are also capable of detecting electrical stimuli. The sense organs and sensory systems transmit information to central sites, primarily, via nervous connections, where interpretation and coordination are carried out.

In most vertebrates vision is the dominant sense, but animals also use the non-visual senses to obtain information about their environment. The non-visual senses are best developed in nocturnal animals or those that inhabit low-light environments. Aquatic non-visual sensory specialists may use mechanosensory, electrosensory or chemosensory systems to gain information about the environment, to communicate with each other, and to guide their activities.

2.2   PHOTIC ENVIRONMENT

The solar radiation reaching the surface of the water is made up of photons of every visible wavelength, together with photons of the infrared and ultraviolet regions of the electromagnetic spectrum. The visible-light photons correspond to colours or wavelengths from violet with a wavelength of 400 nm to red with a wavelength of 700 nm. When light passes through water its intensity decreases, and the loss of intensity varies with colour, i.e. different wavelengths are absorbed by water in differing degrees. Bodies of water that do not contain much suspended organic matter absorb violet and red light much more than they absorb light of intermediate wavelengths, so that the spectral quality of the light changes with depth. Thus, as light passes through the water column it both becomes attenuated, i.e. intensity is reduced, and the spectral bandwidth becomes increasingly restricted.

Photic zones

It is possible to divide the photic environment into zones, depending upon the intensity, or quantity, of light reaching a particular depth; euphotic, dysphotic and aphotic zones can be recognized. The euphotic zone is defined as that area where there is sufficient light to sustain net photosynthetic production. In clear oceanic waters this zone extends from the surface to a depth of about 200 m. In the dysphotic zone, which extends from about 200 m to 1000 m in the oceans, some light is still present, but it is insufficient to allow effective photosynthesis. In the aphotic zone, which encompasses all depths greater than 1000 m, all the surface-derived irradiation has been attenuated, and the only light at these great depths is bioluminescent in origin.

Light transmission and attenuation

In the clearest of waters, maximum transmission occurs at a wavelength of about 460 nm. This means that blue light can penetrate to depths greater than 75–100 m, whereas red and violet light may be fully attenuated at depths of about 25 m. As a result the blue wavelengths dominate in the underwater world of oceanic waters.

Whilst the attenuating and absorbing properties of the water itself may be the only, or most important, determinant of light quality in open oceanic environments, the photic environments of nearshore and freshwater habitats tend to be much more influenced by a variety of abiotic and biotic factors. Amongst the most important abiotic factors influencing the photic environment are the presence of soluble and insoluble, organic and inorganic materials. Such materials are capable of both absorbing and scattering incident light. The biotic influences that are known to affect light quality include bacteria, photosynthetic phytoplankton, macroalgae and higher plants, and animals swimming in the water column.

Lakes, streams and coastal waters may contain significant amounts of both phytoplankton and organic material from decaying plants and animals. In such waters, light of all wavelengths will be more strongly absorbed than it is in clear oceanic waters, i.e. attenuation is more rapid with depth. In addition, the relative colour intensities at each depth will be quite different from what they are in the ocean. The short-wavelength violet and blue light is most strongly absorbed, and green or yellow-green light at wavelengths of 540–560 nm becomes increasingly dominant with depth.

Marshes and swamps contain the same types of light-absorbing organic matter as lakes, streams and coastal waters, but they may also contain high levels of humic material, tannins, lignins and other organic matter indicative of more complete plant decomposition. The combined effects of the increased amounts of dissolved and suspended organic matter can be that nearly all of the light is absorbed within 2–3 m of the water surface. In addition, the maximally transmitted wavelength will be 600 nm or longer, well towards the red end of the spectrum, so the water takes on a reddish or red-brown hue.

In addition to the effects of different types of water on the photic environment, it must also be remembered that there will be changes in both the intensity and spectral quality of the incident radiation on a temporal basis. The intensity of the light reaching the water surface will diminish rapidly over several orders of magnitude at dusk, and will then settle at a level that is dependent upon moonlight and starlight, and the degree of cloud cover. At twilight the sky reddens and the spectrum contains relatively more photons of longer wavelength, but the effects of this on the spectral bandwidth of the underwater illumination are of minor proportions. The daylight and night-time spectra are often found to be similar and the main problem associated with night-time vision underwater would appear to be related to reduced light intensity.

In temperate and polar regions, not only are there pronounced seasonal changes in photoperiod, but fish may also experience marked seasonal changes in light intensities. For example, the irradiance under a seasonal ice cover may be 1% or less than that impinging on the ice surface.

2.3 VISUAL PIGMENTS

In the eye of a fish the photoreceptor cells onto which the incoming light is focused contain photosensitive substances called visual pigments. The visual pigments catch photons of light by absorbing them, and different pigments are capable of maximal absorption of different wavelengths. The pigment *rhodopsin,* for example, transmits red and blue light, whereas its absorption maximum is at a wavelength of 500 nm, in the blue-green region of the spectrum. During the early years of the study of visual pigments, rhodopsin was extracted from the photoreceptor cells of a wide range of fish species and terrestrial vertebrates. An additional pigment named *porphyropsin,* having an absorption maximum at 520–525 nm, was extracted from freshwater fish. There are now known to be visual pigments that have absorption maxima spread over a wide proportion of the visible spectrum, ranging from the blue wavelengths to over 540 nm.

When a visual pigment absorbs photons, the molecules in the pigment undergo structural changes and trigger electrochemical responses in the photoreceptor cells that contain the pigment. The response of the cell is determined by the number of photons absorbed by the pigment, but not by the wavelengths of the photons. That is, once a photon is absorbed, its wavelength no longer affects the output of the photoreceptor cell. Visual pigments absorb light of all visible wavelengths to some extent, but each has specific absorption maxima.

It has become conventional to specify the visual pigments solely by reference to the wavelengths at which they show their absorption maxima. In addition to having a main sensitivity peak, each visual pigment also has a secondary region of heightened sensitivity. These secondary regions of sensitivity, known as cis-peaks or beta-bands, are broadly centred at a wavelength about two-thirds of that at which the visual pigment displays maximum absorption.

Thus, a visual pigment with an absorption maximum at 520–525 nm, within the visible part of the spectrum, will tend to have a cis-peak at around 340–350 nm, within the ultraviolet (UV) range of wavelengths (320–400 nm). Thus, the visual pigment would be expected to be activated, but to a different extent, by wavelengths within both the visible and the UV range. However, since visual pigments are 'blind' to wavelength differences, absorption of energy at any of the wavelengths to which the pigment is sensitive leads to qualitatively indiscriminable signals. The point to be made is that the possession of pigments with such patterns of sensitivity permits detection of wavelengths within the normal visible spectrum and gives rise to the possibility of ultraviolet vision.

Underwater vision

The problems of underwater vision can be summarized as being related to the low light intensities and to the shifts in spectral bandwidth that occur with increasing depth. Thus, fish would be expected to possess photoreceptors having enhanced sensitivity, and improved ability to capture photons, at low light levels. The visual pigments of fish may also be expected to differ depending upon the habitats occupied. Thus, fish living at depth in oceanic waters, where the wavelengths are 470–490 nm, would be expected to possess blue-shifted visual pigments. Many deep-water marine fish do have visual pigments, the *chrysopsins*, or 'visual golds', that have absorption maxima close to the spectral distribution of the ambient light.

Marine fish living closer to the surface tend to have photoreceptors that contain rhodopsins with absorption maxima within the range 494–505 nm, whereas multiple pigment systems may be found in vertically migrating marine species. The vertically migrating species may have both rhodopsins absorbing maximally at about 500 nm, and a chrysopsin which absorbs maximally at a wavelength of about 470 nm. Fish that inhabit shallow bays and coastal waters, where the colour of the water tends towards the yellow-green, have visual pigments with absorption maxima shifted further towards the red end of the spectrum than those of the deep-water species.

The pigments of the coastal-zone species do not seem to follow the spectral distribution of the available light as closely as do the pigments of the deep-water oceanic species. The visual pigments of the shallow-water marine fish have absorption maxima in the range 500–510 nm, even though the local spacelight is in the yellow-green, at wavelengths 525–550 nm.

The visual pigments of some freshwater fish tend to have optima within the range 510–545 nm. This would appear to be poorly matched to the spectral distribution of the ambient light, which tends to be dominated by wavelengths around 600 nm. Most freshwater fish are, however, thought to have trichromatic vision, with the visual pigments having absorption peaks around 455 nm (blue), 530 nm (green) and 625 nm (red).

Several freshwater species, such as some cyprinids, salmonids and percids, have photoreceptors that contain pigments sensitive to UV light, at very short wavelengths (below 390 nm). Rates of attenuation of light having such short wavelengths would be expected to be high, but penetration to depths of a few metres is possible. There are known to be ontogenetic changes in UV sensitivity in some species of salmonids, cyprinids and percids. In these species small juveniles are UV sensitive but during normal development this sensitivity to UV wavelengths is lost. This ontogenetic loss of UV photosensitivity may be related to habitat shifts, in that the fish move from shallow to deeper water as they grow. Because UV wavelengths may not penetrate more than a very short distance into the water column, the retention of UV sensitivity by a fish living in deep water would appear to be of little adaptive significance.

The specific functions of UV sensitivity are not known with any degree of certainty, although it seems that the receptors for short wavelengths, within the UV range, enable the fish to detect the pattern of polarized light. The pattern of polarization depends upon the position of the sun, and the spectral composition of light underwater varies as the sun rises and sets. The intensity of UV light changes considerably during periods of sunrise and sunset, and the UV-sensitive mechanisms involved in the detection of polarization appear to be particularly effective when the light intensity is comparable to that of sunrise and sunset. These times of the day are those at which several species of fish show greatest migratory activity, and it has been suggested that the pattern of UV, polarized light may be used as an aid in navigation.

Sensitivity to the polarization of light can also enhance the contrast of underwater objects, leading to increased ease of detection. Moreover, certain objects may polarize light and reflect polarized light that differs in composition from the background. Zooplanktonic organisms are known to reflect polarized light, so fish that have the ability to detect patterns of polarized light may increase their ability to locate zooplanktonic prey in surface, and relatively shallow, waters. The ontogenetic loss of UV sensitivity seen in the cyprinids and salmonids may, therefore, be associated with changes in feeding habits and habitat shifts. The juvenile fish feed almost exclusively on zooplankton in surface waters, whereas larger fish may consume larger proportions of benthic organisms taken in deeper water.

Whilst the visual pigments of some fish species appear to be well matched to the spectral distribution of the ambient light, the pigments of other species are mismatched to a greater or lesser degree. It has been suggested that the absorption maxima of the visual pigments may be offset from the wavelength of the spacelight in order to maximize visual contrast. When a pigment is matched to the spacelight, there will be a relatively high degree of contrast between a dark object, which deposits relatively few photons on the visual pigment, and the spacelight, which deposits many photons. On the other hand, a bright object will deposit many photons on the pigment, so bright objects will not be easily distinguishable from the spacelight.

If, however, the spectral bandwidth of the light illuminating the object is relatively broad, and the pigment is not matched to the spacelight, the situation will be different. The visual pigment does not have an absorption maximum at the wavelength of the spacelight, so both the spacelight and a dark object will appear dark. This means that a dark object will be indistinguishable from the background. A bright object, however, will continue to deposit many photons on the pigment, and will appear bright against the dark spacelight. Thus, a fish that incorporated both pigments, one matched to and one slightly offset from the wavelength of the spacelight, would be able to distinguish both light and dark objects in the environment.

2.4   THE FISH EYE

The eyes of most species of fish resemble flattened spheres (Fig. 2.1), but in a number of deep-water species the eyes are tubular in form. For fish living in deep water the light comes to them directly from above, and the light rays

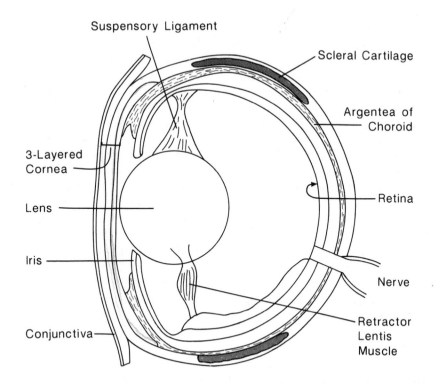

Fig. 2.1 Cross section through the fish eye showing general structure.

will be almost parallel. Little, or no, light reaches the fish from other directions, and the tubular eyes of deep-water fish are usually directed upwards.

The spherical eye has six oculomotor muscles attached to its fibrous outer layer, the sclera. The sclera is also usually reinforced with cartilage or calcified elements. Part of the outer layer, the cornea, is transparent and allows for the passage of light into the eye. In some species, especially those capable of prolonged, fast swimming, this region of the outer surface of the eye is protected by transparent membranous folds of tissue. The refractive index of the cornea is similar to that of the surrounding water (refractive index of about 1.33), and the cornea is almost flat. This means that the cornea contributes little to the focusing of the light.

Focusing

Focusing is carried out by the spherical lens which has a refractive index much higher than that of water (about 1.67). The lens protrudes through the pupil, and may be closely allied with the cornea. This ensures that the eye has a wide field of view.

In the majority of teleosts the pupil is fixed in size, but in elasmobranchs the iris may be capable of contraction. Contractions of the iris result in changes in the aperture of the pupil.

Light impinging on the lens is focused by small back or forward movements of the lens. In teleosts, the lens is pulled backward by a retractor muscle, whereas in elasmobranchs it is pulled forward by a retractor muscle during focusing.

The lenses or corneas of many fish species contain yellow pigments that filter out radiation of violet or ultraviolet wavelengths, but other species have relatively clear lenses that transmit the shorter wavelengths. Relatively clear lenses have been found amongst freshwater fish such as the cyprinids and salmonids which are known to have UV vision, and the least transmissive lenses appear to be found amongst the perciform fishes.

The possession of a non-transmissive lens containing pigment may serve to improve visual acuity. Additional short-wavelength filters may be present in the retinas of some species. The filters consist of coloured oil droplets located within the ellipsoid region of the inner segments of the photoreceptive cone cells.

Light reception

Once the light has transversed the focusing system and the fluid-filled chambers containing the aqueous and vitreous humors, it falls upon the retina and choroidal tissues. The choroid serves the dual functions of nourishing the retina and of either absorbing or reflecting stray light back through the retina.

The retina comprises both receptor cells and a series of neuronal elements. The neuronal elements lie closest to the chamber containing the vitreous humor, and the light passes through this layer of the retina before impinging upon the receptor cells. It is the receptor cells which contain the visual pigments. A pigment epithelium lies outermost, adjacent to the choroid, and cells of this epithelium contain the dark pigment melanin. In the majority of fish species the cell layer lying internally to the pigment epithelium contains two morphologically distinct types of receptor cells, the *rods* and *cones*. The retinas of deep-water fish often contain only rods, and these receptor cells may be extremely numerous in fish living at great depth. Thus, the rods are the cells that lead to increased visual sensitivity under conditions of low light intensity, as occurs in deep water, or in shallow water at night.

Visual sensitivity may be increased by the presence of a mirror, or *tapetum lucidum*, at the back of the eye. This is a common feature of the eyes of nocturnally active species. The reflecting tapetum lucidum appears to take one of three forms in fish species.

A retinal tapetum occurs in the pigment epithelial layer of several freshwater species. The epithelial cells contain mirror-like crystals of guanine, which reflect light to the cells containing the visual pigments. The same cells also contain the dark pigment melanin, which acts to mask the reflecting crystals under conditions of bright light, a process known as occlusion.

Some marine teleost species have a shiny, reflecting tapetum associated with the choroid. Unlike the retinal tapetum of the freshwater teleosts, the choroidal tapetum is not capable of being occluded.

Almost all species of elasmobranchs have a tapetum lucidum within the choroidal tissue layer, but the structure differs from that found in teleosts. The tapetum consists of flattened cells that contain guanine crystals. The reflecting plates are orientated perpendicularly to the incident light. This ensures that the reflection from the plates is highly directional, which appears to be an adaptation to minimize blurring of the image. The combination of a retinal epithelium containing rods, and a reflecting tapetum increases visual sensitivity under conditions of low light intensity.

Fish are, however, capable of changing the effective light intensity impinging on the receptors by several means. The majority of elasmobranchs have a pupil that can be closed or dilated depending upon incident light conditions, and adjustments in the aperture of the pupil will lead to a regulation of the amount of light entering the eye. The retina of the elasmobranchs has a pigment epithelium comprising rod cells, and this, together with the possession of a tapetum lucidum, means that the elasmobranch eye is inherently adapted to dim light conditions, i.e. is 'dark adapted'. The fact that elasmobranchs have the ability to adjust the aperture of the pupil permits closure of the pupil in bright light. This allows the dark-adapted visual system of the elasmobranchs to cope with the higher light intensities of full daylight conditions.

'Light' and 'dark' adaptation

In the vast majority of teleosts there is little or no pupillary control over the amount of light that reaches the retina. In teleosts the adaptation of the eye to light conditions and dark conditions occurs via movements of the visual cells and pigments within the retina – the retino-motor movements.

Long processes extending from the pigment cells of the pigment epithelium run towards and into the cell layer containing the visual receptor cells. Melanin pigment can pass along these processes, either towards or away from the visual receptor cells, the direction of movement depending upon the

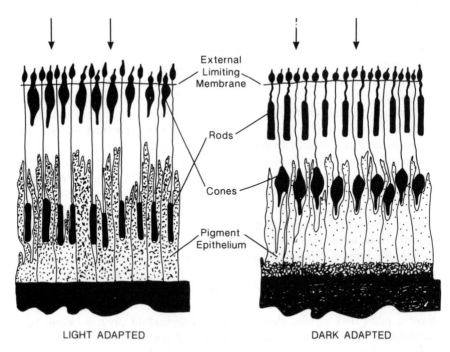

Fig. 2.2 Changes in pigment distribution and in the rod and cone cells of the fish retina during light adaptation and dark adaptation. Arrows represent incident light.

level of incident light. The visual cells, both the rods and cones, are capable of a certain amount of contraction. This enables changes of length to take place, and these changes are also dependent upon the levels of incident light. In the dark-adapted eye, the melanin granules within the pigment cells are drawn back, away from the visual receptor cells. The rods move counter to the melanin in the pigment cells, and this results in the rod cells being fully exposed to the incident light. There is also some movement of the cones, in the opposite direction to the rods, but this movement does not result in them being shielded by the melanin pigment (Fig. 2.2).

An increase in light intensity leads to a migration of melanin granules into the processes of the pigment cells, leading in turn to an increase in the depth of the pigmented layer. Movements of the rods result in the cells entering the pigmented area, where they are protected from exposure to high light intensities. The cones move also, once again in a direction opposite to the rods, so in the light-adapted eye it is only the cones that are fully exposed to the incident light (Fig. 2.2).

Visual sensitivity

The retino-motor movements that lead to light and dark adaptation will tend to follow a diel cycle, but there are several other modifications that influence the visual sensitivities of fish in different light intensities. For example, there may be temporal changes in concentrations of visual pigments related to the seasonal changes in incident light intensities. The relative proportions of the different visual pigments may also change due to the shifts in spectral bandwidths that occur on a seasonal basis.

Retinal epithelia dominated by rods, and the reflecting tapetum lucidum, are also adaptations to low light intensities, but, perhaps surprisingly, a reflecting tapetum is not a universal feature of the eyes of benthic and deep-water species. In many deep-water species the eye is elongate, or tubular. The light-gathering ability of the eye is improved due to the absence of the iris, and the incoming light is focused on the retina at the bottom of the tube by the enlarged lens. The tubular eyes are usually directed upwards, giving binocular vision. Overall, these series of adaptations lead to an increase in light-gathering at the expense of a narrowing of the visual field.

2.5   UNDERWATER SOUND

While sound obeys the same physical principles in air and water, there are certain quantitative differences between sound in these two media. The differences are related to the fact that air and water have different densities and degrees of compressibility. At the sound source there will be alternate compression (rise in pressure) and rarefaction (drop in pressure), and these fluctuations in pressure (sound waves) spread away from the source with the speed of sound.

During wave propagation, the particles of the medium are first compressed. As the particles rebound, or rarefy, they impart their motional energy to neighbouring particles. Thus, during compression and rarefaction each particle moves first in the direction of the wave propagation and then in the opposite direction. Consequently, the pressure wave produced by sound causes individual particles to oscillate along the axis of wave propagation, although the particles themselves are not displaced from their original positions by the passing of the wave. Sound sources will, therefore, provide two types of stimulation – a sinusoidal change in pressure (P), and a back-and-forth motion of particles (expressed as particle displacement or particle velocity, u).

For spherical waves, which are those most likely to be produced by biological sound sources, the ratio between P and u changes as a function of distance from the sound source. The ratio of P to u decreases as the sound source is approached. As the distance from the source increases, the state of a spherical wave-front will come to resemble a plane wave, and then the ratio of P to u remains constant with increasing distance. The distance from the source at which the ratio of P to u becomes constant is known as the near-field:far-field boundary. Thus, for spherical waves the near-field is an area relatively close to the source, in which pressure and particle velocity are out of phase with each other and in which velocity changes at a greater rate than pressure. As the far-field is approached, particle velocity and pressure come together in phase, and their rates of change become more nearly equivalent.

In general, quantitative differences between near-field and far-field effects are not important in the study of acoustic processing by terrestrial animals. This is the case because the near-field, which is related to the compressibility of the medium, extends only a very short distance from the sound source in air, which is highly compressible. In water, which is nearly incompressible, the extent of the near-field increases approximately fivefold over that in air.

The position of the near-field:far-field boundary also depends upon the frequency of the sound. For low-frequency sounds the boundary will be further away from the source than for higher frequencies, and the extent of the near-field shows an inverse relationship with the frequency of the sound. Underwater sounds will generally be of low frequency, within the range 1–400 Hz, with the frequencies produced by animals moving through the water being 3–40 Hz. Given the low compressibility of water, and the fact that the sounds encountered by fish are much lower in frequency than those experienced by terrestrial animals, it should be apparent that near-field effects are important in the processing of acoustic information by fish.

Differences in the characteristics of air and water also give rise to other differences. The speed of sound is 4.5–5 times as great in water as in air, being about 1500 and 330 m s^{-1} in water and air, respectively. Since the frequency (f in Hertz, Hz), wavelength (λ in m) and speed of the sound (c) are related according to $f \times \lambda = c$, it is clear that, for any given frequency, the wavelength of the sound in water will be 4.5–5 times as long as in air. Sound would appear to have much to recommend it as a form of sensory stimulus in the underwater environment. It is transmitted at high speed through turbid water and in darkness, where vision is impossible. Furthermore, unlike a chemical stimulus, sound does not persist. The two stimulatory components of the acoustic signal – pressure and particle velocity (or displacement) – change significantly with distance from the source, the most significant change occurring when the fish move across the near-field:far-field boundary. In the far-field, pressure may be the prevailing component of the sound wave, whereas in the near-field, particle velocity or displacement will be more important. Thus, if fish were able to detect and process both pressure and particle displacement stimuli they could extract a considerable amount of information about their environment from the acoustic sensory system.

The majority of fish species are able to detect both the pressure and the particle displacement components of a sound wave, and this dual ability is attributed to certain mechanical properties of the acoustic system. Particle displacement is sensed by mechanoreceptors, but with the addition of a mechanism for converting pressure into particle displacement signals, the sensory system can also respond to pressure stimuli.

2.6   MECHANORECEPTORS AND THE LATERAL LINE ORGAN

The mechanoreceptors of fish are located in the acoustico-lateralis system, which comprises the inner ear and the lateral line organ. In both the ear and the lateral line, the basic mechanoreceptor is the *neuromast*. In addition to their location within the ear and lateral line system, neuromasts may also be distributed over the body surface of the fish.

The neuromast consists of a group of sensory hair cells surrounded by supporting mantle cells. The hair cells are ellipsoid in shape and have a series of cilia projecting from their apices (Fig. 2.3). Each bundle of cilia consists of a single kinocilium and a group of stereocilia, which may number from 40 to 100 or more. The kinocilium is generally the longest structure in the ciliary bundle, whilst the stereocilia are usually graded in size with the longest of them being adjacent to the kinocilium. The ciliary bundles of the hair cells constituting the neuromast project into a gelatinous matrix, the *cupula*, and the impingement of a stimulus on the cupula will cause the cilia to bend.

The hair cells have a directional sensitivity: if the stereocilia are bent towards the kinocilium, there is an increase in the rate of nervous discharge, whereas a bending of the stereocilia away from the kinocilium results in an inhibition of discharge. The discharge of the hair cell is related to the degree of displacement of the ciliary bundle. In addition, the sensory hair cells are aligned to a common axis within the neuromast, which imparts a directional sensitivity to the system.

Since most of the body tissues of a fish are almost the same density as water, the fish will vibrate in a similar manner to the particles in the water, but there is some differential motion between the fish and the surrounding water. This motion varies along the length of the fish depending upon the distance from the sound source, so there are differential displacements at various points on the body. Consequently, there are advantages in having a long lateral line in which the particle displacement system, consisting of a series of neuromasts, is subjected to differential stimulation. Similarly, by having an elaborate series of canals in different orientations, each with

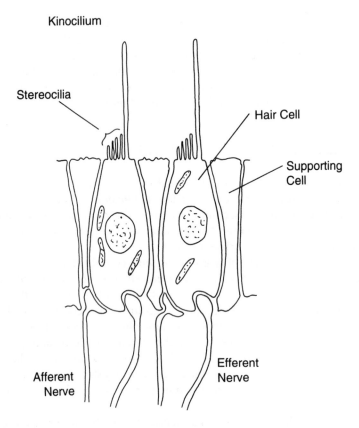

Fig. 2.3 General structure of the hair cell found in the organs of the acoustico-lateralis system.

characteristic axes of sensitivity to particle motion, the fish would be able to detect the direction of the source.

   Typically the lateral line organ consists of series of neuromasts located in canals running around the head and along the trunk. The trunk canal is generally single and runs the full length of the body, whereas the canals on the head may be extensively branched (Fig. 2.4). There is, however, considerable interspecific variability in the morphology of the lateral line system. The lateral line canals, which run below the body surface, open to the exterior at various points via a number of pores. The grouping of the neuromasts into the lateral line organ seems to be particularly well suited to the sensing of the movements of nearby fish, such as conspecifics in a school, the irregular movements of a potential prey, or the approach of a predator.

Fig. 2.4 The general distribution of the lateral line canal system in a fish. The canal running along the length of the trunk is usually single, but the system may be extensively branched in the head region.

The basic neuromast sensory unit is also used in the detection of other types of mechanical stimulation – sound, linear and angular acceleration and gravity. The arrangement of the neuromasts within the inner ear, or labyrinth, is particularly well suited to performing these functions.

2.7 INNER EAR AND OTOLITHIC ORGANS

The inner ear of fish consists of three semicircular canals, a widened area, or ampulla, at the base of each canal, and three sac-like otolithic organs: the sacculus, the lagena and the utriculus (Fig. 2.5). The otolithic organs contain a sensory epithelium, or *macula*, which is made up of a large number of sensory hair cells. In the otolithic organ the gelatinous cupula is specialized to form the otolithic membrane, which encapsulates the calcareous *otolith*.

In teleosts, the otoliths, which have a density of about 3, are hard structures composed of calcium carbonate within a proteinaceous matrix. In the elasmobranchs, on the other hand, the otoliths consist of a gelatinous matrix containing numerous small crystals of calcium carbonate. The otolithic membrane serves to loosely maintain the position of the otolith with respect to the sensory macula, whilst at the same time allowing the two structures to move independently of each other (Fig. 2.6).

The otoliths, which are more dense than the rest of the fish, will have an inertia and so will tend to lag behind the movements of the fish if it accelerates or brakes suddenly. Similarly, the otoliths will be markedly affected by the forces of gravity. Gravitational forces will act on the calcareous otoliths when the fish moves obliquely up or down in the water column (Fig. 2.6).

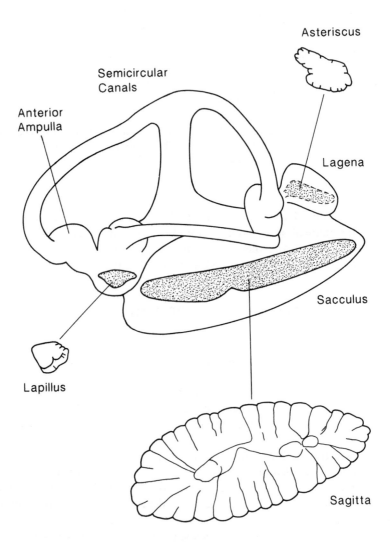

Fig. 2.5 The general structure of the inner ear showing the semicircular canals and otolithic organs of a gadoid fish. In gadoids the sagittae are typically 10–15 mm in length.

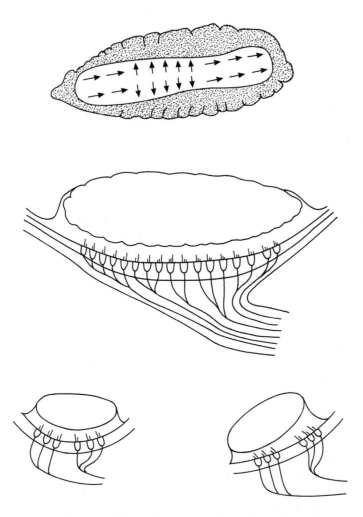

Fig. 2.6 Schematic diagram of the otolithic organ. Upper figure shows the orientation of the hair cells on the macula, arrows indicating different directions of orientation. The centre figure shows the relationship of the otolith to the hair cells of the macula. The effects of various body movements on the spatial relationships between the otolith and the hair cells are shown in the lower figures.

These differential movements of the otolith with respect to the macula will cause the ciliary bundles of the sensory hair cells to bend. This bending will result in either promotion or inhibition of discharge, depending upon the direction in which the cilia are bent.

The neuromasts of the macula are not all orientated in the same direction, there being different patterns of orientation in different parts of the macula. Thus, the sensory cells on the otolithic macula are divided into distinct groups, with all the cells in a particular group being orientated in the same direction (Fig. 2.6). The presence of these differently orientated, and hence directionally sensitive, groups of sensory hair cells on the otolith macula allows detection of different movements of the otolith relative to the macula. Thus, the otolithic organ can provide information relating to positional and directional changes of the fish in the water column, and can also function as an accelerometer.

Most of the fish body has the same density as the surrounding water, and during the passage of a sound wave the oscillatory particle displacements in fish tissues will be similar to those of the water molecules. The neuromast cupulae and hair cells have a density of about 1, similar to that of water, so the sensory cells will vibrate in the same fashion as the bulk of the body tissues. The otolith, is, however, much denser than the remainder of the body and the otolith movements lag behind the rest of the body. Consequently, while the otolith and the sensory hair cells on the macula both move in response to the displacement energy of the sound, the distances moved by the two structures differ. The vibrations, or oscillations, of the body tissues and macula are much greater than those of the otolith, so the fish body can be considered to be vibrating around a relatively static dense otolith. Because the sensory hair cells are embedded within the macula, and the ciliary bundles of the hair cells are in contact with the otolith, the differential oscillations of the body tissues and the otolith provide the movements necessary for stimulation of the sensory cells. Thus, this system provides the mechanism by which a fish can detect the particle displacement component of an acoustic stimulus. The system does not, however, allow direct detection of the pressure component of the sound.

Sound pressure is the dominant stimulus in the far-field, but fish do not appear to have sensory cells that are capable of responding directly to sound pressure. Thus, the pressure stimulus must be converted into particle displacement before it can be detected. Consequently, the detection of both the particle displacement and pressure components of the sound ultimately involve stimulation of the sensory hair cells of the macula of the otolithic organ, resulting in the bending of the ciliary bundles. It has been suggested that one of the functions of the gas bladder is to translate sound pressure into particle displacements, which can then be transmitted to the otolithic organ of the inner ear. In other words, although the inner ear responds to particle displacement, the entire system of gas bladder and otolithic organ can be considered to be pressure sensitive.

2.8   GAS BLADDER AND SOUND DETECTION

The gas bladder appears to respond to sound pressure by pulsating in sympathy with the passing compressions and rarefactions of the sound wave. The pulsations caused by the sound pressure create a secondary near-field within the body of the fish close to the inner ear. The particle displacements so produced are then re-radiated through the tissues to the inner ear where they can stimulate the sensory hair cells of the macula. Thus, the gas bladder may function as a pressure transducer and sound amplifier, but there are marked interspecific differences in the morphological relationships between the gas bladder and the inner ear. It is thought that the closeness of the coupling between the gas bladder and the inner ear is related to hearing ability.

The importance of the gas bladder in the enhancement of hearing ability has been demonstrated by carrying out a series of experimental manipulations. In the Atlantic cod, *Gadus morhua*, for example, deflation of the gas bladder results in impaired hearing. Furthermore, it has been shown that fish possessing gas bladders have better hearing, with respect both to frequency range and to sensitivity, than do fish species that lack gas bladders. Thus, the hearing abilities of the pleuronectids, which lack gas bladders, are generally poorer than those of many other teleosts. The hearing abilities of the pleuronectids can, however, be improved by experimental manipulation in which they are provided with a 'hearing aid' in the form of a small air-filled balloon placed near the head.

In many species of teleost fish, the particle displacements are transmitted from the gas bladder to the inner ear via the body tissues, but in some species there are horn-like extensions to the anterior end of the gas bladder. These extensions bring the bladder and inner ear into close contact. The wall of the skull may also be relatively thin in the region where the two structures come into nearest contact, and this would assist in transmission of the particle displacements.

Other groups of fish have more elaborate specializations linked with sound pressure detection. The cyprinids, ictalurids and their relatives are characterized by having a connecting link between the anterior end of the gas bladder and the inner ear. The link is formed by three small bones – the *Weberian ossicles*, derived from vertebrae – that lie just ventral to the vertebral column. Movements of the walls of the gas bladder appear to cause rocking motions of the Weberian ossicles, and the gas bladder pulsations are thus transmitted to the inner ear without being weakened. Both the frequency range and the sensitivity of hearing appear to be greatly enhanced by the possession of Weberian ossicles. Consequently, the hearing abilities of the cyprinids and ictalurids are amongst the best of any fish species.

The pro-otic bulla

The clupeids (herrings and their relatives) also have excellent hearing abilities. In these fish there are two very fine ducts that extend forward from the anterior end of the gas bladder to connect with the pro-otic bulla in the region of the inner ear (Fig. 2.7). The pro-otic bulla is a bony sphere which is divided into two chambers by an elastic membrane. A fine thread stretches from the membrane across the fluid-filled upper chamber (the fluid is known as perilymph) to connect the membrane to the macula of the utriculus. Thus, any vibration of the pro-otic membrane resulting from changes in sound

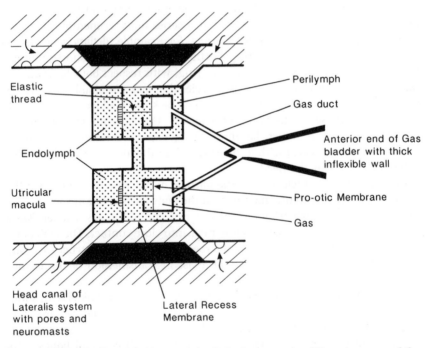

Fig. 2.7 Schematic diagram showing the links between the different organs of the acoustico-lateralis system and the gas bladder in a clupeid fish. Arrows through pores indicate communication channels between the head canal of the lateralis system and the surrounding water. Neuromasts are indicated by half circles on the inner wall of the canal of the lateralis system.

pressure may be transmitted to the sensory hair cells of the macula both via the thread and via movements of the perilymph. Sound pressure detection is achieved by having the lower chamber of the pro-otic bulla filled with gas. This gas responds to the compressions and rarefactions of the sound pressure wave, and thus causes the elastic membrane to vibrate.

The gas bladder does not play a central role in the detection of sound pressure waves because it does not appear that it is vibrations of the gas bladder wall that are transmitted to the elastic membrane of the pro-otic bulla, i.e. sound is not conducted along the fine ducts that connect the gas bladder to the bulla. It appears that these ducts represent a pressure-equalizing system that prevents distortion and rupture of the elastic membrane of the bulla when the fish either dives, or rises rapidly in the water column.

When the fish dives, the increase in hydrostatic pressure will result in compression of the gas in the bulla, and the elastic membrane will tend to bow inwards towards the gas-filled chamber. The gas present in the gas bladder will also be compressed, and since the walls of the gas bladder are compliant, gas will tend to be forced along the fine tubes from the gas bladder to the lower chamber of the pro-otic bulla. Gas flow is enhanced by the fact that the wall of the anterior portion of the gas bladder is thicker, and less compliant, than that of the remainder of the bladder. As gas flows from the bladder to the pro-otic bulla the shape of the membrane will be restored to its original flat form, in which it is most sensitive to stimulation by changes in sound pressure.

If the fish rises in the water column the decrease in the hydrostatic pressure will cause the gas in the bulla to expand, and this will result in the elastic membrane bowing outwards. Gas can, however, flow along the ducts from the bulla into the gas bladder, once again restoring the shape of the elastic membrane. Thus, the connections between the bulla and gas bladder enable corrections to be made for differences in hydrostatic pressure at different depths. Adaptation of the system appears to require about 15–30 s to be completed when the fish moves up and down in the water column.

This system is also unique in the fact that it incorporates a coupling between the pro-otic bulla, the inner ear and the lateral line system. The cavity containing the perilymph not only links with the upper chamber of the pro-otic bulla and the utricular macula, but is also separated from the head canals of the lateral line system by a thin membrane, the lateral recess membrane. Movements of the elastic pro-otic bulla membrane will be transmitted to the lateral recess membrane via the perilymph, and the lateral recess membrane will, therefore, vibrate in sympathy with the bulla membrane. These movements will then be transmitted to the fluid contained within the head canals of the lateral line, leading to a stimulation of the lateral line neuromasts. Thus, this couples the mechanoreceptors of the lateral line to the sound pressure detection system of the pro-otic bulla, thereby leading to further enhancement of hearing abilities.

2.9 ELECTRORECEPTION

A number of species of fish possess electroreceptors which appear to be closely allied to the lateral line organs. These electroreceptors, or *ampullary organs*, are found in the elasmobranchs, some dipnoans and chondrosteans, and a few species of teleosts.

The ampullary organs are characterized by having an epidermal ampulla, containing sensory epithelium, connected to the surface by a long canal (Fig. 2.8). The canal is filled with a highly conductive mucopolysaccharide jelly, and the thin walls of the canal are electrically resistive. Thus, the electrical potential across the receptor is the voltage drop between the base of the canal and its opening. Of the ampullary organs, the ampullae of Lorenzini, found in all elasmobranchs, have been examined in greatest detail. The receptors are located in sacs at the end of long canals that open on to the surface of the body. The canals vary between 5 and 150 mm in length. The receptors are most densely aggregated on the head, on the anterior end of the body and on the pectoral fins. The ampullae of Lorenzini may form extensive networks, and they may be clustered to form a series of rosettes.

The electroreceptors are extremely sensitive, and some elasmobranchs may show behavioural responses when exposed to electrical field strength gradients as low as 5 nV cm^{-1}. The best-documented use of the ampullae of Lorenzini is in prey detection. All animals produce bioelectric fields that are modulated by their ventilatory movements, and elasmobranchs are able to locate their prey by homing in on these bioelectric fields.

The importance of electroreception in prey detection by elasmobranchs has been demonstrated by a series of experiments carried out on the spotted dogfish, *Scyliorhinus canicula*. When presented with a piece of bait (fish flesh) lying on the surface of the bottom substrate the dogfish readily consumed the bait. In this instance the dogfish could have located the bait by responding to visual and chemical, but not bioelectric, cues. The bait could be seen and would be emitting odour, but dead flesh does not produce a bioelectric field. When the experiment was repeated, the dogfish was offered a choice between the bait and an active pair of electrodes (producing a weak electric field) buried in the substrate, the fish attacked the electrodes in preference to the bait. Thus, in the dogfish, the detection of a weak electric field may override both visual and chemical stimuli when the fish are engaged in searching for potential prey.

In addition to prey detection the electrosensory system may also be used for orientation and navigation. The movements of the oceanic currents and those of the animals themselves produce voltage gradients that are within

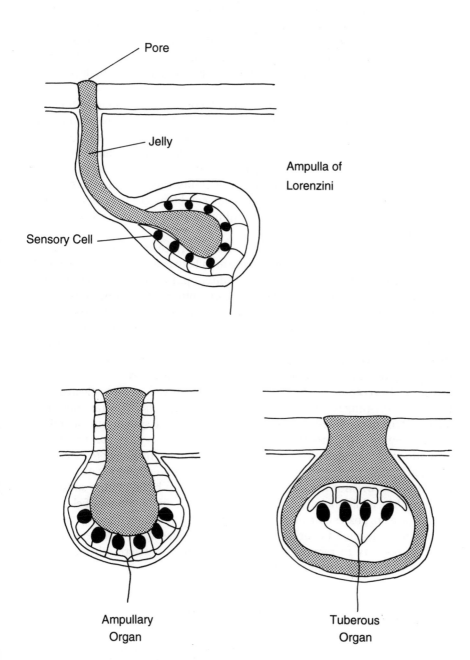

Fig. 2.8 General structures of electroreceptive organs – the ampullary organ, the ampullae of Lorenzini and tuberous organs.

the sensitivity range of the electrosensory system. Laboratory studies have shown that several elasmobranch species are capable of detecting, and using, very weak electromagnetic fields and voltage gradients for orientation purposes. Further studies are, however, needed to show to what extent this information is used in guiding the directed movements made by fish in the wild.

The ampullary organs and ampullae of Lorenzini give long-lasting responses to low-frequency (0.05–8 Hz) stimuli, but some species of fish, notably the weakly electric South American gymnotid and African mormyrid fish, also possess electroreceptors giving brief responses to high-frequency stimuli. The *tuberous organs* consist of epidermal capsules containing sensory epithelium but they lack a connecting canal to the exterior (Fig. 2.8). They are part of a system that combines electroreception with electric organ discharges to provide communication and active electrolocation functions.

The electric fishes have an electric organ, derived from modified muscle tissue, located in the caudal region in the area between and behind the dorsal and anal fins. This organ produces intermittent pulse-like electric discharges. The fish also possess three types of electroreceptors, which provide separate sensory channels for electrocommunication, low-frequency electrolocation and active electrolocation. The ampullary organs are used to detect the bioelectric fields produced by potential prey items, whilst different forms of tuberous organs are used in electrocommunication and in the active electrolocation of inanimate underwater objects and obstacles.

The electroreceptor systems are extremely sensitive and sophisticated, but they are short-range systems with effective receptive distances ranging from a few to tens of centimetres. Despite the fact that these systems are of limited range, they have important sensory roles to play, since they can supply information not provided by other senses. For example, the ability to detect the bioelectric fields generated by potential prey allows prey to be located both in the dark and when it is hidden from view by being buried in the substrate. Fish with the additional ability of both production and detection of electrical stimuli are in possession of a 'radar' system that allows the positions of objects to be located in turbid waters. This enables species possessing this sensory modality to remain active under conditions where other systems, such as vision, are virtually useless.

2.10 CHEMORECEPTION

Chemoreception is an important sensory modality which plays a central role in governing many aspects of fish behaviour, including feeding, reproduction, recognition of individual conspecifics and predators, homing behaviour and orientation. Chemoreceptors of various types occur in numerous locations on the fish body, and three general categories are usually recognized. These categories are olfaction or smell, gustation or taste, and a common chemical sense. These three sensory modalities overlap somewhat, with some substances eliciting reponses by more than one type of receptor. The term 'common chemical sense' is somewhat vague, and probably encompasses the sensory inputs of several different types of receptor, including free nerve endings and the solitary chemosensory cells in the skin. There may also be difficulty in distinguishing between the sensory inputs from the taste bud receptors of the gustatory system and those of the solitary chemosensory cells.

In terrestrial animals the receptors which have highest sensitivity and specificity, and which are distance chemical receptors, are classified as belonging to the olfactory system. Receptors of moderate sensitivity that are stimulated by contact with chemicals in dilute solution are the gustatory receptors, whereas the receptors that are relatively insensitive and have broad specificity are considered to represent the common chemical sense. In fish both the olfactory and gustatory receptors are stimulated by chemicals in dilute aqueous solution, so distinction between these receptor systems must be made on the basis of anatomical and physiological differences.

The greatest aggregations of chemoreceptors are associated with the olfactory and gustatory systems. Olfactory chemoreceptors are restricted to the epithelium of the olfactory rosette located within the nose of the fish, and the olfactory organs are innervated by the first cranial nerve. The gustatory chemoreceptors, or taste buds, have a relatively dispersed distribution in some species. The taste buds are concentrated within the buccal cavity, but gustatory receptors may also be found on sensory barbels close to the mouth, on specialized fin rays of the pectoral fins, and on other parts of the head and anterior regions of the body. The taste buds may be innervated by branches of the VIIth, IXth or Xth cranial nerve, depending upon their location.

Thus, the two major systems of chemoreception – olfaction, or smell, and gustation, or taste – are distinguished from each other by having different detector mechanisms, sensitivities and nervous connections to the brain.

2.11 OLFACTION

The morphology of the olfactory organs is extremely variable, and there are fundamental differences between the olfactory systems of the elasmobranchs and those of the teleosts. In the elasmobranchs, the paired olfactory sacs are usually located relatively close to the mouth on the ventral side of the snout. Each nostril is a single opening, but the opening to the olfactory pit may be divided into two parts by a fold of skin. Thus, there are, effectively, anterior and posterior outlets, with the flow from the latter often leading directly to the mouth. As the fish swims through the water, and as it takes the respiratory current into the mouth, a current of water passes through the olfactory sacs. Thus, in the elasmobranchs, the flow to the olfactory organs is directly influenced by the water currents generated in connection with respiration.

In teleost fish the nostrils are usually on the dorsal surface of the snout and the paired olfactory sacs are located at some distance from the mouth. The size and form of the olfactory sacs varies greatly, usually reflecting the importance of olfaction as a sensory modality for the different species. The nostrils opening to the olfactory sacs are double, so the water that enters and leaves the olfactory sacs is divided into distinct inhalant and exhalant currents. The currents of water enter via the anterior opening and leave via the posterior opening, but there are a number of different ways by which these currents are generated. In some species the olfactory currents are generated passively as the fish swims through the water, but in other fish species the flow of water through the olfactory organs is assisted by the active generation of currents.

In isoosmatic species, the anterior inhalant opening tends to be funnel-shaped, and this probably assists in directing the water flow through the olfactory organ as the fish swims (Fig. 2.9). Water is directed down the funnel and into the olfactory sac. The sides of the funnel and roof of the olfactory sac extend downwards towards the sensitive olfactory epithelium.

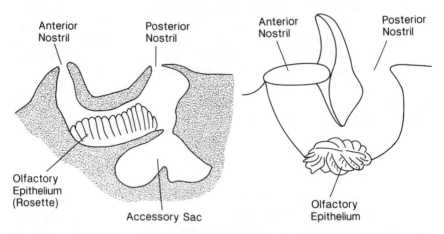

Fig. 2.9 The olfactory organs in isoosmatic and cycloosmatic fish species illustrating the position of the olfactory epithelium (rosette) in relation to the anterior (inhalant) and posterior nostrils. Note the presence of accessory sacs in the cycloosmatic species, and the funnel-like anterior nostril in the isoosmatic species.

These anatomical features ensure that the water current passes over the olfactory epithelium as it flows posteriorly to leave via the exhalant opening. The base of the olfactory sac, in addition to containing the olfactory epithelium, may also be lined with a ciliated epithelium. The ciliated epithelium does not have a sensory function. The beating of the cilia assists in the transport of the water over the olfactory epithelium, ensuring that all parts of the sensory epithelium are exposed to the inhalant current. The beating cilia also ensure that the flow of the water through the olfactory organ is unidirectional.

The olfactory organ of the cycloosmatic species differs from that of the isoosmatic species in that there are accessory sacs closely allied with the olfactory system (Fig. 2.9). These sacs assist in directing water flow over the sensory epithelium. The functioning of the accessory sacs is closely linked to the action of the muscles of the jaw and gill regions. Contractions of the jaw muscles, or the muscular activity that leads to the generation of the respiratory currents over the gills, will lead to volume changes in the accessory sacs. These changes will, in turn, lead to a pumping of water either into or out of the olfactory sac.

In the vast majority of fish species there is no connection between the nostril, and olfactory sac, and the mouth and pharynx. The exception to this general rule is seen in the lungfish, or dipnoans. In the lungfish the nostrils are external anterior nares, and the internal nares open to the mouth. Thus, in the lungfish, there is a direct connection between the nostril and the mouth, as is seen in the terrestrial vertebrates.

Olfactory epithelium

The sensitive olfactory epithelium lines the base of the olfactory sac, but the epithelial tissue is usually raised from the floor of the sac and is thrown into a complex series of folds to resemble a rosette-like structure. The folds of the olfactory rosette vary greatly between species in both direction and number, with the folds, or primary lamellae, serving to increase the surface area of the olfactory epithelium. The olfactory rosette may have an oval, rounded or elongated ovoid form. Primary lamellae are most numerous on elongated rosettes, and the individual lamellae are set at right angles to the longitudinal axis of the olfactory sac. Round rosettes tend to have fewest lamellae, and species with this type of rosette tend to rely little on olfaction as a major sensory modality. Elongated rosettes are usually associated with an acute sense of smell, whereas species with oval rosettes occupy an intermediate position.

The primary lamellae are further subdivided into smaller folds, or secondary lamellae. The chemosensory cells are distributed on these secondary lamellae. The surface of the secondary lamellae comprises the cells of the sensory epithelium interspersed with non-sensory supportive cells. Thus, the sensory cells are not distributed over the entire surface of the olfactory rosette, and three general types of arrangement have been recognized. In some species, such as eels and ictalurids, the sensory cells may have a more or less continuous distribution except for the dorsal parts of the folds, but in other species, the sensory cells have a more limited distribution. In esocids the sensory cells appear to be separated in large areas between the folds, whereas in several cyprinid species the cells occur as small islands of sensory epithelium. Thus, the sensory cells are unevenly distributed in the epithelial tissues of the rosette, but on average there may be 50 000–100 000 cells mm^{-2}.

The sensory epithelium may have a number of different types of receptor cells, these being distinguished according to their structure. The first category consists of the ciliated olfactory sensory cells. These cells have relatively few (four to six) long cilia which take the form of a ring, or crown, on the cell surface. A second type of cell is covered with large numbers of short microvilli, which give the cell surface a brush-like appearance (Fig. 2.10). A small number of species may have a third cell type which has a single thick, rod-shaped process arising from the epithelial surface. Whilst the different categories of sensory cells probably have different functions, i.e. are stimulated by different chemical stimuli, the contributions of the

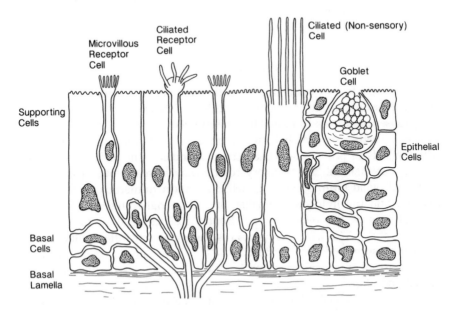

Fig. 2.10 Structure of the olfactory rosette showing the array of different cell types associated with the olfactory epithelium.

individual cell types to the overall olfactory sensitivity and responses are not known.

Responses to olfactory stimuli

Responses of fish to olfactory stimuli have been studied using either behavioural or electrophysiological methods. Behavioural studies often rely on some form of conditioning, either positive or negative, in which the fish is exposed to a chemical followed by a mild electric shock (negative conditioning) or by the presentation of food (positive conditioning). After a few trials the fish will come to associate the presentation of the chemical with the stimulus that follows, i.e. it will have become *conditioned*. Following the conditioning period the fish will, on presentation of the chemical, show behavioural or physiological changes (e.g. in heart rate), in anticipation of the stimulus to follow. Thus, by exposing conditioned fish to chemicals in solutions of different strengths it is possible, by examining the behavioural responses, to gain information about the detection thresholds for the particular chemical substance. When behavioural studies are made on fish there is, at the outset, no guarantee that the fish are using the olfactory sense to detect the chemical, with detection via the gustatory and common chemical receptors also being possible. Thus, in order to ensure that it is olfactory thresholds that are being measured, control experiments must also be carried out on anosmic fish. The fish can be made anosmic either by physically blocking the nostrils or by destruction of the olfactory epithelium.

Electrophysiological methods for the determination of olfactory sensitivity are based upon the direct recording of signals arising from the olfactory epithelium. The signals may be recorded by placing electrodes at different positions on the olfactory rosette or in the nervous tracts leading from the olfactory sacs to the brain. The nerve fibres from the olfactory rosette run in the olfactory nerve to the olfactory bulb, and thence in the olfactory tract to the anterior region of the brain. The length of the olfactory nerve and the location of the olfactory bulb differ markedly between species. In some species the olfactory nerve is short and the bulb is found close to the olfactory organ, whereas in other species the olfactory nerve is long and the bulb is closely allied with the forebrain.

Recordings of activity made directly from the olfactory rosette are known as electro-olfactograms (EOGs), whereas the electrical activities of the nerves are recorded as electroencephalograms (EEGs). When making each type of recording the olfactory epithelium of the fish is bathed in a flow of water that is introduced via the nostril. Different chemical stimuli can be introduced into the inflowing current. EOG recordings can give information about the sensitivity of the olfactory epithelium for different chemical stimuli and also about whether there is a spatial separation, within the rosette, of receptors sensitive to specific types or classes of chemical compounds. Electrical activity in the form of the EEG is usually made by recording from the olfactory bulb, but different individual nerve cells, and regions of the bulb, respond differently to chemical stimulation of the olfactory epithelium.

Thus, within the olfactory bulb, there is some division of labour and different areas of the bulb are responsible for the preliminary filtering, interpretation and transmission of different types of olfactory information. Following initial filtering, the olfactory information is transmitted from the olfactory bulb to the brain via the olfactory tract. Within the tract the nerve fibres may run as distinct bundles, with these bundles running to different regions of the brain. In species in which the olfactory tract is long, recordings made from this part of the olfactory system can provide valuable information about the nature of the olfactory messages conveyed to the various parts of the brain.

For example, EOG recordings made on the olfactory epithelium of the goldfish, *Carassius auratus,* have revealed that there are at least four general olfactory stimulants for this species – amino acids, bile salts, steroids and prostaglandins. Amino acids may act as feeding stimulants, whereas some steroids and prostaglandins may function as sex pheromones used in communication between individuals during the breeding season. These sex pheromones are extremely potent olfactory stimulants, and the detection threshold appears to be within the range 10^{-9}–10^{-12} M.

The olfactory epithelium appears to possess receptors sensitive to the different types of olfactory stimulants, and there may even be specific receptors for the different amino acid series, multiple steroid receptors and different classes of prostaglandin receptor. There is spatial separation of the different sensory cell types, and the nervous connections that leave the olfactory bulb to form the olfactory tracts also appear to carry different sorts of sensory information. Thus, the sex pheromone information appears to be carried exclusively by the medial olfactory tract, whereas information relating to amino acids and various food odours is carried both by the medial and by the lateral olfactory tract.

This separation of sensory information is continued at the level of the brain, with the medial and lateral olfactory tracts projecting to different terminal areas within the forebrain. Once the distinct nerve bundles have been identified and mapped it may be possible to investigate the effects of stimulation of these nerves on the behavioural responses of free-swimming fish. In only very few species are the nervous tracts so distinct that this type of manipulation is possible.

Such experimentation has been carried out on the Atlantic cod in which the olfactory tract is long, and in which four distinct nerve bundles have been identified. Electrical stimulation of each of these nerve bundles results in the fish showing a characteristic behavioural response. For example, in one of the behavioural responses, termed the food-seeking response, the fish adopts a characteristic posture with the head downwards, and the tips of the pectoral fins and barbel in contact with the substrate. The four responses shown following stimulation of the different nerve bundles in the olfactory tract have been categorized as flight, reproductive, snapping (rapid opening and closing of the mouth) and food seeking. In the cod, stimulation of the medial olfactory tract elicited behaviours that were reminiscent of courtship, whereas when the lateral olfactory tract was stimulated certain components of feeding behaviour were evoked. This clearly indicates both that there is spatial separation of olfactory information, and that olfactory inputs are important factors influencing a wide range of behaviours in the cod.

2.12   GUSTATION

The primary sensory organ of the gustatory system is the taste bud. In elasmobranchs the taste buds appear to be restricted to the mouth and pharynx, but in teleosts they may occur on the gill rakers, gill arches, on appendages such as barbels and the pectoral fins, and scattered on the body surface. However, even amongst the teleosts the taste buds are most densely concentrated within the mouth. Here they are often densely packed in the roof of the mouth to form the palatal organ.

The taste bud has a characteristic barrel-like form, and the sensory cells are arranged within the barrel to resemble the segments of an orange. The taste bud comprises three different cell types – the sensory cells, which are elongated and bear ciliary extensions, the supporting cells and the basal cells.

In common with other vertebrates fish are known to have taste buds sensitive to sweet, sour, salt and bitter chemicals. The taste sensitivities of many fish species for acidic taste qualities may be several orders of magnitude greater than those of humans, but the gustatory responses to salts and sugars may be more limited. Compounds such as quinine and strychnine that humans find extremely bitter also elicit gustatory responses in fish. For example, goldfish will reject food pellets that have been soaked in a quinine solution.

Amino acids are the best-studied gustatory stimuli for fish, but the gustatory responses to amino acids differ considerably among fish species. Generally, fish species can be divided into those which display a wide response range, i.e. those that respond to many different types of amino acids, and those which have a limited response range, i.e. fish species that respond to only a few amino acids. Species that show a limited response range include salmonids and some cyprinids, whereas the channel catfish,

Ictalurus punctatus, is amongst the species that respond to a wider range of amino acids.

Carboxylic acids, such as lactic acid and succinic acid, are quite potent gustatory stimuli for some fish species, although salmonids seem to lack gustatory sensitivity to such compounds. Nucleotides stimulate the gustatory receptors of several fish species, and may act as feeding stimulants. Salmonids are sensitive to CO_2 and H^+, there being distinct receptors for both carbon dioxide and protons. The high gustatory sensitivity to CO_2 and H^+ may play a role in the regulation of respiration in the fish because exposure of the palatal organ to CO_2 usually results in an immediate reduction in gill ventilatory movements.

Since the taste receptors of fish may be scattered widely over the body surface, the sensory thresholds of the gustatory system for various chemicals have often been determined in behavioural experiments following destruction or blocking of the olfactory sense. Electrophysiological recordings have, however, been made from nerves innervating distinct concentrations of taste buds, such as those on the barbel or those of the palatal organ. For example, gustatory responses of ictalurids are usually recorded from a branch of the facial nerve (cranial nerve VII) innervating the maxillary barbel, whereas gustatory sensitivities of cyprinids are investigated by recording from the branch of the facial nerve that innervates the lower lip. In other species, including salmonids, gustatory responses are recorded from the palatine nerve innervating the anterior region of the palate and upper lip.

2.13   SOLITARY CHEMOSENSORY CELLS

Solitary chemosensory cells (SCCs) occur in the epidermis of many fish species. They bear a structural resemblance to the sensory cells of the taste buds. The SCCs are defined as being chemosensory cells, embedded within the epidermis, and possessing one or a few apical microvilli. They have connections to primary afferent nerve fibres.

In the majority of species the SCCs are widely scattered over the body surface, and there may be 250–5000 SCCs mm^{-2}. The SCCs almost invariably outnumber those chemosensory cells organized into epidermal taste buds.

In some species of fish there may be high concentrations, up to 100 000 cells mm^{-2}, of SCCs at specific points on the body. In such cases the SCCs may be considered to form specialized chemosensory organs. For example, the free, finger-like rays of the pectoral fins of the gurnards (Triglidae) are well invested with SCCs. These fin rays do not appear to be invested with taste buds, but are used by the fish when probing the substrate in search of food. Thus, it is the SCCs on the pectoral fin rays which are used as the chemical sensors in food-searching behaviour. The nervous signals arising from these SCCs are separate from those of both the olfactory and gustatory systems.

SCCs are also highly concentrated on the anterior dorsal fins of rocklings (Gadidae), but the role of this chemosensory organ does not appear to be linked to food searching and feeding behaviour. The rockling has a series of epidermal taste buds on the paired fins, and the responses recorded from the nerve fibres innervating these fins clearly implicate these taste buds in food search and feeding behaviour. The SCCs of the rockling are relatively irresponsive to classical taste stimuli, such as amino acids and chemical extracts of prey animals. They are, however, stimulated by mucus extracts taken from heterospecific fish, but are not responsive to conspecific mucus or tissue extracts. In the rocklings, therefore, the SCCs may have a role to play in the chemical detection of potential predators.

2.14   CONCLUDING COMMENTS

In the majority of terrestrial vertebrates vision tends to be the dominant sense, but non-visual senses may be better suited to gaining information about the surroundings in low-light environments. The problems of underwater vision are related to the low light intensities, and to the shifts in spectral bandwidth that occur with increasing depth, in aquatic habitats. Nevertheless, many fish species appear to rely on vision as the main sense.

The non-visual senses include mechanoreception, electroreception and chemoreception. All of these alternative sensory systems are used by fish, to a greater or lesser extent, in order to gain information about the environment.

Sound has much to recommend it as a form of sensory stimulus in aquatic environments. Firstly, there is high-speed transmission coupled with non-persistence, and secondly, both of the stimulatory components – pressure and particle displacement – may be used by the fish to gain information. Close to the sound source, i.e. within the near-field, particle displacement is most important, but further from the source pressure is probably the main component.

Within the near-field, sound detection is a function of the direct stimulation of the mechanoreceptors of the inner ear. On the other hand, effective detection of the pressure stimulus within the far-field is probably dependent upon possession of a gas bladder that transduces sound pressure into a secondary near-field. Whatever the system used, however, there is ultimate reliance on the detection of particle displacement by mechanoreceptors, these being based upon the sensory hair cells grouped to form the basic detection unit, the neuromast.

Whilst electroreceptor systems are extremely sensitive they have the disadvantage of being short-range systems with effective receptive distances ranging from a few to tens of centimetres. Nevertheless they can supply information not provided by other senses, and electroreception is particularly useful when other sensory modalities are rendered ineffective. For example, the ability to detect the bioelectric fields generated by potential prey allows prey to be located both in the dark and when it is hidden from view by being buried in the substrate.

The weakly electric fish have the ability to both produce and detect electrical stimuli, and they are, therefore, able to locate the positions of objects in turbid waters. Thus these fish can remain active under conditions where other systems, such as vision, are virtually useless.

Chemical stimuli play a central role in governing most aspects of fish behaviour, including feeding, reproduction and the recognition of conspecifics. Chemicals may be detected by receptors of various types, and chemoreceptors are to be found at numerous locations on the fish body. The major division between the chemosensory systems is, however, that between olfaction and gustation, the senses of smell and taste, respectively. Fish are also known to possess arrays of solitary chemosensory cells (SCCs), and a broad chemoreceptor system given the name the 'common chemical' system.

There is consensus that the sensitivities of the gustatory, SCC and 'common chemical' systems are much lower than those of the olfactory system. For example, the gustatory threshold for certain salts and sugars may be 10^{-5}M, whereas olfactory thresholds for a range of compounds may be 10^{-8}M. Many amino acids can be detected by the olfactory system at concentrations of 10^{-8}M, and detection thresholds for other chemical substances can be as low as 10^{-12} to 10^{-15}M. There are, however, considerable interspecific differences in sensory thresholds, and, following detection, any given chemical substance may elicit a different set of behavioural responses from different fish species.

FURTHER READING

Books

Atema, J., Fay, R.R., Popper, A.N. and Tavolga, W.N. (eds) (1988) *Sensory Biology of Aquatic Animals*, Springer-Verlag, New York.

Bolis, L., Keynes, R.D. and Maddrell, S.H.P. (eds) (1984) *Comparative Physiology of Sensory Systems*, Cambridge Univ. Press, Cambridge.

Coombs, S., Gorner, P. and Munz, H. (eds) (1989) *The Mechanosensory Lateral Line – Neurobiology and Evolution*, Springer-Verlag, New York.

Douglas, R.H. and Djamgoz, M.B.A. (eds) (1990) *The Visual System of Fish*, Chapman & Hall, London.

Hara, T.J. (ed) (1992) *Fish Chemoreception*, Chapman & Hall, London.

Herring, P.J., Campbell, A.K., Whitfield, M. and Maddock, L. (eds) (1990) *Light and Life in the Sea*, Cambridge Univ. Press, Cambridge.

Hoar, W.S. and Randall, D.J. (eds) (1971) *Fish Physiology, Vol V*, Academic Press, London.
Pitcher, T.J. (ed.) (1993) *The Behaviour of Teleost Fishes*, 2nd edn, Chapman & Hall, London.

Review articles and papers

Blaxter, J.H.S. (1987) Structure and development of the lateral line. *Biol. Rev.*, **62**, 471–514.
Crescitelli, F. (1991) The scotopic photoreceptors and their visual pigments of fishes: Functions and adaptations. *Vision Res.*, **31**, 339–48.
Dulka, J.G. (1993) Sex pheromone systems in goldfish: Comparisons to vomeronasal systems in tetrapods. *Brain Behav. Evol.*, **42**, 265–80.
Hara, T.J. (1994) The diversity of chemical stimulation in fish olfaction and gustation. *Rev. Fish Biol. Fisheries*, **4**, 1–35.
Hawryshyn, C.W. (1992) Polarization vision in fish. *Amer. Scient.*, **80**, 164–75.
Jacobs, G.H. (1992) Ultraviolet vision in vertebrates. *Amer. Zool.*, **32**, 544–54.
Kalmijn, A.J. (1982) Electric and magnetic field detection in elasmobranch fishes. *Sci. Amer.*, **218**, 916–18.
Kotrschal, K. (1991) Solitary chemosensory cells – taste, common chemical sense or what? *Rev. Fish Biol. Fisheries*, **1**, 3–22.
Popper, A.N. and Coombs, S. (1980) Auditory mechanisms in teleost fishes. *Amer. Scient.*, **68**, 429–40.
Popper, A.N. and Fay, R.R. (1993) Sound detection and processing by fish: Critical review and major research questions. *Brain Behav. Evol.*, **41**, 14–38.

Chapter three

Foodstuffs and feeding

3.1 INTRODUCTION

The performance of any activity, such as swimming, requires the expenditure of energy, and the fuelling of energy production requires the ingestion of food. The energy sources in the food are organic materials, consisting of carbohydrates, lipids and proteins. These organic molecules consist mostly of carbon, hydrogen and oxygen in varying proportions with minor quantities of other elements such as nitrogen, sulphur and phosphorus. In addition to acting as sources of energy for the performance of work, the organic molecules obtained from food also provide the building blocks for making new body tissues, and act as precursors for the synthesis of biologically active molecules such as enzymes and endocrine factors.

Potential foodstuffs come in many shapes and sizes, and different fish species have become adapted to utilize many different food types ranging from other fish species, to water plants, and even hippopotamus dung. These foodstuffs differ widely in biochemical composition, and in the amounts that must be ingested and processed in order for the fish to obtain a given amount of any specific nutrient (Table 3.1). Thus, fish that feed on detritus may have to process 200 g or more of sedimented material in order to obtain 1 g protein, whereas the consumption of 7–8 g fish prey may supply this amount of protein. With the wide diversity of foods eaten by fish it should come as no surprise that the organs involved in the feeding and digestive processes show considerable variation.

As a generalization fish can be broadly grouped into the following feeding categories:

1. carnivores – eat animal food:
   (a) piscivores – fish eaters;
   (b) benthophages – eat animals living in or on sediments;
   (c) zooplanktivores – eat planktonic animals;

Table 3.1 Summary of the compositions of a number of prey types consumed by fish and an indication of the amounts required to be processed in order for fish to obtain a given amount of nutrient

Prey type	Percentage composition					Energy content (kJ g^{-1})	Weight required per	
	Water	Ash	Protein	Lipid	Carbohydrate		g protein	kJ energy
Sediment + detritus	~98		~0.5	~0.2	1.3	~0.4	~200	~2.5
Algae								
Phytoplankton	~92	~0.8	~2.5	~0.3	~4.4	~1	~40	~1
Macroalgae	~92	~2.5	~0.8	~0.3	~4.4	~0.9	~125	~1.1
Aquatic macrophytes	~94	~0.7	~1.5	~0.8	~3	~1	~65	~1
Oligochaete worms	~85	~2	~11	~1	~1	~3	~9	~0.35
Molluscs (soft parts)	80–85	1.5–3	~10	~2	~2	~4	~10	~0.25
Crustaceans								
Zooplankton	~80	~2.5	~13	~3.5	~1	~4.5	~7.6	~0.2
Crabs	~74	~8	~13	~3	~1.5	~4.5	~7.6	~0.2
Insect larvae	75–85	1–2	10–16	1–5	~2	3–6	~7.6	~0.2
Fish								
Lipid-rich (e.g. clupeids)	~72	~2.5	~14.5	~10	Trace	~7.5	~7	~0.13
Lipid-poor (e.g. gadoids)	~77	~2.5	~17	~3.5	Trace	~5.5	~6	~0.18

    (d) epifauna eaters – scrape or bite prey from stones or rocks;

    (e) parasites;

2. omnivores – eat both plant and animal food;

3. herbivores – eat plant food, either large algae and water plants, or phytoplankton and microalgae;

4. detritivores – eat detritus (sedimented organic matter, bacteria, fungi and encrusting microalgae).

3.2 NUTRIENT CLASSES

Carbohydrates, lipids and proteins make up the major part of the dry weight of the ingested food. These macronutrients can be used directly as fuels (respiratory substrates) or can be stored within the body for utilization at a later date. In addition to the macronutrients, the food will also contain a range of micronutrients. These micronutrients are the vitamins and minerals, which are required to be consumed in small doses if the fish is to survive and grow well.

Carbohydrates

Carbohydrates were originally named so because it was believed that they were hydrates of carbon, that is, $C_x(H_2O)_y$ but it soon became clear that other compounds not showing the 2:1 ratio of hydrogen to oxygen had many of the characteristics and chemical properties assigned to the carbohydrates. For example, deoxyribose ($C_5H_{10}O_4$) has the chemical properties typical of

Table 3.2 Simplified scheme illustrating the classification of the carbohydrates

Monosaccharides	Oligosaccharides	Polysaccharides

Trioses $C_3H_6O_3$
   Glyceraldehyde
   Dihydroxyacetone
 Hexoses
 $C_6H_{12}O_6$
   Glucose
   Fructose
	Disaccharides
	 e.g. Sucrose
	   Lactose
	   Cellobiose
	 Tetrasaccharides
	 e.g. Stachyose
		– Homopolysaccharides
		   (single type of
		   monomeric unit)
		 – Heteropolysaccharides
		   (two or more types of
		   monomeric unit)
		 Storage – Starch
		           – Glycogen
		           – Laminarin
		 Structural – Cellulose
		           – Chitin
		           – Pectins
		           – Gums

the carbohydrates, as have a number of compounds that contain small proportions of nitrogen and sulphur in addition to carbon, hydrogen and oxygen. The carbohydrates are generally grouped into categories depending upon the size of the molecule – monosaccharides, oligosaccharides and polysaccharides (Table 3.2).

Monosaccharides

Monosaccharides are simple sugars that cannot be hydrolysed into smaller units, and the monosaccharides are usually classified according to the number of carbon atoms they possess e.g. trioses $(C_3H_6O_3)$ have three carbon atoms, tetroses have four carbon atoms, pentoses have five and hexoses have six. The monosaccharides are also classified according to their molecular structure. Thus, both glyceraldehyde and dihydroxyacetone are trioses, but they differ in molecular structure. Glyceraldehyde is an aldose, possessing an

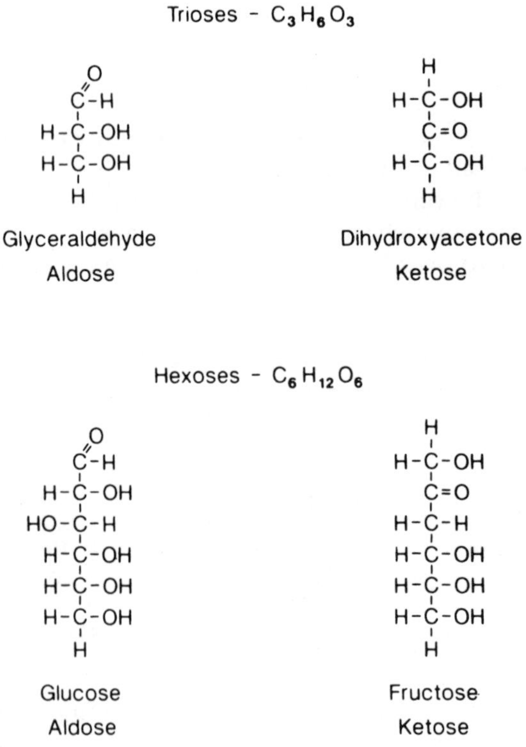

Trioses – $C_3H_6O_3$

```
      O                        H
      ‖                        |
      C-H                   H-C-OH
      |                        |
   H-C-OH                    C=O
      |                        |
   H-C-OH                   H-C-OH
      |                        |
      H                        H

 Glyceraldehyde            Dihydroxyacetone
    Aldose                     Ketose
```

Hexoses – $C_6H_{12}O_6$

```
      O                        H
      ‖                        |
      C-H                   H-C-OH
      |                        |
   H-C-OH                    C=O
      |                        |
  HO-C-H                    H-C-H
      |                        |
   H-C-OH                   H-C-OH
      |                        |
   H-C-OH                   H-C-OH
      |                        |
   H-C-OH                   H-C-OH
      |                        |
      H                        H

   Glucose                   Fructose
    Aldose                    Ketose
```

Fig. 3.1 Structure of triose and hexose monosaccharides indicating the differences between an aldose and a ketose.

aldehyde group (−CHO), and dihydroxyacetone is a ketose, with a keto group (−CO) (Fig. 3.1). The monosaccharide molecule has either an active aldehyde or a keto group and, consequently, these molecules act as reducing substances. They can be oxidized to produce various acids. In the case of glucose, glucuronic acid may be produced on oxidation. The most important sugar acids are the uronic acids, especially glucuronic and galacturonic acids, which are components of large heteropolysaccharide molecules.

An important property of the monosaccharides is their reaction with phosphoric acid. A number of sugar phosphates occur naturally in both plants and animals. The hexose phosphates play a very important role in cell metabolism, with two important compounds being glucose-6-phosphate and glucose-1-phosphate.

If the hydroxyl (−OH) group on carbon atom 2 of an aldohexose is replaced by an amino group (−NH$_2$), the resulting compound is an amino sugar. Examples of amino sugars are glucosamine and galactosamine, both of which occur as components of glycoproteins.

Oligosaccharides and polysaccharides

Oligosaccharides are compound sugars that yield from two to six molecules of simple sugars on hydrolysis, and the term 'polysaccharide' refers to a compound yielding a large number of monosaccharide molecules on hydrolysis. Polysaccharides are often also called glycans, and can be further classified as being either homopolysaccharides or heteropolysaccharides. Homopolysaccharides are condensation polymers of a single type of sugar, with, for example, hydrolysis of glucans yielding glucose. Heteropolysaccharides, on the other hand, are mixed polysaccharides which, on hydrolysis, yield mixtures of monosaccharides and derived products.

In nature, the carbohydrates usually perform either a structural or an energy storage function, and they are usually present in biological materials in the form of long-chain polysaccharides. For example, cellulose is an important constituent of plant cell walls, and the exoskeletons of insects and crustaceans contain varying proportions of chitin as a strengthening agent. The most important storage product of plants is starch whereas in animals the storage carbohydrate is glycogen.

Both starch and cellulose consist of long chains of glucose (a hexose monosaccharide) molecules, the main difference between the two polysaccharides being the nature of the linkages between adjacent glucose molecules, α linkages in starch and β linkages in cellulose (Fig. 3.2). This difference gives rise to differences in physical and chemical properties between cellulose and starch, and also affects the ease with which the two carbohydrates can be digested and absorbed by animals. Two forms of starch may be found in the storage granules of plants − amylose and amylopectin.

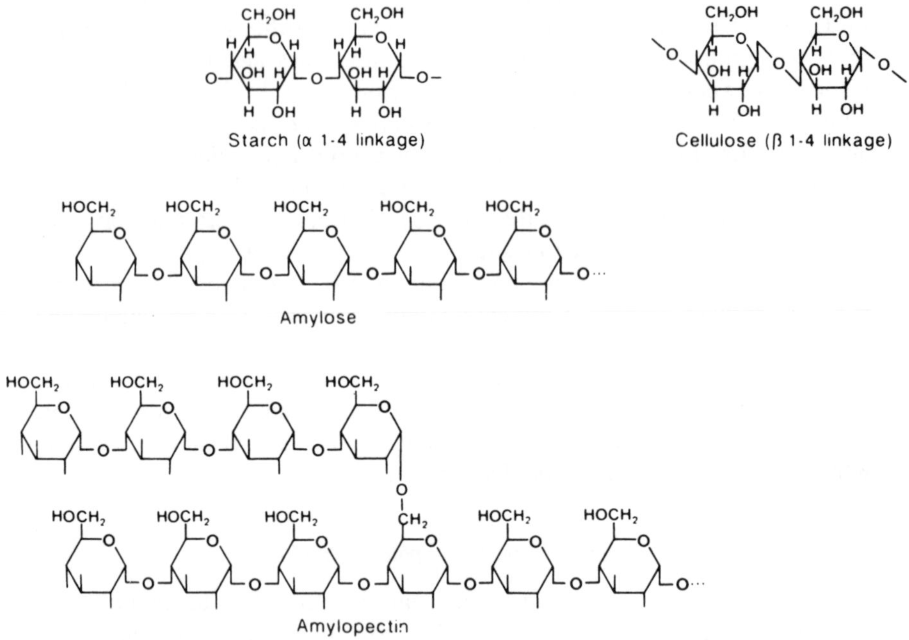

Fig. 3.2 Chemical structures of starch and cellulose, two homopolysaccharides containing glucose as the monomeric unit. Note the different glycosidic linkages between the glucose monomers in the starch and the cellulose.

Amylose is a straight-chained molecule, in which the glucose units are glycosidically linked through carbon atoms at positions 1 and 4 (1,4 linkage), whereas amylopectin is a more complex molecule with side branches. In amylopectin there are both 1,4 and 1,6 glycosidic linkages. The relative proportions of the two molecules in the starch granules differ between plant species and this imparts different properties to the starches of the plants.

Carbohydrates are usually present in much lower proportions in animals than in plants, with the storage carbohydrate, glycogen, being found in the liver and muscle tissues. Glycogen is similar in structure to amylopectin, with the major difference between the two molecules being that glycogen has relatively large numbers of 1,6 linkages. Amylopectin has only one of these linkages for every 30 or so 1,4 linkages.

The fourth common polysaccharide, chitin, differs from cellulose, starch and glycogen in the nature of its monosaccharide subunits. Unlike the other polysaccharides, which have glucose as the basic unit, chitin is composed of long chains of an amino-sugar, N-acetyl glucosamine. This means that the chitin molecule contains atoms of nitrogen in addition to carbon, hydrogen and oxygen. Structurally, chitin is a close relative of cellulose; the hydroxyl (−OH) group at position 2 in the glucose residues of cellulose is replaced by an N-acetylamino (−NHCOCH$_3$) group in the N-acetyl glucosamine units that make up chitin.

Whilst cellulose, starch, glycogen and chitin are the most widespread of the polysaccharides, there are many other glycans that occur in nature. The algae, for example, contain a wide range of glycans as components of their cell walls, and may also have carbohydrates other than starch as the main storage product. For example, alginic acid may be an important component of the cell walls of algal species and can constitute up to 40% of the dry weight of the brown seaweeds (Phaeophyceae). Alginic acid is a linear co-polymer of two uronic acids, mannuronic acid and guluronic acid, linked via 1,4 glycosidic linkages. The relative proportions of the two acids making up the polymer vary between species, and this imparts different properties to the alginic acids extracted from the different brown algae.

The red algae (Rhodophyceae), on the other hand, contain agar and carrageenans as structural elements in their cell walls. These poly-saccharides represent a family of galactans, consisting of a spectrum of molecules having similar, but slightly varying, chemical structure. The molecules are made up of repeating sequences of 1,3 and 1,4 linked units of sulphated or methylated sugar residues. All of these algal polysaccharides are soluble in warm water and, on cooling, they may form firm gels. These colloidal and gelling properties are widely exploited in an industrial context, with algal polysaccharides being used as thickening or gelling agents in the food, pharmaceutical, cosmetic and textile industries.

The storage reserves of algae consist of a range of oils and carbohydrates, the exact composition of the reserves varying between the different phyla. The green algae (Chlorophyceae), for example, tend to store their energy reserves in the form of starch, predominantly amylose, whereas the red algae (Rhodophyceae) accumulate a peculiar form of solid carbohydrate, floridean starch, built up of glucose residues. The brown algae (Phaeophyceae) store their reserves as soluble carbohydrates, such as laminarin, and at least one alcohol, mannitol. Laminarin is a polymer of glucose subunits linked via β 1,3 and β 1,6 glycosidic linkages. A polysaccharide having similar chemical structure, chrysolaminarin, is also found in many species of diatoms.

Lipids

The lipids are a rather heterogeneous class of water-insoluble organic compounds. Biochemical separation techniques yield fractions of lipids that are distinguishable both on the basis of their abilities to dissolve in different solvents, and by the possession of differences in physical and chemical

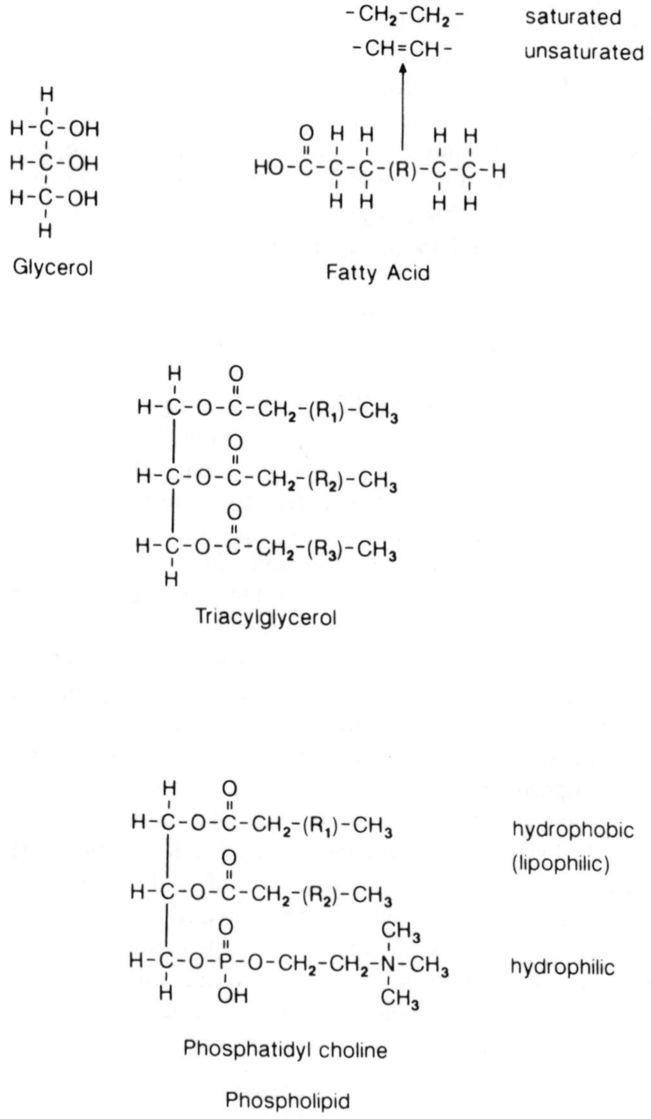

Fig. 3.3 Chemical structures of glycerol, fatty acids, triacylglycerols and phospholipids. (R) denotes the carbon chain of the fatty acid, which may or may not contain double bonds between the carbon atoms. The fatty acids esterified with the glycerol in the triacylglycerols and phospholipids may either be of the same, or of different, configurations.

properties. The class of lipids most frequently encountered in nature is the acylglycerols. The acylglycerols represent the major storage lipids of both plants and animals, and they form the major part of the neutral lipid group. The neutral lipids are formed by esterification of fatty acids with the alcohol, glycerol (Fig. 3.3). Whilst triacylglycerols make up the great bulk of neutral lipids found in nature, mono- and diacylglycerols are also found.

The wax esters are a group of compounds closely related to the acylglycerols. Wax esters, which are formed by the esterification of fatty acids with long-chain alcohols, are found as waterproofing agents on the outer surfaces of fruits and leaves of plants, and on the exoskeleton of insects. Wax esters also appear to be the main lipid storage reserves in some zooplanktonic crustaceans and, hence, may provide an important energy source for a number of fish species.

Fatty acids

Fatty acids differ in the numbers of carbon atoms in the molecule, and in the number and positioning of the double bonds between the carbon atoms. The fatty acids form distinct series, and a shorthand system has been devised for classifying these fatty acid series. For example, the shorthand formula 16:0 represents a fatty acid with 16 carbon atoms and no double bonds, a fatty acid lacking double bonds between the carbons being known as a *saturated* fatty acid. The shorthand formula $18:3(n-3)$ represents an 18 carbon fatty acid with three double bonds, the $(n-3)$ denoting that the first double bond is found in the link between the third and fourth carbon atoms from the methyl end (Table 3.3). Fatty acids with one or more double bonds are *unsaturated* fatty acids. Those having two to four double bonds in the molecule are often termed polyunsaturated fatty acids (PUFAs), and those with more than four double bonds may be described as being highly unsaturated fatty acids (HUFAs). These terms are used to indicate increasing degrees of unsaturation that can occur amongst the fatty acids.

Fatty acids found in terrestrial plants and animals generally have relatively low degrees of saturation and have carbon chain lengths of 14–18, but longer-chain fatty acids, with up to 22 carbon atoms, are commonly encountered in aquatic organisms. The fatty acid components of the wax esters tend to be relatively short-chained saturated fatty acids (14:0 or 16:0) and the alcohols are generally saturated (16:0) or have only one double bond in their carbon atom chains (20:1 or 22:1).

Both the chain length and the number of double bonds (degree of unsaturation) determine the physical and chemical properties of the fatty acids (Table 3.3) and the lipids into which they are incorporated. For example, triacylglycerols, such as pork fat and beef tallow, that contain saturated fatty acids are solid at room temperature. On the other hand, the triacylglycerols

Table 3.3 Examples of fatty acids illustrating the system of classification, and the effects of chain length and degree of unsaturation on melting points

Shorthand formula	Structure	Melting point (°C)
Saturated fatty acids		
16:0	$CH_3(CH_2)_{14}COOH$	63
20:0	$CH_3(CH_2)_{18}COOH$	76.5
24:0	$CH_3(CH_2)_{22}COOH$	86
Unsaturated fatty acids		
18:1 (n–9)	$CH_3(CH_2)_7CH = CH(CH_2)_7COOH$	13.5
18:2 (n–6)	$CH_3(CH_2)_4CH = CHCH_2CH = CH(CH_2)_7COOH$	–5
18:3 (n–3)	$CH_3CH_2CH = CHCH_2CH = CHCH_2CH = CH(CH_2)_7COOH$	–11
20:4 (n–6)	$CH_3(CH_2)_4CH = CHCH_2CH = CHCH_2CH = CHCH_2CH = CH(CH_2)_3COOH$	–49.5

Table 3.4 Examples of fatty acid compositions of fish species from marine (capelin, *Mallotus villosus* and elver of American eel, *Anguilla rostrata*) and freshwater (American eel and perch, *Perca fluviatilis*) environments. Values are presented as percentages of the total fatty acids present. American elvers were sampled whilst entering fresh water from the sesa. Note the relatively high level of the (n–6) series fatty acids in the freshwater fish species, leading to these fish having much lower ratios of (n–3)/(n–6) than the marine species

Fatty acid	Perch	Eel	Elver	Capelin
18:2 (n–6)	3.5	8.9	1.3	1.4
20:4 (n–6)	11.7	4.4	2.9	Trace
20:5 (n–3)	13.4	2.9	5.0	13.8
22:6 (n–3)	25.0	5.4	18.8	11.3
Saturates	24.9	23.3	29.6	28.7
Monoenes	14.4	40.0	30.3	38.8
PUFA	57.7	30.0	36.4	27.9
(n–3)	45.5	13.6	29.1	26.5
(n–6)	15.2	16.4	7.3	1.4
(n–3)/(n–6)	3.0	0.8	4.0	18.9

derived from the seeds of a number of plant species, e.g. soya, sunflower, peanut and rapeseed, contain quite high levels of unsaturated fatty acids and are fluid oils at room temperature. Marine fish oils, which usually contain relatively high proportions of PUFAs and HUFAs, have low melting points and are fluid at room temperature.

Terrestrial and aquatic lipids differ not only in the degrees of saturation of the fatty acids found in the triacylglycerols, but there are also differences in the unsaturated fatty acids typical of food chains in these environments. The (n − 6) series fatty acids are typically found in terrestrial and freshwater environments, e.g. plant seed oils are typically rich in 18:2(n − 6), whilst in marine ecosystems fatty acids of the (n − 3) series tend to dominate. The food chain in fresh water is characterized by 18:2(n − 6), 18:3(n − 3) and 20:5(n − 3) fatty acids. On the other hand, the fatty acid profiles of marine phytoplankton and zooplankton tend to be dominated by 18:3(n − 3), 20:5(n − 3) and 22:6(n − 3) fatty acids, and this pattern is generally reflected at higher levels in the food chain. Thus, when lipids are extracted from freshwater and marine fish species, and the fatty acid profiles compared, there will usually be found to be large differences (Table 3.4). These differences are, primarily, due to the differences in the fatty acid compositions of the prey present in the two environments.

Phospholipids

The vast majority of the lipid extracted from a fish or prey species will be made up of neutral, storage triacylglycerols, but the total lipids will also contain smaller quantities of other lipid classes. The phospholipids are important components of the cell membranes, where they form lipoprotein complexes. The majority of the phospholipids contain nitrogen and phosphorus in addition to carbon, hydrogen and oxygen.

Phospholipids are esters of glycerol, in which two of the alcohol groups are esterified with fatty acids. The third alcohol group is esterified with phosphoric acid, which, in turn, is esterified by a nitrogenous base, the amino acid serine, the sugar alcohol inositol or by glycerol phosphate. The nitrogenous bases may be choline or cholamine (amino ethyl alcohol). The nature of the base or esterification compound provides the basis of the name of the specific families of phospholipids, e.g. phosphatidylcholine, phosphatidylethanolamine, phosphatidylserine and phosphatidylinositol.

The phospholipids combine within the same molecule both the hydrophilic (water-loving) phosphate and nitrogen-containing groups, and the hydrophobic fatty acid residues (Fig. 3.3). These molecules are, therefore, surface active and they play an important role as emulsifying agents in biological systems. Their surface-active nature also explains their function as part of various biological membranes. The major membrane phospholipids are phosphatidylcholine and phosphatidylethanolamine. Phosphatidylinositol, on the other hand, appears to have a number of important roles in the transduction of hormonal signals through biomembranes.

The sphingomyelins are not esters of glycerol, but are made up of fatty acids, phosphoric acids, choline and sphingosine. Since they contain phosphorus, and are closely associated with the phospholipids both in occurrence and in biological function, the sphingomyelins are usually included within the general class phospholipids.

The fatty acid profiles of the phospholipids tend to be quite similar, irrespective of whether a fish comes from a marine or a freshwater environment. The long-chained HUFAs, particularly $20:5(n-3)$ and $22:6(n-3)$, are present in large amounts in the major membrane phospholipids. In phosphatidylinositol, on the other hand, the $(n-6)$ series fatty acid $20:4(n-6)$ is present in large amounts. PUFAs and HUFAs usually account for over half of the phospholipid fatty acids, whereas the triacylglycerols are usually dominated by saturates and monounsaturates.

Proteins and amino acids

Proteins are large organic molecules that contain carbon, hydrogen, oxygen, nitrogen and often sulphur. The elementary composition of most proteins is

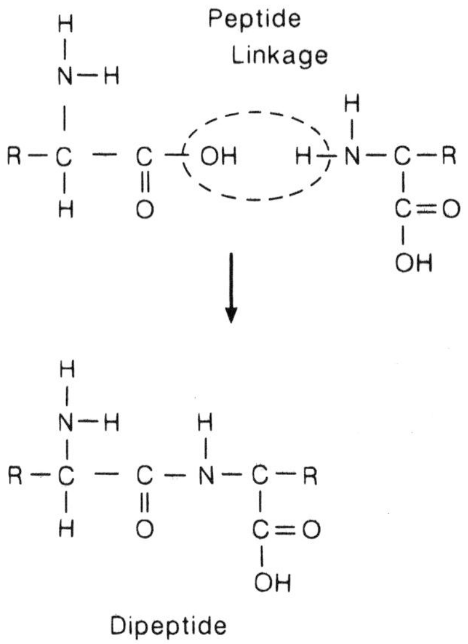

Fig. 3.4 Formation of a peptide linkage between two amino acids, giving rise to a dipeptide. R denotes the side chain of the amino acid.

very similar, approximate percentages being C $= 50–55\%$, H $= 6–8\%$, O $= 20–23\%$, N $= 15–18\%$, S $= 0–4\%$. The fundamental structural unit of the protein molecule is the amino acid. There are 20 naturally occurring amino acids that are incorporated into proteins. These amino acids are joined by peptide bonds to give long repeating units making up the protein molecule (Fig. 3.4).

All of the amino acids have the same basic design, with a carboxyl (–COOH) and an amino (–NH$_2$) group attached to the α carbon atom. The remainder of the amino acid molecule (denoted by R in the basic structural formula shown in Fig. 3.5) can have different configurations. The amino acids can be divided into a number of series depending upon the nature of the R-group:

1. aliphatic series (glycine, alanine, serine, threonine, valine, leucine, iso-leucine);
2. aromatic series (phenylalanine, tyrosine);
3. sulphur-amino acid series (cysteine, cystine, methionine);
4. heterocyclic series (tryptophan, proline, hydroxyproline);
5. acidic series (aspartic acid, glutamic acid);
6. basic series (arginine, histidine, lysine).

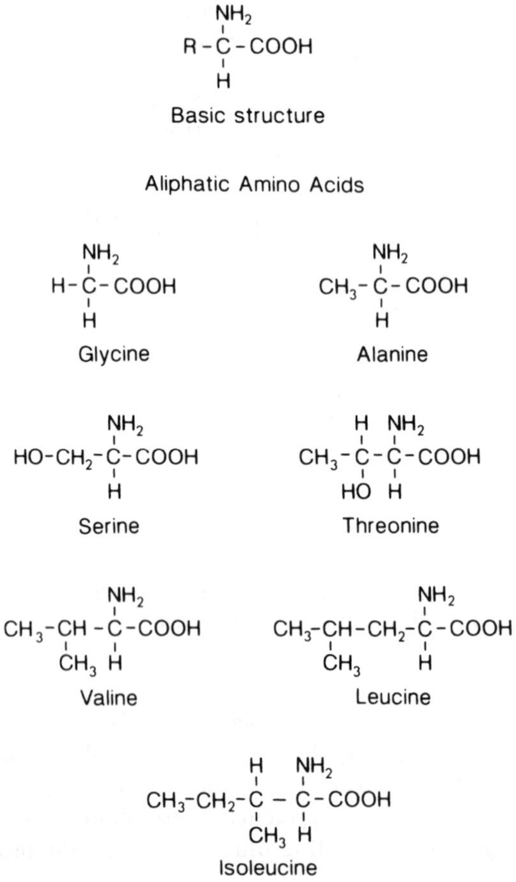

$$NH_2$$
$$R - \overset{|}{\underset{|}{C}} - COOH$$
$$H$$

Basic structure

Aliphatic Amino Acids

$$NH_2$$
$$H - \overset{|}{\underset{|}{C}} - COOH$$
$$H$$

Glycine

$$NH_2$$
$$CH_3 - \overset{|}{\underset{|}{C}} - COOH$$
$$H$$

Alanine

$$NH_2$$
$$HO - CH_2 - \overset{|}{\underset{|}{C}} - COOH$$
$$H$$

Serine

$$H \quad NH_2$$
$$CH_3 - \overset{|}{\underset{|}{C}} - \overset{|}{\underset{|}{C}} - COOH$$
$$HO \quad H$$

Threonine

$$NH_2$$
$$CH_3 - CH - \overset{|}{\underset{|}{C}} - COOH$$
$$\underset{|}{CH_3} \quad H$$

Valine

$$NH_2$$
$$CH_3 - CH - CH_2 - \overset{|}{\underset{|}{C}} - COOH$$
$$\underset{|}{CH_3} \qquad H$$

Leucine

$$H \qquad NH_2$$
$$CH_3 - CH_2 - \overset{|}{\underset{|}{C}} - \overset{|}{\underset{|}{C}} - COOH$$
$$CH_3 \quad H$$

Isoleucine

Fig. 3.5 Chemical structures of the amino acids making up the various series. R in the basic structural formula represents the side chain.

From a nutritional point of view the amino acids can be divided into non-essential and essential groups. Fish and other animals are able to synthesize and interconvert some of the amino acids, but are incapable of the synthesis of others. The essential amino acids are, therefore, those that the fish cannot synthesize *de novo*, whereas the non-essential amino acids are those that can be synthesized from a range of precursor molecules. Studies carried out on a range of fish species, and other animals, have revealed that ten of the amino acids are essential – arginine, histidine, isoleucine, leucine, lysine, methionine, phenylalanine, threonine, tryptophan and valine – and these must, therefore, be obtained via the diet. The amino acid tyrosine

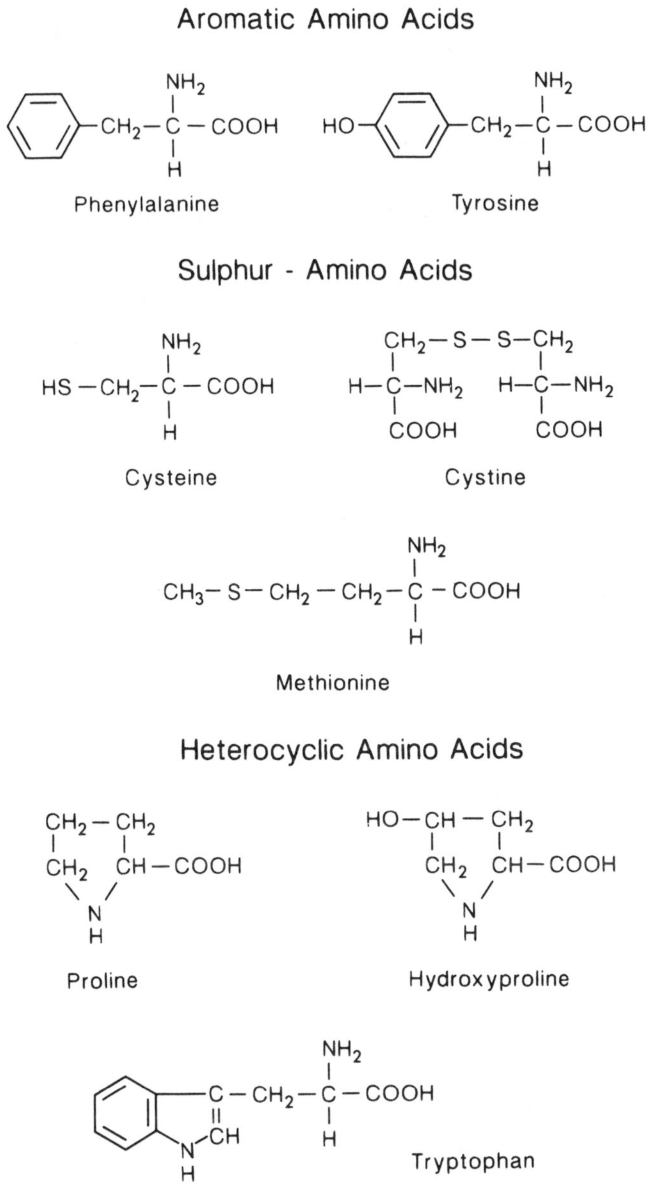

Aromatic Amino Acids

Phenylalanine   Tyrosine

Sulphur - Amino Acids

Cysteine   Cystine

Methionine

Heterocyclic Amino Acids

Proline   Hydroxyproline

Tryptophan

Fig. 3.5 (*cont.*)

Acidic Series Amino Acids

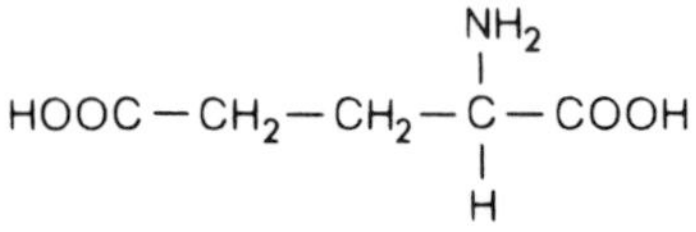

$$HOOC-CH_2-CH_2-\overset{\overset{\displaystyle NH_2}{|}}{\underset{\underset{\displaystyle H}{|}}{C}}-COOH \qquad HOOC-CH_2-\overset{\overset{\displaystyle NH_2}{|}}{\underset{\underset{\displaystyle H}{|}}{C}}-COOH$$

Glutamic Acid 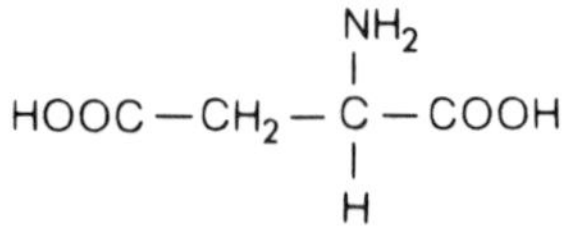   Aspartic Acid

Basic Series Amino Acids

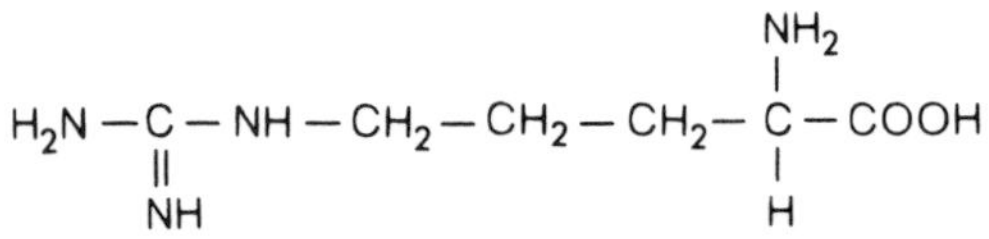

$$H_2N-\overset{\overset{\displaystyle ||}{C}}{\underset{\underset{\displaystyle NH}{}}{}}-NH-CH_2-CH_2-CH_2-\overset{\overset{\displaystyle NH_2}{|}}{\underset{\underset{\displaystyle H}{|}}{C}}-COOH$$

Arginine

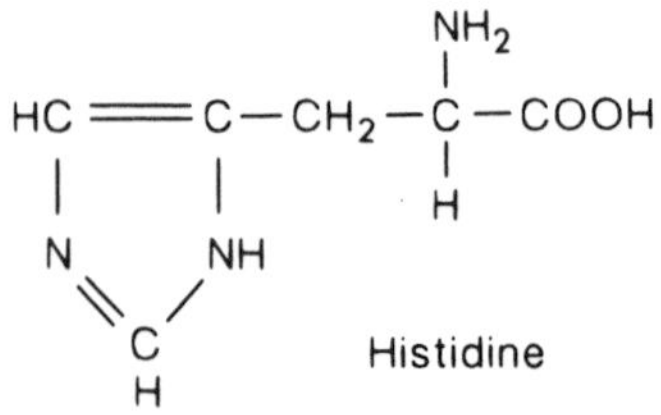

$$HC=\overset{}{C}-CH_2-\overset{\overset{\displaystyle NH_2}{|}}{\underset{\underset{\displaystyle H}{|}}{C}}-COOH$$

Histidine

$$H_2N-CH_2-CH_2-CH_2-CH_2-\overset{\overset{\displaystyle NH_2}{|}}{\underset{\underset{\displaystyle H}{|}}{C}}-COOH$$

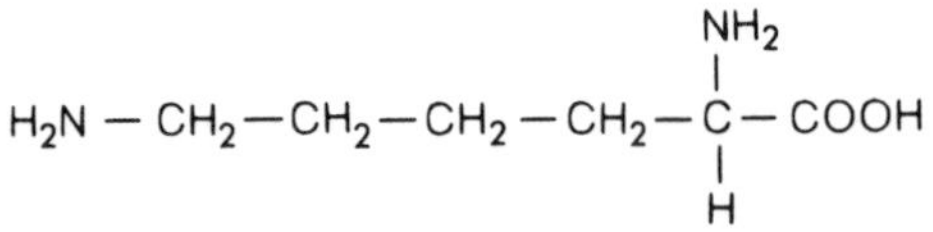

Lysine

Fig. 3.5 (*cont.*)

may be an essential dietary component for some exclusively carnivorous species.

The amino acids can be linked together in chains of different length, and also with differing orders of the amino acids in the chains. Thus it is easy to see that with 20 amino acids available the possible number of combinations is almost limitless. As a consequence, the proteins are extremely hetero-geneous, and they can have markedly different physical properties. The proteins play a range of functional roles within biological materials. They may, for example, act as structural elements, endocrine factors or enzymes.

3.3 ENERGY METABOLISM

Compounds from the three main groups of nutrients (carbohydrates, lipids and proteins) are all capable of being used as fuels in respiratory metabolism. When they are combusted in the presence of oxygen they will all yield carbon dioxide and water, and there will be production of heat. For example, the combustion of glucose can be described by the equation:

$$C_6H_{12}O_6 + 6O_2 = 6CO_2 + 6H_2O \qquad (3.1)$$

with the combustion of 180 g glucose leading to the production of approximately 2800 kJ as waste heat.

The heat produced by the combustion of the different metabolic fuels depends upon their chemical composition. The heat produced by the combustion of carbohydrates is approximately 17 kJ g^{-1} substrate combusted. In the combustion of a saturated fatty acid, such as palmitic acid (16:0):

$$CH_3(CH_2)_{14}COOH + 23O_2 = 16CO_2 + 16H_2O \qquad (3.2)$$

there will be the production of approximately 10 000 kJ as heat, or 39 kJ g^{-1} (10 000/256) palmitic acid metabolized. Combustion of a typical protein leads to a heat production of about 24 kJ g^{-1} protein metabolized.

This combustion produces waste heat, which cannot be utilized by the fish, and what is required is a system by which this heat energy can be chemically bound and harnessed to perform work of various kinds. The way in which the energy is stored is in the form of high-energy phosphate bonds, especially those of the molecule adenosine triphosphate (ATP). The biochemical pathways involved in the production of ATP from adenosine diphosphate (ADP) and inorganic phosphate are summarized in Fig. 3.6. It can be seen that the organic components derived from the different nutrients (carbohydrates, lipids, proteins) enter the citric acid cycle at different positions. These different points of entry will mean that there will be different ATP yields from the different substrates.

The reactions of the cycle can be summarized by considering the fate of glucose as respiratory substrate. The earliest part of the metabolic processing of glucose is termed *glycolysis*, and this results in the production of two molecules of pyruvic acid from each molecule of glucose. At this stage, two molecules of ATP have been produced, and in the absence of oxygen (anaerobic conditions) the pyruvic acid is converted into lactic acid (Fig. 3.6).

Under aerobic conditions the pyruvic acid enters the citric acid cycle where it is further metabolized (Fig. 3.6). The metabolic reactions involving the production of ATP via the citric acid cycle involve *oxidative phosphorylation*. The reactions of the aerobic metabolic pathway, the citric acid cycle and respiratory chain, lead to the production of an additional 34 ATP molecules. Thus, the aerobic metabolism of glucose leads to a total production of 36

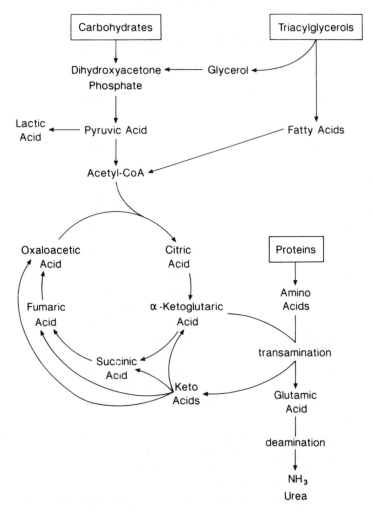

Fig. 3.6 Simplified scheme showing the points of entry of the different respiratory substrates (carbohydrates, lipids and proteins) into the citric acid cycle.

ATP molecules – 2 ATP via glycolysis, plus 34 ATP produced aerobically. This stresses the importance of the citric acid cycle, and the role of oxygen, in the energy storage processes. In the absence of oxygen, under conditions of anaerobic glycolysis the ATP yield is only 2 molecules of ATP per molecule of glucose metabolized, whereas under aerobic conditions the yield is 36 ATP molecules, and considerably more of the energy available from the glucose is harnessed in a form that can be used to produce useful work.

Both lipids and amino acids can enter biochemical pathways that lead to the production of glucose or glycogen (gluco- or glyconeogenesis), so they may enter the respiratory metabolic cycle via indirect routes. It is, however, more usual that they have an alternative route of entry (Fig. 3.6). For practical purposes, it can be considered that there are a relatively small number of metabolic pathways responsible for energy production, and that the different respiratory substrates ultimately enter the citric acid cycle by a limited number of routes. The most important point to be grasped is the central role of the oxidative pathways in energy metabolism, with the energetic yields of such pathways being considerably greater than those involved in the anaerobic metabolism of respiratory substrates.

3.4 PREY DETECTION AND PREDATOR AVOIDANCE

The first major problem in the feeding process is the detection and location of the food. This problem is of minor importance to some fish, such as those feeding on detritus. As far as is known, these fish swim along the bottom sucking in loose surface material, spitting out large particles and swallowing the rest. The majority of fish, however, feed on food which must be actively sought out. Prey detection may involve vision, chemoreception, electroreception or mechanoreception, but the different modes of detection are not mutually exclusive.

Visual detection of prey

Visual detection of objects underwater presents rather different problems from those encountered on land, both because of the absorption of light by the water itself and because of light scattering by the particulate matter suspended in the water column. As light passes through the water it is absorbed and light intensity declines exponentially with increasing depth. The relationship between intensity and depth can be expressed according to the Lambert–Beer law:

$$I_d = I_s\, e^{-e\ d} \tag{3.3}$$

where I_s is the light intensity at the water surface, I_d is the intensity at depth d and e is the extinction coefficient. Light is, however, made up of many different wavelengths and each wavelength is not absorbed to the same extent during its passage through the water. Thus, with increasing depth both intensity and spectral bandwidth are reduced.

In addition, the scattering of light caused by suspended particles will further reduce the light intensity. As light passes down through the water column, suspended particles of fine silt and organic matter will scatter light in all directions including back towards the point of origin, producing a kind of underwater fog effect. The light that is scattered in a horizontal direction is responsible for an even background radiance known as background space-light and it is against this that the fish must detect its prey. Scattering of light that occurs between the prey and the predatory fish gives rise to a veiling brightness. This, along with the scattering of the light emanating from the prey, causes a severe decrease in both the sharpness and contrast of the visual image with distance.

Visual pigments and prey detection

Both light intensity and spectral bandwidth vary greatly depending upon water type and depth, so there are wide ranges of photic environments in marine and freshwater habitats. There may be close correlations between the peak absorption wavelength of a fish visual pigment and the peak wavelength of light to which the fish is exposed in its natural habitat, but most fish possess multipigment visual systems which provide an efficient detector system under a variety of conditions. In deep water, for distant objects and for objects darker than the background spacelight, visual pigments with absorption maxima matched to the background light are most efficient. In the case of bright objects viewed in shallow water, however, pigments most sensitive to wa-velengths slightly offset from the maximum transmission of the water give best results for detecting contrast between the object and background light. Thus, a system consisting of pigments matched to, and slightly offset from, the wavelength of the background spacelight is likely to provide the best detector system for all-round visual performance in shallow- and midwater fish.

The distribution of visual pigments has been examined in relation to the hunting strategies of some predatory fish species. The mahi mahi, or dol-phinfish, *Coryphaena hippurus*, which attacks its prey horizontally in shallow (down to 15 m) water, will detect its prey against a blue horizontal back-ground. This species possesses three visual pigments with absorption maxima at wavelengths of 521, 499 and 469 nm. The pigments have ab-sorption maxima matched to the horizontal and upwelling spectrum, and offset towards longer wavelengths. This multiple pigment system tends to maximize the visibility of targets that are darker and brighter than the blue horizontal background spacelight against which they are detected. A similar multipigment system is found in the rainbow runner, *Elagatis bipinnulatus*, which is also a diurnal pelagic species that hunts close to the surface.

In contrast to the mahi mahi and rainbow runner, the skipjack tuna, *Katsuwonus pelamis*, most other tunas, and the wahoo, *Acanthocybium solandri*, attack their prey from beneath and will, therefore, see the prey as a dark silhouette outlined against the bright downwelling light. Skipjack tuna have a single visual pigment which has an absorption maximum at 483 nm. This closely matches the wavelength of the background spacelight for all directions, including upwards. A number of other species employing this attack strategy also appear to have monochromatic vision.

Several fish species that feed on active prey adopt the strategy of attacking obliquely from below, and this has been observed for Atlantic cod, *Gadus morhua*, saithe, *Pollachius virens*, and some salmonids. This strategy has the advantage of giving increased visibility of the prey when viewed against the bright downwelling light. At the same time, a predator attacking from below is less easily detected than one attacking horizontally. A problem for a predator that attacks horizontally is that, being larger than the prey, it will, other factors being equal, be detected by the potential prey before it can see the prey. On the other hand, whilst a predator that attacks prey from below sees a clear silhouette of the prey against the downwelling light, the prey species has difficulty in detecting the predator approaching from a region of low illumination. This mode of attack is likely to be particularly effective under relatively low light conditions (e.g. at twilight), where only in the downward direction is there sufficient light for vision. Thus, a number of predatory fish species appear to concentrate their hunting and feeding bouts to the periods around dusk and dawn.

Camouflage

The difficulties involved in prey detection underwater are increased by various camouflage adaptations shown by potential prey species. Many zooplanktonic species, including pelagic fish larvae, are difficult to see due to their transparency and the matching of the colour of the reflecting parts of the body to that of the background spacelight. For example, although midge larvae, *Chaoborus* spp., have a much larger body size than water-fleas, *Daphnia* spp., they are less easily detected by planktivorous fish. The midge larvae have very little pigment in their bodies and transmit over 90% of the incident light whereas daphnia may contain red fat droplets within the body and transmit only 60–90% of the incident light. In addition, midge larvae spend much time stationary in the water column using hydrostatic organs to maintain position, whereas daphnia hold position by swimming. The swimming movements of daphnia consist of an upward 'hop' followed by a slow drift downwards in the water column, so the swimming movements of daphnia appear very jerky. These different movements of the two potential prey species contribute to conspicuousness, with moving prey tending to attract the attention of potential predators more readily than prey that remains stationary.

Larger potential prey species, such as fish, do not have transparent bodies but fish that swim in mid-water camouflage their bodies by means of countershading and silveryness. Many fish are dark-coloured or mottled, particularly in the dorsal region of the body, and this makes detection from above difficult.

The sides and ventral aspect of the body usually have a silver sheen. These silver-coloured sides have mirror-like properties that play an essential role in underwater camouflage. The silver colour results from reflection of light from thin guanine crystals in either the scales or the skin. A flat mirror standing vertically in water will not be detected when viewed obliquely since an observer will see light reflected from the mirror but will be unable to distinguish this from light which would have reached the eye in the absence of the mirror. Although the sides of fish are not flat and the scales follow the body contours the reflecting guanine crystals are orientated vertically and hence function in the same way as a mirror.

The fish cannot be camouflaged in this way when viewed directly from below since it will be silhouetted against the downwelling light. There are two solutions to this problem: the fish can have the most ventral part of the body very narrow, as in the clupeids and other small pelagic fish species, or the fish may possess special light-producing organs (photophores) on the ventral body surface. The light produced by the photophores can be used to mimic the bright downwelling light.

Other species have body colours that provide a camouflage protection enabling them to remain unseen when concealed amongst vegetation, rocks, or when lying on the substrate. This camouflage is often in the form of a mottling of dark and light blotches or stripes, which serves to disrupt the body outline. For example, the patterns of spots and blotches on the body surfaces of the pleuronectids enables them to remain effectively concealed when lying on, or partially buried in, the substrate. Mottled body coloration may also serve to camouflage predators such as the Eurasian perch, *Perca fluviatilis*, and pike, *Esox lucius*, which remain well hidden when lurking amongst underwater vegetation. In other species the camouflage is more elaborate, with the development of outgrowths from the body or fins that may resemble small projections of rock or the fronds of underwater plants.

Physical protection of prey

Whilst camouflage represents one method by which predation can be avoided, physical protection of the body represents a viable alternative. Thus, a well-armoured body decked with dermal teeth, or the possession of sharp fin rays, represent effective alternative forms of protection against predation.

Representatives of the boxfish family, Ostraciidae, have their bodies encased within an inflexible bony armour, and the porcupinefish, *Diodon holocanthus*, is decked with spines. Several species have some of the fin rays of the dorsal and pectoral fins elongated and stiffened to form protective spines. The threespine stickleback, *Gasterosteus aculeatus*, for example, has three stiff dorsal spines that can be locked into position when the fish is attacked by a predator such as a pike. The spines may create problems for the predator by making the prey difficult to grasp and swallow. Even if the predator is successful in grasping the prey, the prey may still be able to make an escape when the predator relaxes its grip in order to make manipulative movements in preparation for swallowing.

Additional protection against potential predators can be achieved by having poison glands associated with protective spines. Combinations of poison glands and spines are found in the weevers, *Trachinus draco* and *Trachinus vipera*, which have venom glands at the base of the spiny dorsal fin and a strong venomous spine on the gill cover.

Chemical detection of prey

Many fish are very sensitive to chemical stimuli in the water and some species can detect amino acid concentrations as low as 10^{-9}M. The use of chemoreception to detect food has the advantage that chemical stimuli can be used in conditions of poor visibility, but because such stimuli are persistent they may give information as to where potential prey has been rather than where it is at present. The chemoreceptor systems used in prey detection consist of the olfactory (smell) and gustatory (taste) systems, with the olfactory sense being the more sensitive.

Predator species may not respond to the full range of chemicals emanating from a prey but may only be attracted by a few amino acids and other organic compounds, such as betaine, nucleotides and nucleosides. There is evidence that water-borne ATP can stimulate feeding behaviour in some carnivorous species, whereas ADP and AMP may have little effect, or may even be inhibitory. ATP is found universally in biological material, but it is a highly labile compound that is readily broken down to ADP and AMP. Thus, ATP present in the water would indicate freshly damaged tissue, whereas ADP and AMP may be indicative that degradation has taken place.

It is probable that the various predatory species detect and respond to different chemical stimuli emanating from prey, and there are unlikely to be any chemical mixtures that act as universal feeding stimulants or attractants. Since predators may be sensitive to only certain chemical stimuli it is possible that different predatory species can have different chemical pictures of any given prey species.

The olfactory chemoreception system and the responses shown to chemical stimuli may vary in complexity. In some fish, the system may register only the presence of prey odour, as seems to be the case in some thunnids. Thunnids, which hunt largely by sight, appear to rely on olfactory cues only insofar as they need to locate the general presence of prey, and once prey is detected the final search and location is via visual clues.

A number of species of fish respond to chemical stimuli from food by swimming against the current until the source of the stimulus is found.

These fish do not appear to be able to detect small changes in the concentration of the chemicals emanating from the prey. Other species of fish, such as the ictalurids, appear to have olfactory systems that can detect relative concentrations of the chemical stimulus. These fish can, therefore, locate the source of an odour in still water by swimming up the concentration gradient. Even when the olfactory system is destroyed, some ictalurids can still locate prey by means of chemical stimuli. The prey is detected using chemoreceptors and taste buds located on various parts of the body. In the ictalurids these receptors are particularly concentrated in the barbels, but are also found on other areas of the body surface.

Several species of fish that search for prey hidden in mud and sand have sensitive barbels in the region of the mouth. These barbels may be used both to detect tactile stimuli and in chemoreception. Some species, such as red hake, *Urophycis chuss*, and gurnards, *Trigla* spp., have chemoreceptors located on specialized fin rays which are used for probing the substrate in search of prey.

Protection of prey by chemical means

Many species of fish rely, to a greater or lesser extent, on chemical senses in locating their prey, and a number of potential prey types rely on chemical protection in order to deter predators. A number of species, ranging from algae to insects and fish, either selectively accumulate or produce noxious chemicals which are released once they are attacked or captured by a predator. For example, many algae of tropical reef areas retain secondary metabolites within their growing fronds and these are thought to deter a number of fish species.

Both terrestrial and aquatic plants produce a range of *secondary metabolites* that may act as feeding deterrents for grazing animals. Alkaloids, a family of nitrogen-containing compounds, are commonly produced by terrestrial plants, but the secondary metabolites produced by algae tend to be halogenated organic compounds (i.e. compounds that contain chlorine or bromine), or polyphenolic tannins. For example, the green seaweed, *Avrainvillea longicaulis*, produces a brominated diphenylmethane derivative which deters feeding in a wide range of coral-reef fish species, and the high phenolic content of the temperate zone brown alga, *Fucus gardneri*, appears to act as an effective feeding deterrent. Not only do the polyphenolic tannins present in the tissues of this brown alga deter feeding by herbivorous fish, such as the monkeyface prickleback, *Cebidichthys violaceus*, but the secondary metabolites may also interfere with the digestive processes of the fish leading to reductions in digestive efficiency.

The secondary metabolites produced by plants do not function as universal deterrents and such plants may be subject to attack by several grazing species. In a number of cases the grazers themselves may take advantage of the fact that the plants produce secondary metabolites by accumulating these compounds within their own body tissues. Thus, several species of invertebrate grazers do not produce any deterrents themselves, but store and accumulate the chemicals ingested in their plant food in such concentrations that they can deter potential predators. For example, the gastropod mollusc, *Costasiella ocellifera*, feeds on *Avrainvillea longicaulis* and sequesters the secondary metabolites produced by the alga into its own tissues. These compounds then act as an effective defence against predation on the gastropod by fish. The defensive chemicals employed by molluscan species appear to be diet-derived, with the greatest number of recorded examples coming from the opisthobranch molluscs. Both the herbivorous sea hares and the carnivorous nudibranchs use diet-derived feeding deterrents which they sequester from their food organisms: algae, sponges, corals, bryozoans or tunicates.

Arthropods, including insects and crustaceans, are common prey for many fish species. Some arthropods are, however, capable of producing chemical deterrents that inhibit the feeding activities of fish predators. Aquatic beetles of the family Dytiscidae, for example, are rejected as food by many predatory fish, presumably because they produce and release a range of noxious or distasteful chemicals. Dytiscid beetles have a prothoracic gland which discharges onto the body surface, and the secretions produced by this gland contain a mixture of steroids. Addition of the steroid mixture or individual steroid components to otherwise attractive food pellets has been shown to reduce food consumption by the bluegill sunfish, *Lepomis macrochirus*, with the fish often mouthing the food pellet before rejecting it. Thus, the chemical deterrents of this family of aquatic beetles appear to act upon the gustatory receptors of potential predators, resulting in the sensation of a noxious or unpleasant taste. The free-living larvae of the pea crab, *Pinnotheres ostreum*, also appear to produce chemical deterrents that act via the gustatory sense of the predator, because the larvae are frequently rejected after having been captured by predatory fish.

The pelagic hyperiid amphipod *Hyperiella dilatata* does not appear to produce any deterrent chemicals, nor does it seem to sequester secondary metabolites from its food. This amphipod is a common prey item found in the stomach contents of a number of plankton-eating Antarctic fish species. The amphipod may adopt a novel form of chemical defence by associating with the pelagic pteropod *Clione limacina*. The pteropod is not eaten by fish predators, since it appears to possess some form of chemical deterrent. The amphipods seem to exploit the chemical defences of the pteropod by capturing an individual pteropod, grasping it close to the dorsal abdominal segments, and then carrying it for a prolonged period. The carrying of a pteropod appears to provide the amphipod with an effective defence, since, in laboratory studies, amphipod–pteropod pairs were almost invariably rejected by predatory fish, whereas amphipods that were not carrying pteropods were consumed readily.

Irrespective of the source of the foul-tasting deterrent substances, they seem to act via the gustatory system and result in the predator spitting out the prey rather than swallowing it. A number of the species that accumulate the chemicals from their food, and some of those that produce the deterrent chemicals themselves, also have distinctive warning, or *aposematic*, body markings. Potential predators soon learn that such coloration represents an unsavoury item that should be avoided rather than attacked.

Thus, aposematic coloration, or the performance of an aposematic behavioural display, can serve to warn a predator that the prey is unpalatable. This form of advertisement would seem to be advantageous for the prey, since even for a prey that is distasteful it may be best to avoid capture or intimate interaction with predators. Amongst fish species, the midshipman, *Porichthys porosissimus*, which has a toxic spine, appears to use an aposematic display in order to warn predators that it is both potentially dangerous and unpalatable. In response to an attack or close approach by a predator the midshipman flashes its photophores, and this usually leads the predatory fish to abort its attack.

Several fish species rely on chemical defence but do not seem to have this linked with either warning coloration or a specific aposematic behavioural display. The noxious chemicals are usually localized in venom glands associated with spines or elongated sharp fin rays, or the chemical deterrents may be distributed amongst specialized cells in the skin. Thus, the chemicals may be released following capture of the fish by a predator, but either before it has received any injury or following only slight surface damage. Consequently, many fish will survive attack by a predator without having received serious injury.

Thus, the skin secretions of a number of fish species appear to have repellent or toxic properties that may be of deterrent value. For example, the skin secretions of the marine catfish, *Arius thalassinus*, have a number of toxic effects, and the secretions released by the boxfish, *Ostracion lentignosus*, and the soles, *Pardachirus pavoninus* and *Pardachirus marmoratus*, are known to deter potential predators.

Several species of puffers, Tetraodontidae, and goatfish, Grammistidae, are also known to produce and secrete chemicals with toxic and deterrent properties. The puffer fish toxin, *tetrodotoxin*, has been isolated from representatives of a number of animal groups, including octopus (Mollusca), crabs (Arthropoda) and salamanders (Amphibia). The toxin is known to be produced by a range of bacterial species. The wide distribution of tetrodotoxin in bacteria may point to a bacterial origin for this metabolite. Tetrodotoxin-producing bacteria have been isolated from the guts of some fish species. In addition, the skin of the puffer, *Fugu poecilonotus*, is known to contain *Pseudomonas* spp. bacteria that are capable of producing the toxin.

When coho salmon, *Oncorhynchus kisutch,* are exposed to extracts taken from the skin of either largescale suckers, *Catostomus macrocheilus,* or northern squawfish, *Ptychocheilus oregonensis,* they appear to become distressed. This may indicate that the skin secretions of both the sucker and squawfish species contain some toxic components, but some other form of interspecific chemical alarm signalling can not be ruled out in this case.

Chemical communication between potential prey

The chemical substances released by some fish species when attacked and injured do not serve to deter the predator but act as a warning to conspecifics. In other words, these chemicals function as alarm signals. In order for an alarm signal system to be effective it must include a series of elements: recognition of the hazard or danger by the sender, signal generation and release by the sender, reception of the signal by the receiver and the performance of the appropriate behavioural actions or 'avoidance response' by the receiver. In practice, the alarm signal could consist of an auditory signal (an 'alarm call'), a visual display, or could be chemical in nature. Amongst fish species chemical signalling has been extensively studied, with the alarm substance or pheromone (*Schreckstoff*) system found in the ostariophysan fish having received most attention.

Alarm substance and flight responses

The skin of the ostariophysan fish, which include cyprinids, ictalurids and characinids amongst their numbers, is characterized by having distinctive epidermal club cells which contain the alarm substance. The alarm substance has not been completely identified with respect to its chemical composition, but the compound hypoxanthine-3(N)-oxide is probably one of the most important signal components. The fish do not actively secrete the alarm substance into the water on identification of a hazard, or at the approach of a predator, but the club cells must be damaged or broken before the alarm substance is released.

Thus, although the fish do not actively release the alarm substance, they do produce a specific pheromone in fragile epidermal cells. The infliction of a minor injury by a predator will be sufficient to rupture some of the club cells causing the release of the alarm substance into the surroundings. The alarm substance is detected by neighbouring fish using the olfactory, rather than the gustatory, sense. Following detection of the pheromone the fish will perform a behavioural response or 'flight reaction' that is species specific.

In a small, schooling cyprinid species, such as the European minnow,

Phoxinus phoxinus, the chemical alarm substance will be released from the club cells when a fish is captured by a predator, such as a pike, *Esox lucius*, and the skin is lacerated. The very low concentrations of alarm substance in the water are detected by conspecifics using the olfactory sense, and these fish then respond by displaying a fright reaction. The school breaks up as individual fish rapidly seek cover amongst vegetation or close to the substrate, and having found a refuge the fish then remain stationary for some period of time. Once a single, or a small number of, fish have started to display the fright reaction, the warning of imminent danger may be rapidly transmitted to other members of the school via visual and mechanical alarm signalling in addition to chemical. Thus, some of the rapid turning movements and 'skitters', sudden darting motions, that occur in the fright reaction repertoire of the minnow may act as visual signals to other members of the school, and the mechanical disturbances made by these movements may also be detected by the lateral line organs of conspecifics.

The type of behavioural reaction displayed may vary among populations of a given species, and individuals within a given population may not include all components of the flight reaction within their own repertoire. For example, dashing and skittering often precedes tight schooling and reduced activity in the behavioural responses of cyprinids to alarm substance. Dashing movements are not, however, shown by every individual that reduces activity, and some individuals even seem to be unresponsive to the alarm substance.

Some fish may respond to alarm substances produced by other species, but the sensitivity is generally lower than to the chemicals produced by conspecifics. Generally, the more distant the taxonomic relationship between the species, the lower is the sensitivity to the heterospecific alarm substance.

There may also be seasonal, and/or sexual, differences in production of, and responses to, alarm substance. For example, in the zebrafish, *Brachydanio rerio*, males seem both to have fewer epidermal alarm substance cells, and to show lower response thresholds than the females. In other species, such as the fathead minnow, *Pimephales promelas*, and the pearl dace, *Semotilus margarita*, the number of epidermal club cells may become reduced during the breeding season, when abrasion and mechanical damage to the skin caused by courtship and spawning activities could lead to the release of alarm substance. Even though the production of alarm substance may be drastically reduced at this time of the year the fish may still respond to exposure to the pheromone by showing the characteristic fright reaction.

The detection of alarm signals can lead to a variety of effects beyond the immediate behavioural responses characterized by the flight reaction. For example, following exposure to alarm substance, foraging behaviour and feeding activity can be suppressed for some days, or the fish may not return to forage in the area where they were exposed to the alarm pheromone.

The chemical alarm signalling system involving the epidermal club cell appears to occur only amongst the ostariophysan fish, but since the ostariophysans account for almost 30% of all known teleost species, and about 70% of those found in fresh water, this alarm system is extremely widespread. Alternative systems of chemical signalling exist in a range of other fish species, including certain percids, cottids and gobies. As with the ostariophysan system, the release of the alarm substance seems to require some form of injury or mechanical damage to the skin, rather than active secretion of pheromone. The alarm pheromones of the percids, cottids and gobies differ, however, from those of the ostariophysans, both in their chemical nature and in their sites of production and storage. The non-ostariophysan species do not have epidermal club cells, but the alarm pheromones appear to be produced in large sacciform or vacuolated cells that are distributed throughout the epidermal layer of the skin. It is, however, as yet uncertain whether production and storage of the alarm substance is restricted to these cell types.

Chemical labelling of predators

Whilst alarm pheromones may act as effective signals they do not usually have a protective function preventing an attacked or captured individual from being consumed by predators. Since the faeces of predators contain undigested parts of their prey, and predator urine and mucus secretions may contain various metabolites, it is possible that predators may become chemically labelled after having consumed particular prey types. Consequently, the possibility exists that specific chemical cues emanating from predators could be used by prey organisms in order to detect and avoid areas of potential danger. For example, fathead minnows have been shown to exhibit a fright response following exposure to water that had contained a pike that had previously fed on conspecific minnows. The fright response was not, however, shown when the minnows were exposed to water that had held a pike fed on green swordtails, *Xiphophorus helleri*, a heterospecific fish prey. Thus, it appears that the predatory pike had become labelled with fathead minnow alarm pheromone, and traces of the label emanating from the pike were sufficient to indicate that it posed a potential danger to conspecific fathead minnows.

Chemical labelling of predators may not, however, be restricted to those species, such as the ostariophysans, that are known to produce alarm pheromones. For example, juvenile coho salmon significantly reduce activity when exposed to chemical cues emanating from mergansers, *Mergus merganser*, feeding on conspecifics, and juvenile brook trout, *Salvelinus fontinalis*, have been shown to avoid water conditioned by Atlantic salmon, *Salmo salar*, fed on fish prey (goldfish, *Carassius auratus*), but not water conditioned by salmon fed on mealworms. Thus, these juvenile salmonids may be capable of using chemical cues to detect, and avoid, predators that have been feeding on both conspecific and heterospecific fish prey. Whilst the salmonids are typically considered to use vision in both prey detection and predator avoidance, the ability to detect predators by chemical cues would be particularly beneficial in turbid or cloudy water when vision becomes impaired.

Electroreception of prey

Any animal that has a body fluid concentration different from the surrounding medium will create a weak electric field due to the exchange of ions that occurs over the body surface. In addition, nervous impulses are also very weak electric currents. Compared with other signals, electric energy has much to recommend it as a form of sensory stimulus. The conduction of electrical stimuli is rapid in water, but unlike visual stimuli electric stimuli have the advantage that they are not diminished in turbid water. Electrical signals have advantages over chemical ones, both with respect to speed of transmission and also in that they are non-persistent.

The discovery that the spotted dogfish, *Scyliorhinus canicula,* is highly sensitive to weak electric fields suggested that electrosensitivity could be used to detect prey. Laboratory studies showed that prey could be detected by means of the bioelectric field it produced and that the electric sense of the dogfish was associated with the ampullae of Lorenzini (see also Chapter 2).

Some teleost species are capable of electroreception, and the ampullary receptors are widely distributed on the body of electrosensitive teleosts. Amongst the elasmobranchs, the electroreceptors appear to be concentrated around the head in sharks, and over the upper and lower surface of the enlarged pectoral fins of skates and rays.

Mechanoreception and prey detection

The acoustic and lateral line systems enable fish to respond to vibrations in the water, and vibrating objects such as other swimming fish are detected using the hair cell mechanoreceptors of the inner ear and lateral line. Projections (kinocilium and stereocilia) of the hair cell are enclosed in a gelatinous cupula and are bent as the cupula is mechanically deformed by vibrations. This bending of the hair cell projections gives rise either to an increase or to a decrease in the nerve impulse frequency, depending upon the direction in which the projections are bent. The lateral line receptors in most fish lie in canals on the head and in a canal extending along the body. The receptors are distributed throughout this canal system and by comparing the responses of the hair cell receptors from different parts of the system, the fish may be able to use the receptors to locate both the direction and the distance to the source of the vibrations.

The types of vibrations produced by the erratic swimming of an injured fish will differ from those produced by a fish which is swimming normally. Predators are able to distinguish between different patterns of vibrations, and thereby use the mechanoreceptor systems to detect potential prey.

3.5   FEEDING BEHAVIOURS

Once prey has been located it must be captured, and the first problem to be overcome is that of getting the food item into the mouth. Fish display a wide range of feeding behaviours, and there are numerous different feeding strategies and associated adaptations involved in food capture and ingestion. For example, when fish eat insects or small aquatic organisms the prey is generally captured and swallowed whole, but for fish feeding on plants, or large prey items, it is more usual that bites are taken from individual food items.

The major behavioural and morphological adaptations displayed by a given species are obviously related to the main categories of food consumed, but there may also be some flexibility in the types of feeding behaviour shown by individual fish. For example, when American eels, *Anguilla rostrata*, are feeding on small prey they rely on suction to draw individual food items into the mouth, but when prey is larger, the eels may attempt to break the food down into smaller portions by biting and tearing. If the food item is particularly large and tough, the fish may resort to a third type of behaviour, known as spinning. In this type of behaviour the eel grasps the food in the mouth and then rotates rapidly about its longitudinal axis, and in this way breaks small pieces from the large food item. The investments of time and energy required for the performance of each of these feeding behaviours are markedly different. Suction feeding is the least energy-consuming and spinning requires the greatest investment. Consequently, when an eel encounters a large food item it will first attempt to break it into pieces by biting and tearing, and only if this is unsuccessful will the fish then resort to employing the more costly spinning behaviour.

Carnivorous species – macrophagous feeding

Many species of *piscivorous* fish eat their prey whole, but others tear and bite the prey into smaller pieces before swallowing it. The fish prey of the piscivores is usually quite large, being perhaps 40% of the length or 5–6% of the volume of the predator. As fish are active prey, predators have developed a number of methods of prey capture and each of these requires different morphological and behavioural adaptations.

When engaged in *active hunting* a predator will locate and capture prey by swimming over long distances, and a fusiform body shape is advantageous for fish displaying this type of feeding behaviour. Piscivores of this type are usually surface or midwater hunters feeding on pelagic, planktivorous fish which occur in shoals or schools. Examples of predators feeding in this manner include barracudas, *Sphyraena* spp., and several species of thunnids and scombrids.

Predators that attack potential prey by *stealth*, or sneaking, are usually well camouflaged and they approach the prey species slowly, by swimming amongst submerged weeds or roots. The prey is captured by the predator swimming rapidly over a very short distance (often less than a few metres). The ideal form for fish that hunt using this strategy is a relatively elongate and flexible body. The body should have similar depth along its complete length, and the median fins should be placed well back close to the caudal fin, as exemplified by the pike.

Lie-and-wait predators also tend to be well camouflaged, and they may have a specialized body shape to aid concealment in crevices or for burrowing in sand. For example, the conger eel, *Conger conger*, is elongated and hides in crevices, whilst the groupers, *Epinephalus* spp., are dorsoventrally flattened and have dark, mottled coloration. This makes the fish difficult to detect as they lie concealed amongst rocks. When prey species come within range the predator makes a sudden dash from its hiding place and snaps up the prey before it can escape.

A refinement of the lie-and-wait strategy may involve the *luring* of prey into the vicinity of the predator. Thus, prey is attracted into the area around the predator by means of a lure which is usually some part of the body adapted to resemble a worm or other such edible morsel. For example, in the anglerfish, *Lophius piscatorius*, the anterior ray of the dorsal fin is free and has a fleshy tip resembling a worm. The fin ray can be moved to and fro, and this movement attracts prey species into the striking range of the predator. Stargazers, *Uranoscopus* spp., lie buried in sand with their eyes and upturned mouth exposed. Within the mouth is a membranous process of the respiratory valves. This can be moved about, to resemble the movements of a small worm. Other species may have fleshy outgrowths on the head and in the region around the mouth. These serve to attract potential prey to within striking distance of the predator, which has the remainder of its body well concealed from view.

Since many piscivorous fish eat their prey whole, the mouth gape of these species is large and there are usually well-developed jaw teeth used in prey capture. The jaw teeth of piscivores are usually conical and backwardly pointing to prevent the escape of the prey. Some species, such as the anglerfish do not have jaw teeth that are rigidly fixed, but the teeth are hinged to the jaw by means of ligaments. The hinge ligament is arranged such that the teeth fold inwards as the prey enters, but the teeth lock on the jaw so

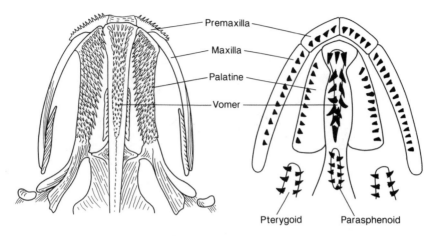

Fig. 3.7 Schematic diagram showing the bones in the 'upper jaws' of fish that may bear teeth (right), and an illustration of the location of teeth in the 'upper jaw' of the pike, *Esox lucius,* (left).

they cannot be bent outwards. This ensures that the prey remains entrapped within the mouth. Teeth are not, however, restricted to the jaws. The teeth that occur on the palate of piscivorous species function both to grip the prey and to lacerate the skin which exposes the underlying flesh (Fig. 3.7).

Suctorial feeding by microphagous species

Fish that feed upon small prey tend to have a narrower gape than the piscivores, and the prey is generally sucked into the mouth. In suctorial feeding it is advantageous to have a tubular mouth in which the premaxilla is loosely hinged to the cranium. The premaxilla will then be able to swing forward as the lower jaw is depressed and the mouth opens (Fig. 3.8).

The sequence of events involved in suctorial feeding is as follows.

1. The fish swim along with mouth closed and the opercular and buccal cavities in a contracted condition.
2. The tubular mouth opens by depression of the lower jaw and the swinging forward of the maxilla and premaxilla. The buccal and opercular cavities expand and there is a drop in pressure. The suction pressure created enables the fish to suck in the food item.
3. Once food has been sucked in the mouth is closed. The buccal and opercular cavities are still in an expanded condition – obviously if they were contracted before the mouth closed the food would be blown out of the mouth.

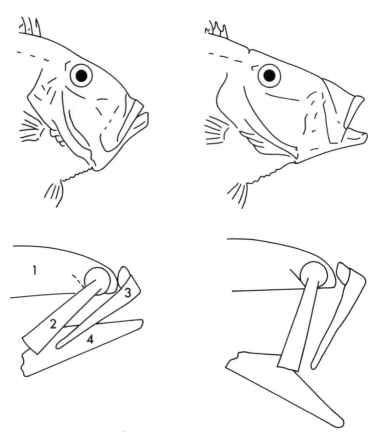

Fig. 3.8 Jaw structure in suctorial feeders. Note the loose connection between the premaxilla (3) and the cranium (1), which enables the premaxilla to rotate as the mouth opens due to the movements of the lower jaw (4) and maxilla (2). In many species of suctorial feeders the premaxilla will both rotate and slide outwards and downwards as the mouth opens.

4. The buccal and opercular cavities are contracted and the excess water is forced out via the opercular openings.

Benthophages

Suctorial feeding is usually adopted when fish take prey from the bottom or when they pluck small worms or insect larvae from the leaves of underwater plants. Thus the *benthophages* usually feed by suctorial feeding, but the juveniles of a number of pleuronectids may also feed by grazing on the exposed siphons of burrowing bivalves, such as *Tellina* and *Mactra* spp., and the tentacles of tube-dwelling worms, e.g. *Lanice* spp.

The adult pleuronectids, on the other hand, feed on a range of benthic prey species, which are taken into the mouth by suction. The jaw teeth are generally small and chisel-like, but the fish also have sets of teeth located in the pharyngeal region. The pharyngeal teeth of the pleuronectid species vary, reflecting the differences in food organisms eaten by the adult fish.

The diet of the plaice, *Pleuronectes platessa*, includes large numbers of bivalve molluscs, and the pharyngeal teeth of plaice are blunt cones used for crushing the bivalve shells. The pharyngeal teeth of the dab, *Limanda limanda*, are relatively poorly developed, and are incapable of crushing the shells of bivalves. The diet of the dab consists, largely, of crustacean species with much weaker exoskeletons than the bivalve molluscs. The European flounder, *Platichthys flesus*, has a more varied diet, consisting of worms, crustaceans and small bivalves, and the pharyngeal teeth are capable of crushing animals having relatively weak exoskeletons. In this species the lower pharyngeal teeth form a triangular plate and the upper pharyngeal teeth consist of blunt cones.

Planktivores

Suctorial feeding is also usually adopted by *planktivorous* species when feeding on larger zooplankton in mid-water. These prey are located visually and are drawn into the mouth by a combination of suction and the forward swimming movements of the predator. When feeding on smaller prey at relatively high density the predator may not orientate towards individual prey but will gulp several prey into the mouth simultaneously by opening the mouth to full gape and sucking at the same time. These small prey organisms are prevented from escaping through the opercular openings by fine filamentous gill rakers which act as traps (Fig. 3.9). At very high prey concentrations these gill rakers may be used as filters. The fish does not suck but merely swims forward with its mouth open, taking in both plankton and water. The water passes through the buccal and opercular chambers and out via the opercular openings. During respiratory movements, in which water passes from the buccal to the opercular chambers, the gill filaments are usually held close together creating a high resistance to water flow. During filter feeding increased rates of water flow through the chambers are achieved by the gill filaments being held slightly apart by muscles at their bases. The food organisms carried in the water currents impinge on the gill arches and may be trapped by the filamentous gill rakers.

Most planktivorous species feed on zooplankton, but there are a small number of species that have gill rakers fine enough to trap phytoplankton. For example, the Peruvian anchoveta, *Engraulis ringens*, and menhaden,

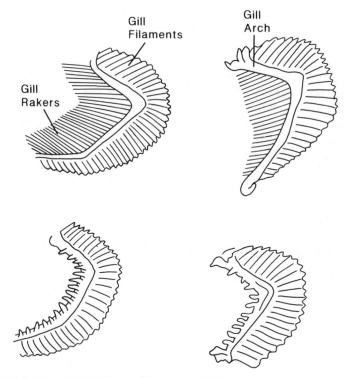

Fig. 3.9 Structure of the gills in fish species consuming small prey, such as zoo-plankton, compared with the gills of macrophagous species. Note the fine, long, tightly-packed gill rakers on the gill arches of the planktivorous species (upper figures). The gill rakers of macrophagous species (lower figures) tend to be blunter and more widely-spaced.

Brevoortia spp., can collect particles as small as 13–16 μm. Certain species of tilapia are also capable of filtering small particles of phytoplankton from the water.

Some fish species have special epibranchial organs that seem to be involved in the collection of phytoplankton. These epibranchial organs are pockets in the pharyngeal region, that are lined with mucus-producing cells. In the thread-fin shad, *Dorosoma pretense*, the material filtered by the gill rakers appears to be funnelled towards the epibranchial organs where it is gathered in the mucus. Epibranchial organs are also found in the milkfish, *Chanos chanos*, a species which feeds on a mixture of organic detritus, micro-organisms, algal and other plant material (known as *lab-lab*, in the areas of South East Asia where the fish occurs). It seems that the function of the epibranchial organs is to gather and compact food particles into balls of mucus before they are swallowed.

Biting and crushing of prey

Neither the jaw teeth nor the pharyngeal teeth of species that feed on zooplankton, phytoplankton and suspended organic matter are particularly well developed. Teeth are often restricted to the jaws and consist of fine bristle-like appendages which can be used for sorting and filtering small particles. This stands in marked contrast to the teeth of fish which feed on epifaunal species or on bivalves and coral. These fish species often have highly specialized teeth and jaws adapted for removing animals from rocks, crushing the shells of bivalves and snails or for breaking pieces from coral.

The jaw teeth may be fused to form a powerful beak as in the parrotfish, or scarids, the beak being used to break off hard pieces of coral. The wrasses, or labrids, which are closely related to the parrotfish, do not have beaks but have strong conical jaw teeth used for breaking shells of snails and bivalves. The jaw teeth of the wolffish, *Anarhichas lupus*, are similar in form to those of the wrasses and are used for breaking the shells of molluscs and echinoderms. The teeth on the palate are more rounded and flatter, being used for crushing the prey. Parrotfish, wrasses and the related pile perch, *Rhacochilus vacca*, also possess a specialized set of pharyngeal teeth, termed the pharyngeal mill, used for grinding and crushing the prey (Fig. 3.10).

Relatively well-developed pharyngeal teeth are also found in a number of *omnivorous* and *herbivorous* fish species, with the form of the teeth being related to the composition of the diet (Fig. 3.11). Many plants have tough cellulose cell walls which must be broken down before the organic cell contents can be released. Amongst the cyprinids, the species that consume the greatest proportions of plant material have the flattest, most molariform pharyngeal teeth, whereas those species which tend to feed more on insect larvae, zooplankton or benthic organisms have pharyngeal teeth that are either conical or elongate in form. The grass carp, *Ctenopharyngodon idella*, may be almost exclusively vegetarian, feeding on the leaves of aquatic plants and terrestrial plants submerged by flooding. The pharyngeal teeth of this species have blunt dorsal edges with marked serrations which are used to grind and masticate the plant material.

Fish feeding almost exclusively on plant material are comparatively rare in temperate waters, but fish having a high proportion of plants in their diets are more common in tropical and subtropical regions. In many of these environments the majority of the plant material consumed usually consists of algae, which may contain relatively large amounts of readily available extracellular organic material. Higher plants, which have little extracellular storage material and tough cellulose cell walls, may also be consumed, but in lower proportions. In freshwater habitats, many of the plants consumed are encrusting algae which the fish scrape from the rocks using their ventrally placed mouths with short bristle-like teeth, or horny toothless jaw plates.

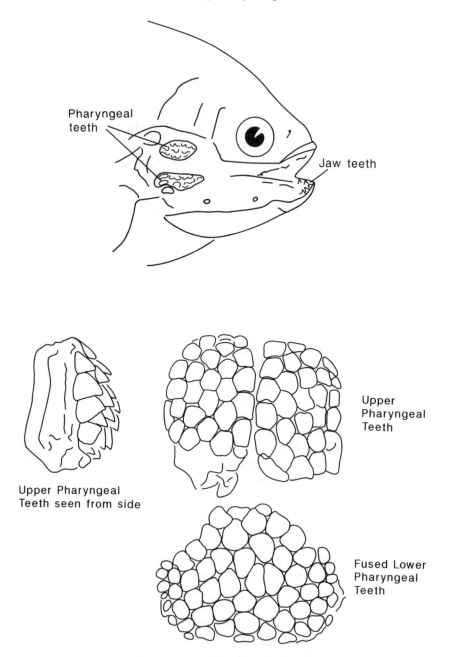

Fig. 3.10 Diagrams indicating the localization and structure of the pharyngeal teeth in the pile perch, *Rhacochilus vacca.*

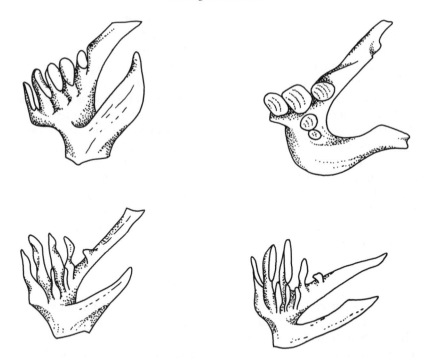

Fig. 3.11 Pharyngeal teeth in cyprinids. Note that the pharyngeal teeth of species that feed on benthic molluscs and plant material are more rounded and molariform (upper figures) than are those of species that consume greater proportions of insect larvae and crustaceans (lower figures).

Mechanical breakdown of this plant material may be continued in the pharyngeal region using the pharyngeal tooth plates.

A number of fish species of tropical coral reefs consume algal material, but many species of algae deter potential consumers either by having calcareous supportive structures, or by retaining secondary organic metabolites as feeding deterrents. Algae are similar to terrestrial plants in that they produce terpenes, aromatic compounds, amino acid derivatives and polyphenolics as secondary metabolites. They differ from terrestrial plants in that they incorporate halogens (e.g. chlorine, bromine) into their secondary metabolites, and do not produce alkaloids (nitrogen-containing compounds) that are common in some terrestrial plant species. The compounds produced and retained are often species specific, and do not have a universal deterrent action. For example, a compound that may be effective in deterring a range of fish species may be completely ineffective against invertebrate grazers, and vice versa.

Some scarids feed on calcareous algae, using the powerful beak to break off pieces of the algal fronds, and then the ingested material is ground to a fine homogeneous mass by the pharyngeal mill. Certain fish species are not deterred from consuming algae, despite the retention of secondary metabolites, and others promote the growth of palatable species of algae by engaging in 'gardening' activities. The fish removes encrusting algae, debris and small sedentary animals from certain rock faces within its feeding territory. These cleared areas form the gardens, on which algal mats can develop. By regularly removing unpalatable species and debris, the fish can ensure the establishment of a monoculture of palatable algae, and these are often species that are otherwise underrepresented on the reef. By cropping different gardens in turn the fish can ensure that there is fairly rapid regrowth of the algae, whilst at the same time the fish has an insurance against the overgrazing of its food source. Obviously these algal gardens may prove to be attractive to other potential grazers and the feeding territory must be actively defended against intruders.

Parasitic feeding habits

Parasitism is perhaps the most unusual feeding habit found amongst fish species, it being a form of carnivory in which the 'predator' has a smaller body size than the prey. There are relatively few parasitic fish species, and the parasitic species tend to show either very specialized morphological or behavioural adaptations. For example, some species of gobies and blennies that are specialized to feed on the scales and flesh of larger fish have body forms and colours that mimic those of 'cleaner' fish. Certain labrids and gobies feed on the ectoparasites of larger species, and these cleaner fish may establish distinct stations which are visited at regular intervals by large fish. The cleaner fish perform a useful function for the large fish in ridding it of its ectoparasites. However, mingling amongst the cleaner fish may be small numbers of mimics. The parasitic mimics, in contrast to the cleaners, do not feed on ectoparasites, but dart in amongst the cleaners and take bites of scales and flesh from the large fish waiting at the cleaning station.

3.6   ONTOGENIC CHANGES IN PREY SELECTION AND HABITAT SHIFTS

Fish display a wide range of feeding habits, and it is rare for fish to specialize in one particular prey category throughout their entire life cycle. Thus, it is most usual that fish show distinct ontogenetic changes in feeding habits and prey selection. For example, species that are piscivorous as adults are usually planktivorous as larvae and juveniles, and there is a gradual change in feeding habits as the fish increase in size. This is, to a large extent, obviously governed by the changes in mouth size and gape that occur as the fish grows. A small fish would be unable to capture, subdue and swallow a large prey item. On the other hand, a large individual would be required to expend large amounts of time and energy in searching for the large numbers of very small prey that would be needed in order to meet its energy requirements and ensure adequate rates of growth. This has led to the development of the idea that prey of different sizes, or types, will give the best return per unit effort for fish of different sizes.

The optimal food particle size (defined as that which results in the fish displaying the greatest rate of growth) is, therefore, expected to change during ontogenetic development. For example, when juvenile Atlantic salmon are fed on pellet feeds the optimal size of the food particles increases from about 1 mm diameter for fish of 4 cm in length to 5 mm for fish that are 20 cm long. By definition, when fish are provided with food that differs in size from the optimal they will display suboptimal rates of growth. When allowed a free choice of pellet sizes salmonids are often found to select food particles that are close to the optimum in size. If these findings obtained under laboratory conditions are extrapolated to natural fish populations, it may be predicted that, if conditions permit, (1) the fish would select prey sizes and types that result in optimal growth, and (2) prey sizes, and types, selected would change in step with fish growth and increased size.

Fish will rarely be exposed to conditions where they have a free choice of prey covering all sizes, and the question arises, when should the fish change from one prey to another when there is only a limited choice? This question can be examined simply by considering factors that may influence the point at which a fish will switch prey types when given a choice between only two different types.

Figure 3.12(a) illustrates the rates of growth of fish of different sizes when fed on either prey A or prey B, these two prey types being of different sizes. Prey A is relatively small and is the optimal prey size for small fish (S_A), i.e. fish that differ from S_A in size will display suboptimal rates of growth when they feed on prey A. Prey B is larger than prey A, and cannot be consumed by very small fish. This prey gives optimal growth rates for fish of size S_B.

When the growth relationships of fish of different sizes feeding on either prey A or prey B are plotted on common axes (Fig. 3.12(a)), the two growth rate curves intersect at a specific point. At the point of intersection, a fish of size S_1 would grow equally well when feeding on prey A and prey B. Fish smaller than S_1 grow better when consuming prey A than they do when feeding on prey B. Fish larger than S_1 display more rapid rates of growth when feeding upon prey B. Consequently, under conditions when there is a choice between these two prey types, the fish should change from feeding on prey A to feeding on prey B when it reaches a size of S_1. If the two prey types occur in different environments (e.g. prey A is a pelagic zooplanktonic species

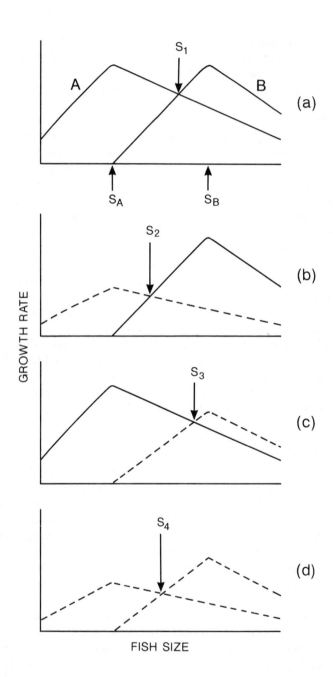

of open-water areas, and prey B is a benthic species associated with aquatic vegetation), a change in prey type will also mean that the fish moves from one habitat to another.

The condition described above relates to a situation where a fish is given a choice between two prey types, but it is also possible to envisage a situation where a number of species may exploit the same types of prey and competition between fish species may arise. If there are several species that exploit prey A the competition that arises may mean that it is less profitable for the fish to feed on this prey type. In other words, competition results in the growth rates of fish exploiting prey A being depressed in comparison with the rates displayed in the absence of competition (Figs 3.12(a) and (b)). Under conditions of competition for prey A it will be more profitable for the fish to switch to feeding upon prey B at a smaller size (Fig. 3.12(b): S_2) than under conditions of no competition (Fig. 3.12(a): S_1).

Increased exploitation and greater competition between species for prey B would make it more profitable for the fish to delay the change from feeding on prey A to feeding on prey B until a larger body size was achieved (cf. Fig. 3.12(c): S_3 with Fig. 3.12(a): S_1). From this it should be clear that both prey selection and the sizes at which the fish switch from one prey to another will be influenced both by the relative sizes of the available prey and by the amount of competition between species for these different prey types (e.g. S_1, S_2, S_3 or S_4 in Figs 3.12(a–d)).

A good example of how competition between species can influence prey selection, shifts in habitat and rates of growth is given by the results of a study carried out on three North American species of sunfish – bluegill (*Lepomis macrochirus*), pumpkinseed (*Lepomis gibbosus*) and green sunfish (*Lepomis cyanellus*) – reared in small freshwater ponds. The available habitats and prey types could be broadly classified as being open water with zooplankton, bottom sediments with benthic prey such as small worms, molluscs and insect larvae, and aquatic vegetation in the shallower fringes of the ponds, with insects and a range of invertebrate prey types.

Fig. 3.12 Theoretical considerations relating to the size at which fish should switch between different prey types in order to maintain most rapid rates of growth. Fish of size S_A and S_B will display the highest rates of growth on prey types A and B respectively. (a) Under conditions of no competition for prey, prey type B becomes more profitable, i.e. promotes greater growth, than prey A once the fish has reached the size S_1. Thus, the fish would be expected to switch from consuming prey A to consuming prey B once they reach a size of S_1. (b) When there is severe competition for prey A this prey becomes less profitable and the fish should switch to prey B at a smaller size, S_2. (c) When there is competition for prey B, but not prey A, the fish should delay switching prey until a larger body size is reached, S_3. (d) When there is competition for both prey types the switch should occur when prey B becomes more profitable than prey A, i.e. when the fish reaches a body size S_4.

When held in single-species groups, fish of all three species were most frequently observed in shallow water close to vegetation. The major part of the diet comprised prey organisms that lived either on or close to the aquatic plants. On the other hand, when fish from all three species were held together, there were marked changes in the habitats occupied, the prey types consumed and the rates of growth displayed by individuals.

The pumpkinseed seemed to be relatively little influenced by the effects of competition with the other sunfish species. Rearing in mixed groups resulted in only minor changes in dietary composition, i.e. prey associated with aquatic vegetation dominated the diets of pumpkinseed held in both single-species and mixed-species groups. Furthermore, growth rates were little depressed by rearing pumpkinseed in mixed-species groups.

The other two species of sunfish were more strongly affected by competition, and they showed marked changes in habitat and prey selection. Under conditions of competition the green sunfish consumed a greater proportion of benthic prey than when reared alone, and the bluegill showed a shift in habitat from the pond fringes to open-water areas. The diet of the bluegill became increasingly dominated by zooplanktonic organisms, there was a marked decrease in the average size of the prey and growth rates of the bluegill were considerably depressed in the mixed-species groups. Thus, these results give a clear indication of the importance of interspecific competition in governing prey selection and habitat choice.

An additional factor that has been demonstrated to have marked influences on feeding behaviour, prey selection and choice of habitat is the risk of predation. In the presence of a predator there will often be observed to be changes in the behaviour of fish. The fish will become more wary and, in the short term, may cease searching for food, and will seek shelter. Long-term exposure to predation risk may, however, also influence prey and habitat choice. Under conditions where a specific area is both highly profitable with respect to the availability of food resources, and dangerous because of the risk of predation, there must be some balance to be made between the profit to be gained by foraging in the area and the chances of being captured and killed by a predator.

Results from numerous studies carried out with potential prey species – European minnows, *Phoxinus phoxinus*, and other small cyprinids, sunfish, *Lepomis* spp., threespine sticklebacks, *Gasterosteus aculeatus*, and some gobies, *Pomatoschistus* spp. – show that there is a marked reduction of foraging activity in the presence of a predator, and this may also be accompanied by a move to a less profitable habitat. Thus, the fish move from good feeding areas to areas where predation risk is reduced, but at the cost of poorer feeding conditions. Consequently, the fish may avoid being predated upon but will suffer a depression in growth rate. Similarly, movements into areas with poorer food resources and changes in prey selection have been observed for the juveniles of some Pacific salmon species, *Oncorhynchus* spp., when they are exposed to a high predation risk. The juvenile salmonids usually search for food in open-water areas and feed pelagically on zooplankton, but in the presence of predators the fish move to areas where there are denser growths of aquatic vegetation, providing hiding places but poorer feeding conditions. Fish may, therefore, be expected to select habitats and prey based upon the following criterion: select areas where the relationships between predation risk and feeding rates are minimized; but problems arise in trying to express these factors in terms of a common currency. Nevertheless it should be clear that both the timepoint and body size at which a fish will show a change in habitat will be dependent upon the interplay between predation risk and prey availability. In addition, it must be remembered that the profitability of a given prey type may be influenced by interspecific competition, which further increases the difficulties in making accurate predictions about prey selection and the compositions of the diets of fish populations.

3.7 CONCLUDING COMMENTS

The potential foodstuffs available to fish species come in many different shapes and sizes, and they may also differ markedly in nutrient composition. Different fish species have become adapted to exploit these foodstuffs, and there are herbivorous, omnivorous and carnivorous species, with each displaying a particular set of morphological and behavioural adaptations.

The sensory modalities used during prey detection and capture include vision, electroreception, the mechanical and the chemical senses. Some species use both olfaction and gustation in the detection of prey, but others will tend to rely upon gustatory responses only in the final stages of prey acceptance or rejection. Thus, the gustatory (taste) system appears to be divisible, both anatomically and functionally, into two distinct, though interrelated subsystems. The extraoral part of the facial taste subsystem, which is innervated by cranial nerve VII, seems to be used to detect chemical stimuli at a distance. This subsystem may, therefore, serve a food-localization function during foraging. On the other hand, the oral gustatory subsystem seems to perform a discriminative function leading to the selection and ingestion, or rejection, of prey.

Potential prey species may use a range of methods to avoid being detected, attacked and consumed by predators – crypsis, chemical and physical protection, and an array of communication mechanisms – and the behaviour of the prey may change substantially when under threat of predation.

The majority of fish species commence life as carnivores, with the larvae and juveniles of many species feeding on small zooplanktonic organisms.

There are, however, usually pronounced ontogenetic changes in both the types and sizes of prey consumed, with, for example, fish that have been zooplanktivores as juveniles becoming either herbivores or piscivores as adults. The ontogenetic changes in dietary habits may also involve the fish in making habitat shifts. For example, there may be a change from feeding in open water on zooplankton as a juvenile to the occupation of habitats associated with aquatic vegetation if the diet of the adult fish contains greater proportions of insects and benthic organisms.

FURTHER READING

Books

Cowey, C.B., Mackie, A.M. and Bell, J.G. (eds) (1985) *Nutrition and Feeding in Fish*, Academic Press, London.

Halver, J.E. (ed) (1989) *Fish Nutrition*, Academic Press, London.

Hara, T.J. (ed) (1992) *Fish Chemoreception*, Chapman & Hall, London.

Hoar, W.S., Randall, D.J. and Brett, J.R. (eds) (1979) *Fish Physiology*, Vol. VIII, Academic Press, London.

Jobling, M. (1994) *Fish Bioenergetics*, Chapman & Hall, London.

Keenleyside, M.H.A. (1979) *Diversity and Adaptation in Fish Behaviour*, Springer-Verlag, Berlin.

Pitcher, T.J. (ed) (1993) *The Behaviour of Teleost Fishes*, 2nd edn, Chapman & Hall, London.

Review articles and papers

Duffy, J.E. and Hay, M.E. (1990) Seaweed adaptations to herbivory. *BioScience*, **40**, 368–75.

Hay, M.E. and Fenical, W. (1988) Marine plant–herbivore interactions: The ecology of chemical defense. *Ann. Rev. Ecol. Syst.*, **19**, 111–45.

Horn, M.H. (1989) Biology of marine herbivorous fishes. *Oceanogr. mar. Biol. ann. Rev.*, **27**, 167–253.

Huntingford, F.A. and Torricelli, P. (1993) Behavioural Ecology of Fishes. *Mar. Behav. Physiol. (Special Issue)*, **23**.

Lima, S.L. and Dill, L.M. (1990) Behavioural decisions made under the risk of predation: a review and prospectus. *Can. J. Zool.*, **68**, 619–40.

McFall-Ngai, M.J. (1990) Crypsis in the pelagic environment. *Amer. Zool.*, **30**, 175–88.

Scheuer, P.J. (1990) Some marine ecological phenomena: Chemical basis and biomedical potential. *Science*, **248**, 173–77.

Smith, R.J.F. (1992) Alarm signals in fishes. *Rev. Fish Biol. Fisheries*, **2**, 33–63.

Werner, E.E. and Gilliam, J.F. (1984) The ontogenetic niche and species interactions in size-structured populations. *Ann. Rev. Ecol. Syst.*, **15**, 393–425.

Chapter four

Vascular transport and gaseous exchange: the circulatory and respiratory systems

4.1 INTRODUCTION

The circulatory, or cardiovascular, system of fish consists of a relatively simple single circuit (Fig. 4.1). The blood is driven through a series of four different types of vessel – arteries, arterioles, capillaries and veins – by the heart, which acts as a pump. This system is the main transport route of the body, not only for nutrients, metabolites and waste products but also for respiratory gases and the blood-borne endocrine factors. For example, the blood serves to transport oxygen from the gills, or other respiratory organs, to the tissues, and transports carbon dioxide in the opposite direction. Nutrients absorbed from the gut are transported in the blood to the liver, and from there to the other organs of the body. Similarly, the waste products of metabolism are transported from the tissues to their sites of excretion via the circulatory system.

4.2 MAJOR VESSELS OF THE CIRCULATORY SYSTEM

The heart pumps blood into the ventral aorta. The ventral aorta blood pressure will generally be within the range 30–50 mmHg, depending upon species and the activity of the fish. (The SI unit of pressure is the Pascal (Pa), but many physiologists prefer to use mmHg (or Torr). 1 mmHg = 1 Torr = 133.3 Pa.) The ventral aorta passes anteriorly and divides into a series of afferent branchial arteries running to the gills (Fig. 4.2). From each of the

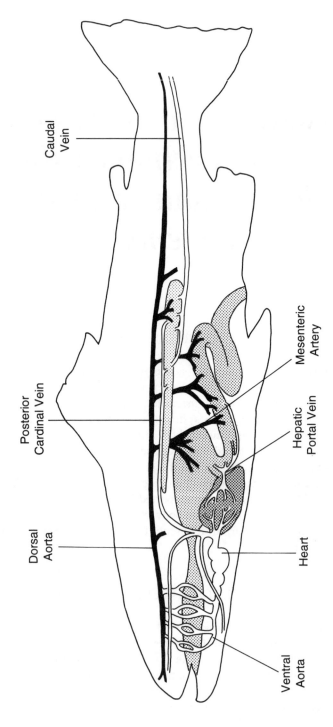

Fig. 4.1 General overview of the circulatory system in a fish showing the major vessels.

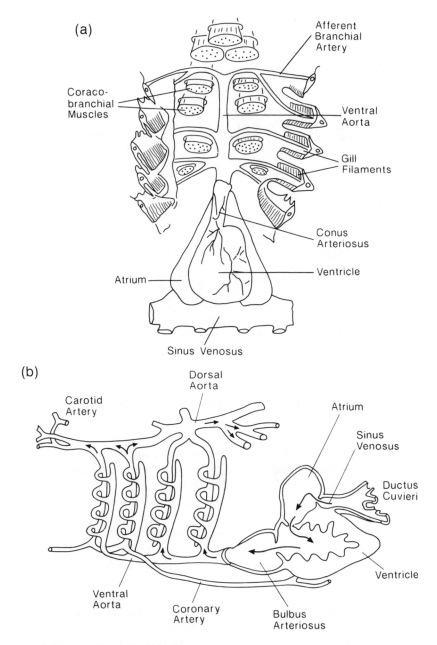

Fig. 4.2 General structures of the hearts of (a) an elasmobranch (ventral view) and (b) a teleost (lateral view), also showing the ventral aorta and major vessels of the branchial region. Arrows show direction of blood flow.

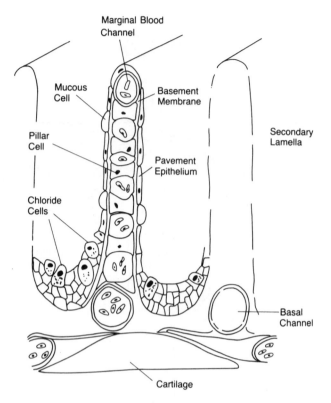

Fig. 4.3 Cross section of secondary lamellae on the gill filament showing the blood channels.

branchial arteries there arise a number of filamentar arteries which carry blood down the gill filaments to the arterioles (afferent lamellar arterioles) that lead to the secondary lamellae (Fig. 4.3).

Once it has flowed through the lamellar channels, the oxygenated blood is collected by a series of efferent blood vessels (efferent lamellar arterioles and efferent filamentar arteries) which then coalesce. On each side of the body there is a branching of the larger vessel, with one branch running anteriorly towards the head and the other branch running posteriorly towards the abdominal cavity (Fig. 4.2).

These branches, or aortae, of each side may fuse both anteriorly and posteriorly, forming a ring, the circulus cephalicus, but the exact anatomical arrangement tends to vary with species. In the brown trout, *Salmo trutta,* and roach, *Rutilus rutilus,* for example, the circulus cephalicus is formed by the fusion of the lateral aortae arising from the two most anterior efferent branchial arteries on each side of the body. The two posterior efferent branchial arteries on each side do not link with the circulus cephalicus, but join a single main vessel to form the dorsal aorta. In the whiting, *Merlangius merlangus*, however, the two lateral aortae do not unite posteriorly until they reach the anterior end of the body cavity. Thus, in this species the circulus cephalicus receives the blood from all the efferent branchial arteries.

The vessels that run anteriorly supply blood to the organs of the head and give off branches to the jaws, eyes and brain, whereas the vessels that run posteriorly eventually fuse to form the dorsal aorta, the major artery of the body. The gills exert a resistance to blood flow and there is a difference of 25–35% in blood pressure between the ventral and dorsal aortae. The blood pressure in the dorsal aorta is lower than in the ventral aorta, and dorsal aortal blood pressure is generally within the range 25–35 mmHg.

Vessels of the abdominal region

The dorsal aorta runs backwards below the spine, through the body cavity to the abdominal region. It gives rise to median arteries to the stomach, liver, intestine and gonads (e.g. mesenteric artery to the viscera, including the gas bladder) and paired segmental arteries to the myotomes, fins and kidneys (e.g. subclavian arteries that arise from the anterior end of the dorsal aorta and run to the pectoral fins, and the renal arteries supplying the kidneys). At the hind end of the body cavity the dorsal aorta passes into the haemal canal and continues to the caudal region (Fig. 4.4).

In the haemal canal, which is the canal formed by fusion of a series of ventral vertebral processes into haemal arches, the dorsal aorta runs alongside the caudal vein. The haemal canal is not a completely sealed tube and the aorta gives off segmental branches supplying blood to the muscles in the caudal region. The muscles in this region are particularly well invested with slow, aerobic, 'red' fibres, and it is imperative that these fibres receive a good supply of oxygen and respiratory substrate via the circulatory system if they are to function effectively.

Venous return from the caudal region

Blood from the muscles in the caudal region is collected in segmental veins, and the blood then enters the caudal vein. The caudal vein, which is contained within the haemal canal, is protected by the rigid vertebral haemal arches. This rigid protection serves to insulate the caudal vein from the backwardly moving pressure waves generated by the contractions of the myotomes during swimming. These pressure changes are, however, harnessed to enhance venous return and do, therefore, play an important role in circulating the blood.

Muscular contraction associated with swimming is thought to compress

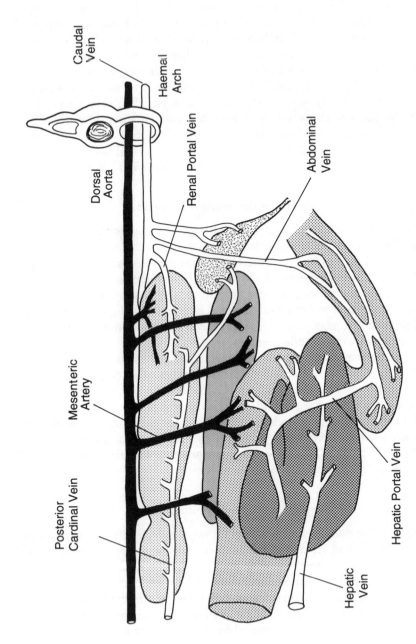

Fig. 4.4 Schematic diagram showing the major blood vessels in the abdominal cavity.

the capillaries and veins lying within the myotomes, thereby forcing the blood along the segmental vessels. At the points of entry of the segmental veins into the caudal vein there are a series of valves which serve to prevent backflow of blood. The pressure generated by muscular contraction forces the blood past the valves and into the caudal vein. This, in turn, can lead to an increase in caudal vein blood pressure, and thereby lead to the flow of blood being directed towards the heart.

Thus, this system of the haemal arch and valved veins can be considered to form a pumping mechanism for returning the blood from the tissues to the heart (the *haemal arch pump*), with the driving force for the pump being generated by the same muscular contractions that serve to propel the fish forwards when it is engaged in swimming.

The swimming movements may also provide the driving force for an additional venous pump that returns blood to the caudal vein (the *caudal pump*). The cutaneous veins run posteriorly and drain into a marginal vein, which lies superficially just beneath the skin in the caudal region. The marginal vein is linked to deeper-lying vessels, the caudal sinuses, via a series of short, valved connections. These sinuses lie close to the elements of the caudal fin skeleton directly beneath the caudal vein. The blood drains from the sinuses to the caudal vein through valved connecting vessels. When the muscles that deflect the tail fin contract, the caudal sinuses will alternately enlarge and then be stretched and compressed. Thus, the sinuses will be filled with blood draining from the cutaneous veins during expansion, and when the direction of the sweep of the caudal fin changes, the stretching and compression of the sinuses will force the blood into the caudal vein. Backflow is prevented by the series of valves.

A caudal pump of this type, which is found in many elasmobranchs, will only function when the fish is swimming, but in some species special adaptations of the muscles allow the caudal pump to continue to function in the absence of a full deflection of the caudal fin. In a resting fish, simultaneous contraction of the muscles on both sides of the body can be used to send rippling waves of compression down the caudal fin. These waves force the blood from the cutaneous circulation into the caudal sinuses and past the series of valves into the caudal vein.

Caudal pumps, sometimes called caudal hearts, are also present in a number of teleost species, but the anatomical arrangement tends to differ from that of the caudal sinuses of the elasmobranchs. The teleost caudal pump does not consist of two caudal sinuses, but is usually a two-chambered structure, with one chamber lying each side of the caudal fin skeleton. One of the chambers receives all of the blood draining from the cutaneous circulation, and this blood then passes to the other chamber via an opening in the wall between the two chambers. The second chamber connects with the caudal vein via an opening guarded by a valve. In this arrangement of the caudal pump the blood is driven by the alternate compression of the two chambers that results from the contraction of the small skeletal muscles that are closely allied with the vertebrae in the region of the pump. Compression of the first chamber forces the blood into the second, and then the compression of the second chamber results in the blood being forced into the caudal vein.

Venous return from the abdominal region

The caudal vein collects blood from the cutaneous circulation, the posterior myotomes and from the remainder of the caudal region. As the caudal vein passes through the body cavity it receives additional drainage from segmental veins, but the blood is not returned directly to the heart. The caudal vein divides into a pair of renal portal veins, and these further subdivide to form a series of capillary networks surrounding the kidney tubules. Additional branches arising from the caudal vein may run to the urinary bladder, the rectum and the hindermost part of the intestine (Fig. 4.4).

The blood from the capillaries surrounding the kidney tubules is collected and drains into the posterior cardinal veins for return to the heart. The blood that reaches the bladder from the caudal vein via the bladder vein is also returned to the heart via the posterior cardinal vein route. The posterior cardinal veins of the left and right sides of the body initially run alongside, and partially within, the kidneys as they pass forward through the body cavity. The posterior cardinal vein of each side then joins with the anterior cardinal vein of the same side, to form the Ductus Cuvieri (or Cuvierian vein).

Vessels in the anterior region of the body

The vessels that run anteriorly from the gills carry oxygenated blood to the jaws, eyes, brain and other organs of the head, and the blood is returned to the heart in a series of veins. The anterior cardinal veins arise from just above the gills. They receive blood from the branchial veins, and from the region of the eye. There are valves guarding the junctions between the branchial veins and the anterior cardinal veins. The openings of the anterior cardinals into the Cuvierian veins are also guarded by valves.

The walls of these blood vessels are relatively thin, and are readily compressed by the muscular activity associated with the ventilatory movements required for driving water over the gills. Thus, in this venous return system some of the power produced by muscular contraction for gill ventilation is harnessed in order to pump blood from the anterior region of the body to the heart.

From the point of fusion between the anterior and posterior cardinal veins, the Cuvierian veins lead down into the sinus venosus, but before entering the sinus they receive drainage from the jugular and hepatic veins (in some species of elasmobranchs the hepatic veins open directly into the sinus venosus). The jugular veins return the blood from the jaw region, whilst the hepatic veins carry the blood that has passed through the liver.

The blood that is returned to the heart via the hepatic veins is, predominantly, that which reached the viscera via the mesenteric arteries. All of the blood that flows to the stomach and intestine is collected in a series of veins which coalesce to form the hepatic portal vein. The hepatic portal vein runs from the gut to the liver, and en route is joined by a large vein draining the spleen. The hepatic portal vein also receives the venous return from the gas bladder. The hepatic portal vein then enters the liver where it subdivides to supply blood to the numerous hepatic sinusoids. The drainage from the sinusoids is collected in the hepatic vein and the blood is then returned directly to the heart.

4.3 CHARACTERISTICS OF THE BLOOD VESSELS

The circulatory system contains four types of vessels – arteries, arterioles, capillaries and veins – each of which performs specific functions and each of which has distinct structural properties and peculiarities. The arteries serve to distribute the blood to all parts of the body, with minimum loss of pressure. The shorter arterioles have a role in reducing the pressure to a sufficiently low level that it will not cause damage to the thin delicate vessels that form the capillary networks. The capillaries invest the organs and tissues, providing an extensive system for the exchange of nutrients, respiratory gases and metabolic waste products between the tissues and the blood. The veins both collect the blood draining the capillary networks for return to the heart, and act as a reservoir of relatively large volume in which blood may be stored.

Arteries

The *arteries* have quite large diameter and there is relatively little change in diameter along the length of the vessels. Thus, both the pressure and the velocity of the blood flow remain fairly constant as the blood moves along these vessels from the heart towards the periphery.

The arterial wall is typically muscular and consists of three layers. The innermost layer is a single layer of thin, flat endothelial cells. This is usually surrounded by a much thicker layer consisting of elastin and smooth muscle fibres, interspersed with a small number of collagen fibres. The smooth muscle fibres are capable of contraction, and can therefore contribute to driving the flow of blood along the artery. When the elastin fibres are stretched and released they tend to return to their original length. Thus, this layer of the arterial wall has considerable elastic and contractile properties. The collagen fibres are more resistant to stretching than the elastin fibres and they are fibres having great tensile strength. The outermost layer of the arterial wall comprises a layer of collagenous tissue which imparts strength to the wall.

There may be differences in structure between the walls of the arteries in the different parts of the circulatory system, for example, between the ventral and dorsal aortae. The wall of the ventral aorta, the vessel carrying the blood from the heart to the gills, tends to contain a large number of elastin fibres. These may be arranged into a series of distinct layers, and the layer of collagenous tissue overlying the elastin fibres is relatively thin. By contrast, the wall of the dorsal aorta is densely collagenous and relatively inelastic. The structure of the dorsal aorta wall reduces the risk of there being any marked changes in vessel diameter and volume arising from sudden changes in blood pressure and flow.

Arterioles

The *arterioles* are, as their name suggests, small arteries, having an internal diameter of 250–300 μm or less. These blood vessels are also relatively short in length. The walls of the arterioles contain comparatively little elastic tissue, but have several layers of smooth muscle fibres. The contraction of these fibres leads to changes in the muscular tension developed in the walls.

The narrow diameter of the arterioles means that they have a relatively high resistance to blood flow, and this, in turn, leads to there being a large drop in pressure along the length of the arteriole. For example, according to the Poiseuille equation, the resistance (R) to flow is calculated as:

$$R = 8 \times l \times \eta \times \pi^{-1} \times r^{-4} \tag{4.1}$$

which means that the resistance to flow (R) in a tube of length (l) depends upon the viscosity (η) of the fluid and is inversely proportional to the fourth power of the radius (r^4). Thus, a reduction in radius of only 16% in a vessel would lead to a doubling in the resistance to flow. The relationship between resistance (R), blood flow rate (Q), in volume per unit time, and pressure difference (δP) will be described by Ohm's law:

$$\delta P = Q \times R \tag{4.2}$$

so for vessels in which there is a high resistance to flow (R) there must be a large drop in pressure along the length of the vessel for the rate of flow to be maintained constant.

Capillaries

Blood at reduced pressure flows from the arterioles to the *capillaries*, which are thin-walled vessels of small diameter (5–10 μm). The capillary wall, which is usually 1–2 μm thick, consists of a thin layer of endothelial cells lying over a basal lamina. The thinness of the capillary walls permits rapid diffusion of gases and nutrients from the blood to the surrounding tissues.

The ability of the capillary network to function as a diffusion and exchange system is also dependent upon the transit time of the blood through the network. The longer the time the blood takes to pass through the system the greater is the likelihood that gas and nutrients will diffuse across the capillary wall.

The volume of blood (ml s^{-1}) flowing through the successive parts of the vascular system (aortae, arterioles, capillary networks) is equal. In other words, the same volume of blood flows through the dorsal aorta, per unit time, as flows through all the arterioles or all the capillary networks of the circulatory system. On the other hand, the velocity of the blood flow (cm s^{-1}) along any given vessel depends upon the total cross-sectional area of the vessels making up the particular part of the circulatory system. The total cross section of a capillary network is much greater (by an order of magnitude or more) than that of the arterioles immediately preceding it, so the velocity of flow (cm s^{-1}) along the capillaries is much slower than along the arterioles. Thus, the velocity of flow is fast in the dorsal aorta, intermediate in the arterioles, and slow in the capillaries. Slow flows mean long transit times, allowing effective diffusion, and diffusion is also enhanced by the very large total cross-sectional areas of the capillary networks.

Veins

The *veins* collect the blood and return it to the heart. They are vessels of quite large diameter, and they offer very little resistance to the flow of blood. By the time the blood has passed through the capillary networks very little of the pressure imparted to it by the heart remains, so the veins can be considered to represent a low-pressure system.

In order to assist the return of the blood to the heart this pressure is augmented by a variety of venous pumps which exploit the movements of adjacent muscles and tissues (e.g. the haemal arch pump, the caudal pump and the branchial venous pump). The action of these pumps is enhanced by the presence of valves (*ostial valves*) at the points where the tributary segmental veins join the larger longitudinal veins. These ostial valves enforce a one-way flow from the capillary networks, by preventing backflow of the blood from the longitudinal veins into the segmental veins.

The walls of the veins are generally quite thin and the vessels are easily dilated to their full size by the relatively small pressures existing within the low-pressure system. The walls of some of the veins contain very little elastic tissue, no muscle fibres, and are little more than thin-walled channels lined with endothelial cells.

As the longitudinal veins (anterior and posterior cardinal veins, and the hepatic vein) approach the heart they expand to form large thin-walled channels, or *sinuses*. These function as reservoirs of blood that provide the blood to fill the atrium of the heart each time the atrium becomes enlarged due to the cardiac suction generated by the contraction of the more muscular heart ventricle.

4.4   THE HEART

The heart is a muscular pump, the contractions of which serve to propel the blood around the body. It is located ventrally and anteriorly in the body, close to the midline between the pectoral fins. The heart is enclosed within a membranous bag of tissue, known as the *pericardium*. The pericardium has an important accessory role to play during the filling of the heart from the blood sinuses.

The major tissue of the heart is the cardiac muscle which consists of a meshwork of interconnected contractile myocyte cells. The myocytes are small (diameter 2–10 µm) cells rich in mitochondria and the oxygen-binding pigment myoglobin, something which attests to the intensity of the aerobic metabolic activity in these cells. Interspersed amongst the myocytes are collagenous fibres which serve to strengthen the tissue. The ratios of fibrous material to contractile myocytes vary between the different structural layers making up the walls of the heart chambers.

The chambers of the heart are arranged in series, with the thin-walled *atrium* opening into the thickly muscled *ventricle*. These are the two true chambers of the heart, but they are closely associated with the *sinus venosus*, and the *conus arteriosus* or the *bulbus arteriosus*. These four chambers form a single series, and the fish heart usually appears to have an S shape (Fig. 4.2(b)).

The blood that returns from the body initially flows into the sinus venosus, a relatively thin-walled sac-like structure which opens into the atrium. In some species of fish a layer of cardiac muscle extends out to invest the sinus venosus, but in others this layer is either absent or is reduced to a few scattered myocytes. The wall of the atrium contains thin layers of cardiac muscle, the contraction of which generates sufficient pressure to fill the more muscular ventricle, with which the atrium connects via the atrio-ventricular valve.

The ventricle, which generally makes up 65–85% of the weight of the heart, is a thick-walled muscular chamber. It is this chamber that generates the pressure required for driving the blood along the ventral aorta to the gills, and eventually to the organs and tissues. The ventricular wall is often divided into an outer compact layer and an inner spongy tissue layer. The compact layer (compact myocardium) consists of alternate layers of muscle fibres arranged at right angles to each other, to give sheets of muscle that run longitudinally and circumferentially around the ventricle. Simultaneous contraction of these muscle fibre sheets is extremely effective in generating the pressure required to propel the blood. The compact myocardium is particularly well developed in active pelagic species such as the thunnids. This layer is invested with its own blood supply, and a capillary network runs throughout the compact myocardium.

The blood is supplied to the compact myocardium via the coronary arteries, which arise from the vessels surrounding the second and third gill slits. The capillary network supplies oxygenated arterial blood to the muscle cells of the compact myocardium, but the capillaries do not penetrate far, if at all, into the spongy layer of the ventricular wall.

As its name suggests, the spongy layer consists of large numbers of small interconnecting chambers surrounded by muscular tissue to form a spongy mass. Thus, the blood flowing through the ventricle passes through these chambers and becomes intimately associated with the layers of muscle fibres. The spongy layer lacks a blood supply from the coronary capillary network, and the myocytes must, therefore, obtain their supplies of oxygen and nutrients from the blood flowing through the ventricle.

It must be remembered that this is venous blood, which, under certain circumstances such as when the fish is engaged in strenuous swimming activity, may be low in oxygen (low oxygen tension, P_{O_2}). Thus, at high activity levels, the ability of the muscle of the spongy layer of the ventricle to generate power could be limited by low oxygen availability in the blood being returned to the heart. It should, therefore, come as no surprise that the most active species amongst the teleosts (e.g. thunnids and scombrids, salmonids, and clupeids) have a well-developed compact myocardium with an independent blood supply. The more sedentary species, such as the pleuronectids, some cyprinids and percids, either lack or have much reduced compact layers in the ventricular wall.

Immediately anterior to the ventricle, and lying between it and the ventral aorta, is a fourth chamber enclosed within the pericardium. In the cartilaginous fish, and some osteichthyans, the fourth chamber, known as the conus arteriosus, consists of a barrel-shaped tube with walls made up of layers of cardiac muscle fibres overlying a sheet of elastic fibrous tissue. The inner surface of the wall of the conus arteriosus is invested with sets or rings (two or three sets, depending upon species) of valves, each set consisting of three valves which project into the lumen of the vessel. The lowest set of valves lies at the junction between the ventricle and the conus, and the uppermost set is located at the point where the conus arteriosus joins the ventral aorta. The valves allow the pressure within the conus and ventral aorta to be maintained, and prevent backflow of blood from the ventral aorta to the ventricle when the ventricular muscle begins to relax and ventricular pressure falls.

As with the conus arteriosus of the elasmobranchs, there is a ring of valves separating the teleostean bulbus arteriosus from the ventricle of the heart. The bulbus arteriosus of the teleosts differs, however, in structure from the conus of the elasmobranchs in that the wall is much more elastic, and it is not invested with cardiac myocytes. The main layer of the wall consists of circumferentially running (radial) elastic tissue interspersed with smooth muscle fibres. This means that the bulbus is highly distensible.

According to the LaPlace relationship, the tension (T) developed in the wall of a cylindrical vessel is dependent upon the transmural pressure (P_t) (the pressure in the vessel minus the pressure surrounding the vessel) and the vessel radius (r):

$$T = P_t \times r \tag{4.3}$$

which means that P_t is inversely proportional to the radius of the vessel. Thus, as a vessel expands, and the radius increases, there is expected to be a decline in transmural pressure, but this does not seem to occur when the bulbus arteriosus becomes distended.

The reason for this seems to be related to the special structure of the bulbus. In the wall of the bulbus arteriosus there is a series of septae, containing longitudinal fibres, which protrude into the lumen. Expansion of the bulbus will lead to a stretching of both the radial and the longitudinal elastic fibres in the wall. It is the unique septate structure of the wall, with longitudinal fibres, that seems to allow pressure to be maintained within the vessel despite the increased radius that results from stretching.

The main role of the bulbus arteriosus appears to be as a depulsator. Thus, the bulbus serves to even out the flow of blood during the course of a cardiac cycle, and reduces the level of pulsatile flow of blood along the ventral aorta to the gills.

The four chambers – sinus venosus, atrium, ventricle, and conus, or bulbus, arteriosus – are all enclosed within the pericardium, a two-layered membranous sac. The inner layer of the pericardium is intimately allied with the surface of the heart, whilst the outer layer is attached to the skeletal and muscle elements in the vicinity, and receives support from them. Thus, the walls of the pericardium are made semirigid due to their attachments to the skeletal structures, and they cannot, therefore, easily follow the changes in the volume of the heart within. Thus the walls of the pericardium tend to resist collapse during the phases of the cardiac cycle when heart volume is reduced. This plays an important role in creating a suction pressure for the filling of the atrium from the blood sinuses and sinus venosus.

4.5   THE CARDIAC CYCLE

The efficient operation of the heart is dependent upon the sequential activation of the four chambers, but since cardiac activity is myogenic there must be tissue at some site within the heart itself that initiates, or generates, the heart beat. Once a heart beat has been initiated, coordination is required to ensure not only that the chambers are activated in sequence, but also that the muscular layers in different regions of each specific chamber exhibit synchronous, or near-synchronous, activation.

The heart beat is initiated by the action of pacemaker cells or tissue, and this sets up a wave of electrical activity which spreads throughout the myocardium. The location of the pacemaker tissue appears to differ from one fish species to another. In a number of species the pacemaker cells are located in the junction between the sinus venosus and the atrium, whereas in others the pacemaker tissue may be found in the walls of either the atrium or the ventricle.

Once the heart beat has been initiated it is important that the electrical impulses do not spread at such a rate that they initiate muscular activity in the different chambers prior to the arrival of the blood to be pumped. This appears to be achieved by partial isolation of the different chambers from each other, with the direct passage of electrical activity only being possible at specific points. The fibres occurring at these bridging points between chambers ensure that there is a delay in the conduction and spread of the electrical activity from one chamber to the next. In contrast to the low conduction velocity desirable between chambers, conduction velocity within chambers should be high in order to ensure that electrical activity reaches all parts of the myocardium very rapidly. There are certain fast-conducting tracts and fibres within the ventricle of fish species, and these result in the spread of electrical activity to all parts of the ventricle within the space of 50–100 ms.

Electrocardiogram

From the above description, it should be clear that the mechanical phases of the cardiac cycle are preceded by a series of electrical events. This electrical activity can be recorded as the *electrocardiogram* (ECG), which gives a graphical representation of changes in electrical potentials. The ECG record is represented as a time trace showing the electrical activity as a series of deflections or waves (denoted as P, Q, R, S and T waves). It is conventional that

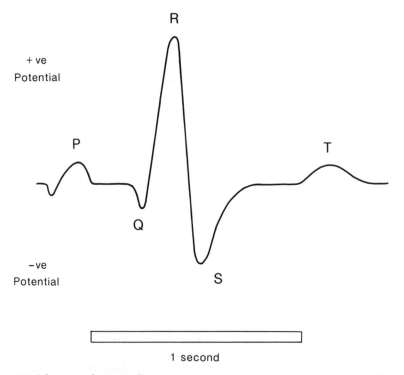

Fig. 4.5 Schematic diagram illustrating an electrocardiogram (ECG). The ECG is a graphic representation of the changes in electrical potentials that occur as a result of cardiac activity. The ECG trace indicates the different waves, or phases, of activity, denoted as P, Q, R, S, and T waves.

a positive potential is denoted by an upward deflection, and a negative potential by a downward deflection (Fig. 4.5). There are clear links between the ECG and the mechanical events of the cardiac cycle. The ECG can give information about heart rate, the rates of spread and decay of the electrical impulses, and the cardiac rhythm.

Cardiac output

The ECG does not, however, give any information about pumping efficiency, blood pressure, stroke volume or cardiac output. The *stroke volume* refers to the amount of blood pumped per cardiac cycle, and the *cardiac output* relates to the volume of blood pumped within a given time interval, e.g. ml min^{-1}. Thus, cardiac output is represented by (stroke volume × heart beat frequency), and any factors that influence either the stroke volume or the frequency at which the heart beats will affect cardiac output. Such factors may include water temperature, the oxygen content of the water and the activity level of the fish.

When data are required for comparative purposes, most physiologists attempt to determine baseline levels for cardiac output using fish at rest. Cardiac outputs generally lie within the range $10-100$ ml min^{-1} kg^{-1}, with scombrids and thunnids being at the top end of the range, salmonids intermediate, and more sluggish species such as the spotted dogfish, *Scyliorhinus canicula*, Atlantic cod, *Gadus morhua*, and sea raven, *Hemitripterus americanus*, being towards the lower end of the range.

In the vast majority of fish species, changes in cardiac output are brought about more by changes in stroke volume than by increases and decreases in heart rate. Resting heart rates in many species lie within the range $30-60$ bpm (beats per minute), and maximum heart rates may be only $20-30\%$ greater than this. This upper limit to heart rate is probably imposed by the supply of O_2 to the cardiac muscle. High heart rates will require rapid diffusion of O_2 to the myocardium. In many fish species the rate of diffusion may be limited due to low concentrations of myoglobin in the muscle and by the absence of a separate coronary circulation.

In the thunnids, on the other hand, which have high cardiac levels of myoglobin and a well-developed coronary circulation, heart rates at high activity levels may be two- to threefold as high as those at rest, e.g. an increase from $90-120$ bpm to $250-260$ bpm. It appears, therefore, that, in tunas, changes in cardiac output may be governed more by changes in heart rate than by changes in stroke volume. Thus, the majority of fish increase cardiac output by increasing stroke volume to a greater extent than heart rate (volume-modulated cardiac pumping), whereas the thunnids, in common with mammals, birds, reptiles and amphibians, increase cardiac output via increases in heart rate (frequency-modulated cardiac pumping).

Muscular contractions during the cardiac cycle

The changes in electrical activity, recorded as the ECG, will be followed by contractions of the muscular tissues in the different chambers of the heart. This will result in a series of changes in both volume and pressure.

When the blood volumes contained within two chambers are in open communication with each other, the pressure profiles will display a common pattern of changes. The chamber that is contracting, and expelling blood, will, however, have a slightly higher pressure than the chamber into which the blood is flowing. When the valves between chambers are closed, the chambers are isolated from each other and the pressure profiles of neighbouring chambers are at liberty to deviate from each other. Thus, recordings of the pressure changes within each chamber during the course of the cardiac cycle will give revelations about when the chambers are in contact, and blood is flowing between them, and about the times at which the valves between the chambers open and close (Fig. 4.6).

The cardiac cycle can be considered to commence in the sinus venosus and atrium, with the electrical activity denoted by the P wave of the ECG, being

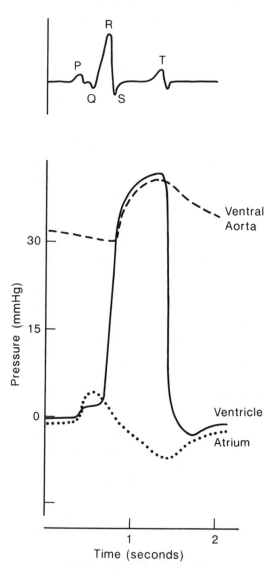

Fig. 4.6 The events of the cardiac cycle, showing both the ECG and the pressure changes that occur in the different chambers of the heart, and in the ventral aorta.

followed by contractions of the muscles in the atrial wall. The muscular contractions cause the atrium to be compressed towards the roof of the ventricle, and this results in a slight rise in pressure (2–5 mmHg) (Fig. 4.6). Thus, blood flows through the open atrio-ventricular valve and into the ventricle.

This flow results in both an increase in ventricular pressure and a passive stretching of the muscle fibres in the ventricular wall. At about the same time, electrical activity occurs in the ventricle (the QRS wave) and this is shortly (0.1–0.3 s) followed by muscular contraction of the ventricle. These muscular contractions lead to a sharp rise in pressure, and the closure of the atrio-ventricular valves prevents backflow into the atrium.

The pressure increase eventually becomes sufficiently high to force the blood into the conus, or bulbus, arteriosus and ventral aorta (Fig. 4.6). The end of ventricular contraction is indicated by the T wave of the ECG. Following the cessation of muscular contraction, there is a rapid fall in pressure. Backflow from the ventral aorta is prevented by closure of the semilunar valves guarding the opening of the ventricle.

There is then a steady decline in the pressure within the ventral aorta as blood passes through the gills and into the efferent vessels taking blood to the body tissues. This completes the cycle, and further blood flow requires the initiation of a new cycle in the atrium, but in order for this to occur the atrium must be refilled with blood.

The filling of the atrium occurs at the same time that the ventricle is contracting. Atrial pressure is very low at this time. As blood is expelled from the ventricle there is a reduction in volume of this chamber, but the attachment of the heart to the semirigid pericardium prevents any marked reduction in total volume. In other words, the reduced volume of the ventricle is, to a large extent, compensated for by an increase in atrial volume resulting from the pull of the pericardial attachments. An increase in atrial volume results in a reduction in pressure (or the generation of a suction pressure) (Fig. 4.6), which assists in drawing blood from the large venous sinuses and sinus venosus. Thus, atrial filling is aided both by the pressure generated by the venous pumps and by a suction pressure arising from within the pericardial sac.

The muscular contractions and pressure changes that occur during the course of the cardiac cycle provide the primary propulsive force for blood flow. The blood that leaves the heart passes first through the gills and is then transported via the major vessels to the systemic circulation.

4.6 GASEOUS EXCHANGE: THE RESPIRATORY GASES

Respiration can be defined as the sum of the processes by which the respiratory gases oxygen (O_2) and carbon dioxide (CO_2) pass from the environment to the tissues, and vice versa. Thus, respiration encompasses the movement of the gases between the environment and the circulatory system at the gas exchange organs, and is also concerned with the transport of the respiratory gases between the respiratory organs and the various internal organs and tissues.

Physiochemical phenomena will dominate any discussion of respiration because O_2 and CO_2 have very different properties with respect to both their rates of diffusion and the physical form in which they occur in various respiratory media and body fluids. Thus, the processes of transport and elimination of CO_2 are not simply a reversal of those seen in connection with the uptake of O_2 at the respiratory surface and transport to the tissues.

Two important variables governing respiration will be the concentrations of the respiratory gases and their partial pressures. The *concentration* refers to the quantity of a particular gas and the *partial pressure* is an expression defining the proportion of the total atmospheric pressure of a gas mixture that is contributed by a particular gas. In other words, the partial pressure exerted by a gas is directly proportional to its fractional concentration in the gas mixture. For example, if a gas mixture is composed of 40% O_2 and 60% nitrogen (N_2) and the overall gas pressure is 500 mmHg, the O_2 partial pressure (P_{O_2}) is $500 \times 0.4 = 200$ mmHg. Most importantly, it must be realized that the partial pressure of a particular gas depends upon its fractional concentration in the overall gas mixture and the barometric pressure, but not on the absolute quantity of gas present.

Gas solubility

Gases are soluble in water, and if water is brought into contact with a gas atmosphere, some of the gas molecules will enter the water and go into solution, i.e. the gas will become dissolved in the water. Eventually an equilibrium will be reached when an equal number of gas molecules will enter and leave the solution per unit time. The amount of gas that is then dissolved depends upon the pressure of the gas in the gas phase. (The dissolved gas exerts no measurable pressure and it is, therefore, somewhat misleading to use the term partial pressure when referring to dissolved gases. The term *tension* is usually preferred, with the tension of a gas in solution being defined as the pressure of the gas in the gas mixture with which the particular solution is in equilibrium. In practice, the terms partial pressure and tension are often used interchangeably.) In addition to the pressure of the gas in the gas phase, factors such as the nature of the gas itself, the temperature, and the presence of other solutes in the water will also have a marked influence on solubility.

Two coefficients have been employed to express the results of solubility measurements with gases. The first is known as the *absorption coefficient*, and this may be defined as the volume of gas (in millilitres STPD) dissolved in 1 l of water when the pressure of the gas is 1 atm (760 mmHg). The amount of gas is expressed as the volume this gas would occupy if the dry gas were at 0°C and 1 atm pressure; this is designated standard temperature and pressure dry (STPD). Thus, if v_0 is the volume of gas dissolved reduced to STPD, V is the volume of the solvent, and P_G is the partial pressure of the gas in atm, then the absorption coefficient is given by $v_0/(V \times P_G)$. By rearranging this to

$$v_0 = \text{absorption coefficient} \times (V \times P_G) \qquad (4.4)$$

it can be seen that the amount of gas dissolved in a given volume of solvent will depend upon the partial pressure of the gas. If the gas partial pressure is doubled, then twice as much gas will be dissolved.

In other words, a gas will dissolve in any given solvent according to its partial pressure in the gas phase, but there will be marked differences in gas concentration between the two media. Thus, if water is at gaseous equilibrium with a 40% O_2:60% N_2 mixture overlying it, and the total partial pressure is 500 mmHg, the P_{O_2} of the gas mixture dissolved in the water will be 200 mmHg even though the concentration of the O_2 dissolved in the water is far lower than the concentration of the O_2 in the gas mixture.

The second coefficient used in describing the solubility of gases is the *coefficient of solubility*. This may be defined as the volume of the gas taken up by a unit volume of the solvent under a particular set of temperature and pressure conditions. In other words, if v is the volume of the dissolved gas, measured at temperature T and a partial pressure P_G atm, then $v = (T \times v_0)/(T_0 \times P_G)$, where T_0 is 0°C i.e. 273 K. A re-arrangement gives

$$v_0 = (v \times P_G \times T_0)/T \qquad (4.5)$$

which shows that the solubility of a specific gas will decrease as temperature increases. For example, under equilibrium conditions between atmospheric oxygen (P_{O_2} within the range 154–158 mmHg) and fresh water the solubility of O_2 decreases from about 8 ml l^{-1} at 10°C to 5.6 ml l^{-1} at 30°C.

The solubility of gases will also be influenced by the presence of solutes in the water, and an increase in dissolved solids will reduce the solubility of dissolved gases. Dissolved gases will not, however, affect the solubilities of other gases. In other words, any gas will dissolve according to its own partial pressure in the gas phase, independent of the presence of other gases. The fact that dissolved solids influence gas solubility means that the solubilities of atmospheric gases are lower in sea water than in fresh water, the difference in oxygen solubility between the two media being approximately 20%.

In addition it must be realized that the solubilities of different gases are markedly different. For example, if the solvent is water, and temperature is held at 15°C, the solubility of oxygen will be 34 ml O_2 l^{-1} water when the gas pressure is 1 atm. Corresponding solubilities for CO_2 and N_2 are about 1020 and 17 ml l^{-1} water, respectively. Thus, CO_2 is 30 times more soluble in water than O_2, whereas N_2 is only half as soluble as O_2.

Finally, it is important to distinguish between both the coefficient of solubility and absorption coefficient, and the *capacitance coefficient*. The capacitance coefficient may be defined as the increase in concentration that results from each unit increase in partial pressure of a particular gas. It must be realized that the capacitance coefficient includes not only the gas that is physically dissolved in the solvent, but also chemically bound gas.

In water, for example, where there are only very small amounts of chemically bound molecular oxygen, the coefficient of solubility and the capacitance coefficient for O_2 will be almost identical. In the same water, however, carbon dioxide can be present as dissolved CO_2, as the bicarbonate ion (HCO_3^-), and as carbonate (CO_3^{2-}), so the capacitance coefficient is higher than the coefficient of solubility. In body fluids, where both carbon dioxide and oxygen may be chemically bound in large quantities, the capacitance coefficients are very much higher than the coefficients of solubility. Thus, when a solvent contains large amounts of a gas in a chemically bound form, it is possible for the gas to diffuse down its partial pressure gradient and, at the same time, up a concentration gradient. In respiration this is of particular importance in connection with the transfer of oxygen from the water to the blood.

4.7   RESPIRATORY ORGANS

In the vast majority of fish species the main organ of respiration is the gills, and the gill area may represent 60–75% of the total body surface area in adult fish. The gills are, however, not well developed in embryonic and larval fish, and the surface area of the gill buds is comparatively small.

During the earliest part of development the larval fish rely on diffusion of gases across the general body surface to meet tissue oxygen demand. This is possible because of the large surface-to-volume ratio of the larvae, and also because there are no scales or other tissues that would hinder gaseous diffusion. In the larvae, diffusion over the skin may account for 85–90% of the gaseous exchange, and in some species oxygen uptake may be enhanced by the presence of a 'red layer' running almost completely around the body just beneath the skin. This red layer is rich in the muscle pigment myoglobin, which is capable of binding oxygen. Consequently, it is suspected that this red layer has an important respiratory function in the larvae.

In adult fish the skin may account for up to 20% of the oxygen uptake. In adults the skin obtains much of its oxygen, and in some species all of it, by direct diffusion from the water. It is an open question whether all the oxygen taken up over the skin is used locally or whether it also makes a contribution to supplying the oxygen demand of other tissues.

Whilst the vast majority of fish species are unimodal water-breathers relying on the gills for extracting oxygen from the water, a number of species are bimodal, that is, they can breathe in both air and water. A small number of these bimodal species use the gills when respiring in both air and water, but the majority of bimodal breathers use the gills when respiring in water, and some other, respiratory organ (air-breathing organ – ABO) when breathing air.

4.8 AIR-BREATHING FISHES

The majority of the bimodal species occur in tropical or subtropical swamps or freshwater lakes where they may regularly experience oxygen-deficient (hypoxic) conditions. There are, however, some bimodally breathing fish species in temperate-zone habitats. The bimodally breathing fish may be either *obligate* or *facultative* air-breathers, with most being the latter. Facultative air-breathers will breathe using the gills under normoxic conditions and only switch to breathing air when the oxygen content of the water falls below some critical level. Obligate species, on the other hand, breathe air irrespective of the levels of dissolved oxygen in the water.

There is a range of tissues that can be modified to have an air-breathing function. These include epithelial tissue in the buccal, opercular and

Table 4.1 Air-breathing organs (ABOs) used by a range of bimodally breathing fish species

Organ used for aerial respiration	Fish	Comment
Gills/skin	*Hypopomus, Anguilla, Periophthalmus*	Facultative air breather
Mouth/opercular cavity		
Pharyngeal epithelium	*Electrophorus*	Obligatory
Diverticula	*Clarias, Anabas*	Facultative
	Ophiocephalus	Obligatory
Gastrointestinal tract		
Stomach	*Eremophilus, Ancistrus*	Facultative
Intestine	*Hoplosternum*	Obligatory
	Misgurnus	Facultative
Gas bladder	*Arapaima, Lepidosiren, Protopterus*	Obligatory
	Amia, Lepisosteus, Neoceratodus	Facultative

pharyngeal regions, the stomach and intestine, and the gas bladder (Table 4.1).

The European eel, *Anguilla anguilla*, is capable of moving quite large distances over land whilst crossing from one water body to another. Most of these overland excursions occur at night through damp grass and moist vegetation. When the eel is out of water it keeps the gill cavity filled with air, and some oxygen is taken up over the gills. The air in the gill cavity is replaced at regular intervals, but the oxygen removed from this air is insufficient to meet the entire oxygen demand. Oxygen taken from the air is supplemented with oxygen removed from the gas bladder, and the skin also plays an important role in respiration when the fish are on land.

When the eels are in water over 80% of the total oxygen uptake is via the gills, but in air this falls to 30–35%, and the amount of oxygen taken up over the skin increases to over 50% of the total. Even though both the gills and the skin function as respiratory organs in air, the amounts of oxygen taken up are usually insufficient to meet demand. When the fish are held out of water for periods of several hours there will be a substantial accumulation of lactate due to the anaerobic metabolism of respiratory substrates.

Mudskippers of the genus *Periophthalmus* also rely upon the skin and gills when respiring in air. In at least one species, *Periophthalmus vulgaris*, the opercular chambers, which act as an air store during aerial respiration, are highly modified into sac-like structures bounded by highly vascularized and folded epithelium. The mudskippers are common in muddy estuarine areas and mangrove swamps of tropical regions, but some of the closely related gobies of temperate regions may also spend some time out of water, and/or be capable of breathing air when exposed to hypoxic conditions.

Some air-breathing fish use a combination of the gills and a heavily vascularized buccal cavity to extract oxygen from air held in the buccal and opercular cavities. The swamp eel, *Synbranchus marmoratus*, for example, not only switches to aerial respiration whilst making terrestrial forays in search of prey, but also breathes air when exposed to hypoxic conditions in the stagnant swamp waters in which it lives. Whilst the gills may be the primary respiratory organ, the walls of the gill chambers are highly vascularized and certainly make some contribution to gas exchange. Thus, in the swamp eel the ABO consists of large numbers of highly vascularized minute papillae spread throughout much of the buccopharyngeal mucosa, giving the appearance that the gill chambers are lined with a fine red velvet carpet.

In the obligate air-breathing electric eel, *Electrophorus electricus*, of South America, it is the highly vascularized oral cavity that functions as the ABO. In this species there is a multiple folding and papillation of the buccal epithelium. The papillations project into the oral cavity from both the floor and the roof of the mouth, and thus form extensive vascular surfaces for aerial gas exchange. In the electric eel the ABO may be the primary respiratory organ because the gills of adult fish are reduced in size and seem to have become almost vestigal.

A variety of branchial chamber modifications occur in air-breathing fish. For example, the gouramies and snakeheads, e.g. *Anabas, Macropodus, Trichogaster* and *Ophiocephalus* spp., possess a labyrinthine suprabranchial organ that functions as an ABO. The organ takes the form of elaborate, highly vascularized plates or tufts which are derived from branchial tissue. The ABOs lie within the dorsally expanded opercular cavity. Air taken in through the mouth passes through the labyrinth where oxygen is absorbed into the blood. As fresh air is taken in, the old air is expelled from the labyrinth and is then forced out via the openings between the body wall and the gill covers.

In some species of bimodally breathing fish, parts of the gastrointestinal tract, notably the stomach or portions of the small intestine, have become modified to serve a role in respiration. The part of the gut functioning as an ABO becomes highly vascularized and this facilitates gas exchange between the blood and the swallowed air. Air is then voided either via the anus or via the mouth. Members of the families of armoured catfishes (Callichthydae and Loricaiidae) that inhabit fresh waters of Central and South America offer typical examples of gastrointestinal air-breathers.

The fish gas bladder may have originated to function as an ABO, but in most extant species the bladder serves primarily in buoyancy control, sound production or sound detection. Even amongst the species in which the gas bladder functions as an ABO there is a great variety of structural complexity. In some species, such as the bowfin, *Amia calva*, the bladder remains essentially sac-like although the bladder wall is more highly vascularized than in those species which use the bladder solely for buoyancy control.

Several species, including the obligate air-breathing pirarucu, *Arapaima gigas*, of South America, have gas bladders in which the inner wall has become subdivided by a series of septa. The greatest septation is seen in the gas bladders of the dipnoans, or lungfish. In the lungfishes, internal septa, ridges and pillars effectively divide the gas bladder into a series of smaller lateral compartments that open into a central cavity. Even within the lungfishes, however, there is much variation, with the single 'lung' of the Australian lungfish, *Neoceratodus fosteri*, being less highly divided than the paired 'lungs' of the African, *Protopterus* spp., and South American, *Lepidosiren paradoxa*, species.

Bimodally breathing fish possessing ABOs display modifications to the cardiovascular system compared with those seen in unimodal species. The circulatory systems of the bimodal fish differ in a number of respects from those seen in the unimodally breathing fish in order to incorporate a series of vessels that allow a shunting of blood flow through the ABO during aerial respiration (Fig. 4.7). When breathing air, bimodal breathers will use the

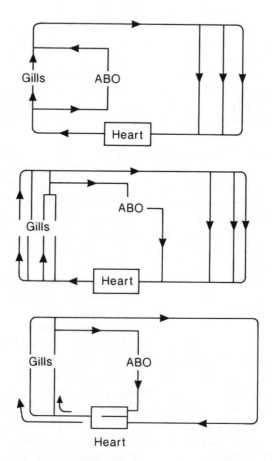

Fig. 4.7 Simplified schematic diagrams of modifications to the circulatory system displayed by bimodally breathing fish using different types of air-breathing organ (ABO). In species in which the ABO consists of modified gills, the buccal mucosa or chambers extending from the opercular cavity the ABOs are in parallel with the gills so that blood enters directly into the systemic circulation [upper figure]. In the majority of bimodally breathing species the blood that has passed through the ABO must also flow through the gills before entering the systemic circulation. There are modifications to the circulatory system when the gas bladder functions as an ABO, as in some teleosts [middle figure], and in the lungfish [lower figure]. Note the partial separation of the blood flow in the lungfish.

ABO for oxygen uptake, but the blood will flow from the ABO to the gills, and the gills will be used for carbon dioxide excretion and ion exchange.

4.9 GILLS AND GAS EXCHANGE

Compared with air, which contains 200 ml oxygen per litre, the oxygen content of water is very low (0.04–12 ml l^{-1}). This means that fish must pass more of the medium over the respiratory surface, for any given oxygen gain, than do terrestrial animals that breathe air. Furthermore, water has a higher density and viscosity than air, and therefore more work is required in ventilation (pumping the medium over the respiratory surface). The liberation of energy for this work uses oxygen and it is estimated that approximately 10% of the oxygen uptake in fish is used in pumping water across the gills.

Each gill consists of two sets of filaments attached to the gill arch, and the gills are supported by the skeleton of the branchial arches (Fig. 4.8). Bony fish have four branchial arches and each of these has a series of filaments which form a double V-shaped row. Each filament is supported by a gill ray. The surfaces of the filaments are thrown up into a number of secondary folds which increase the surface area for gaseous exchange. The gill filaments are richly supplied with blood from the ventral aorta via the afferent branchial arteries which divide repeatedly to supply each of the secondary lamellae.

Ventilation of the gills

The flow of water across the gills is generally maintained by active ventilation. There are several features of gill ventilation and perfusion that ensure that fish gills are extremely effective organs for the purpose of gaseous exchange.

The essential feature of the countercurrent exchange system observed in the gills is that the flows of water and blood are in opposite directions. This ensures that blood partly loaded with oxygen meets water that has had little oxygen removed from it. Therefore, a fairly constant gradient of oxygen tension is maintained between the blood and water throughout the passage over the gills. This ensures that a high level of oxygen transfer can be achieved (Fig. 4.9).

When a teleost fish breathes, it opens its mouth and water is drawn into the buccal cavity. After passing over the gills, the water leaves the opercular cavities through the slits which appear when the gill covers (opercula) are expanded and come away from the side of the fish. This discontinuous flow of water into and out of the system may give a false impression of the water flow across the gills themselves. In practice, despite the fact that the flow is

(a)

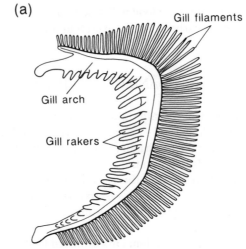

Gill filaments

Gill arch

Gill rakers

(b)

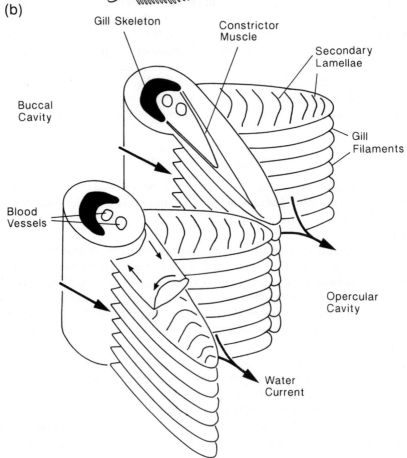

Gill Skeleton

Constrictor Muscle

Secondary Lamellae

Buccal Cavity

Gill Filaments

Blood Vessels

Opercular Cavity

Water Current

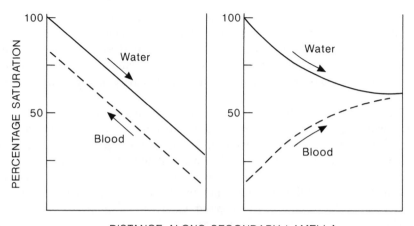

DISTANCE ALONG SECONDARY LAMELLA

Fig. 4.9 The importance of the countercurrent flow of water and blood at the fish gill for the efficient transfer of gases between the two media. Note that parallel flow (right) results in a far less efficient transfer than does countercurrent flow (left).

pulsatile, water flows over the gills almost continuously throughout the entire breathing cycle.

Bucco–opercular pump and ventilation

The mechanism which makes continuous water flow possible involves the operation of two muscular pumps which are slightly out of phase with each other. In the teleost fish, the pumping actions caused by changes in the volumes of the buccal and opercular cavities are produced by muscular action. There are two major phases in the cycle, one in which the opercular pumps are active, and the other when the buccal pump forces water across the gills. The two transition phases take up only one-tenth of the entire cycle, and contribute little to it.

During the inspiratory phase the buccal cavity expands and water enters through the mouth. At the same time, the opercular cavities expand, but water cannot enter through their external openings because a thin membrane around their outer rim functions as a valve.

During expansion of the opercular cavity, the hydrostatic pressure becomes less than that in the buccal cavity and water is drawn across the gills. In other words, the opercular cavity acts as a suction pump.

Fig. 4.8 Schematic diagrams showing (a) the gross structure of the fish gill, and (b) the directions of blood flow (small arrows) and water flow (large arrows).

During the phase of decrease in volume, which also starts in the buccal cavity, the pressure becomes greater than in the external medium. The mouth is closed to prevent backflow of water. The increase in pressure within the buccal cavity is greater than that in the opercular cavities and water continues to pass over the gills, with the buccal cavity acting as a pressure pump.

Throughout almost the entire cycle, there is an excess pressure tending to force water across the gills from buccal to opercular cavities. There is a brief period, however, when the pressure difference is reversed, but the inertia of the water is such that any reversal in water flow is unlikely. In this way a continuous flow of water is maintained over the gills in a direction opposite to that of the blood flow through the lamellae and a very high percentage (75–80%) of the oxygen in the inspired water can be removed (Fig. 4.9).

This double pump system is rather complex and functions under the control of the series of muscles that serve to alter the volumes of the buccal and opercular cavities. Expansion of the buccal cavity takes place mainly in the ventral direction following contraction of the large sternohyoideus muscle which runs from the cleithrum of the pectoral girdle to the hyoid bone. Lateral expansion of the buccal cavity is brought about by contraction of another muscle, the levator arcus palatini. Contraction of this muscle also initiates the expansion of the opercular cavity.

Movement of the operculum is largely brought about by contraction of the dilator operculi muscles, whereas ventral and lateral expansion of the opercular cavity is brought about by separation of the branchiostegal rays. These latter movements appear to be largely produced by contractions of the hyohyoideus muscles.

Whilst this is probably a reasonably accurate description of the major movements and effector muscles influencing the buccal and opercular pumps, it is by no means a complete account. The entire pumping cycle almost certainly involves a complex series of interactions between relatively large numbers of muscles and skeletal elements, each of which has something to contribute to the cycle.

Amongst the teleosts there are several variations on the basic theme, with the relative contributions of the buccal and opercular pumps differing between species. The modifications to the cycle appear to be largely linked to the lifestyles of, and habitats occupied by, the different species. Amongst the teleost species which swim more or less continuously, the buccal pressure pump appears to be well developed, but the more sluggish, bottom-dwelling species are often more reliant upon the opercular suction pump.

Thus, bottom-dwellers, such as gurnards, *Trigla* spp., dragonets, *Callionymus* spp., and pleuronectids, tend to have enlarged opercular cavities. There may be large numbers of supportive skeletal rays in the opercular region and the opercular openings may be relatively narrow. In the dragonets, for example, the buccal pump makes only a very minor contribution to gill ventilation and the water is drawn over the gills due to the suction pressure created by the gradual expansion of the opercular cavity. Then, during a brief contraction phase, the water is forced out of the cavity through the narrow opercular openings.

In flatfish (e.g. pleuronectids) which lie permanently on one side and may be almost completely buried in the substrate when resting on the bottom, other problems arise. The gills are equally developed on both the left and right sides of fish such as the plaice, *Pleuronectes platessa*, and Dover sole, *Solea solea*, and it seems certain that water is pumped through the opercular cavities of both sides. Because the fish often lie buried, there is a potential danger that sand, or other particulate material, could enter the opercular cavities and cause damage to the delicate gill membranes. In these species, however, there does not seem to be any reversal of the differential pressure between the buccal and opercular cavities. In addition, active control of the opercular valves probably prevents the entry of any current carrying sand particles.

Ventilation volume and ventilation frequency

As fish become increasingly active they will require more oxygen. Consequently, the rate of water flow over the gills must be increased. The increase in the water flow over the gills (ventilation volume) is achieved by small increases in breathing rate (ventilation frequency) and large changes in the stroke volume of the respiratory pumps.

As a consequence of increased water flow rates, the time of contact between water and gill tissue is reduced from 250 ms in a resting fish to about 30 ms at high swimming speeds. Despite the increase in water flow, the amount of oxygen removed from the water passing over the gills (oxygen extraction efficiency) is not drastically reduced. One of the factors contributing to the maintenance of oxygen extraction efficiency is lamellar recruitment. In a resting fish, only about 60% of the secondary lamellae are perfused with blood, but as the fish begin to swim, more of the gill lamellae are recruited and supplied with blood. Lamellar recruitment serves to increase the gill area available for oxygen uptake and gaseous exchange.

Ram ventilation

The pumping of a dense, viscous medium, such as water, over the gills is energy consuming, and as the activity of the fish increases then progressively more energy is used by the muscles driving the respiratory pumps. Thus, the energetic costs of ventilation increase with increasing activity and some species adopt an alternative ventilatory strategy at high swimming speeds.

Active species such as striped bass, *Morone saxatilis*, and bluefish, *Pomatomus saltatrix*, and several species of salmonids use cyclic ventilatory movements, involving the branchial muscles, when at rest or swimming at low swimming speeds. At higher swimming speeds they adopt *ram ventilation*, open-mouthed swimming which uses the swimming musculature to force the water over the gills. This alternative ventilatory mode allows the fish to dispense with the buccal and opercular pumps at high swimming speeds, and yields energetic savings.

The savings are, however, less than the total energetic expenditure required to drive the respiratory pumps. The adoption of the ram ventilatory mode requires the fish to swim open-mouthed with the opercula flared. This means that there is some disruption of the body profile presented to the water, and this leads to increased turbulence. Consequently, there are increased swimming costs for fish that adopt ram ventilation, but nevertheless there are energetic savings to be made compared with using the ventilatory pumps at all swimming speeds.

Some of the most active pelagic fish species (tunas and mackerels) are highly dependent upon ram ventilation, and usually swim continuously in order to ensure the passage of water over the gills. It is sometimes said that the Atlantic mackerel, *Scomber scombrus*, must swim continuously in order to maintain water flow over the gills and will suffocate once swimming ceases – this is not true because mackerel can use the branchial pump system for ventilation and they are, therefore, not totally dependent upon ram ventilation.

Gill structure and gaseous exchange

Gill ventilatory mechanisms are related to the lifestyle of the fish, and there are corresponding variations in gill structure. For example, the number of gill filaments tends to be greater in active species than in benthic species.

Each gill filament carries numerous secondary lamellae at right angles to the long axis and it is these lamellae which constitute the respiratory surface (Fig. 4.8). The total numbers of lamellae possessed by different fish species are obviously dependent upon both the numbers of filaments, and the counts of lamellae per unit length of filament. Bottom-living and sedentary species usually have 10–20 lamellae per mm, most species lie within the range 15–30 per mm and highly active species may have 40 or more lamellae per mm filament. Thus, active and sluggish species differ greatly with respect to gill area for gas exchange. For example, thunnids have a gill area of 1500–3500 $mm^2 g^{-1}$ body weight, scombrids over 1000 $mm^2 g^{-1}$ and other active species range from 500 to 1000 $mm^2 g^{-1}$, whereas the majority of fish species have gill surface areas of 150–350 $mm^2 g^{-1}$ body weight.

The oxygen is taken up from the water by diffusion across the gill lamellae, and the importance of the gill dimensions can be seen by considering Fick's equation:

$$M = (P_W - P_B) \times A \times K \times d^{-1} \tag{4.6}$$

where M is the rate of diffusion. $(P_W - P_B)$ is the oxygen partial pressure difference between the water and the blood: it is this difference which drives the diffusion of the gas over the gill tissue. K, Krogh's diffusion constant, describes the diffusion characteristics of the tissue, A is the gill area in mm^2 and d is the length of the diffusion pathway (the thickness of the lamellar wall separating the blood and the water). From the Fick equation it follows that oxygen uptake is aided by having a large P_{O_2} difference between the blood and the water, a large gill area and thin walls to the secondary lamellae (Fig. 4.3). Not only do active species have a greater gill surface area than sluggish species, but the lamellar walls also tend to be thinner. For example, thunnids have lamellar walls about 0.5–1.0 μm thick, most species lie in the range 2–4 μm and the less active species have lamellar walls up to 10 μm or more thick.

4.10 TRANSPORT OF RESPIRATORY GASES

Once the oxygen has been absorbed at the respiratory surface it is transported around the body, to the tissues, in the blood. A certain amount of oxygen is dissolved in the blood plasma but the oxygen-carrying capacity of the blood is increased by the presence of a respiratory pigment, *haemoglobin*, contained within the red blood cells. Haemoglobin is a metalloporphyrin, that is, it is a combination of haem (an iron porphyrin) with globin, a protein. The globin protein units differ between the haemoglobins of different fish species and this imparts different properties to the haemoglobins of the various species. Indeed, many species possess more than one type of haemoglobin, differing in their abilities to bind and transport oxygen. The types of haemoglobins present in the blood may also change during the life of the fish. Some salmonids, for example, may have up to 18 different haemoglobins during their life cycle, and it is not uncommon for fish blood to contain five to ten different haemoglobins at any one time.

The ferrous iron atom in the haem can bind to one molecule of oxygen, this being termed oxygenation of the haemoglobin (sometimes the term 'oxidation' is used, even though the binding reaction is reversible and non-destructive). The haemoglobins are tetrameric, i.e. have four chains, so each haemoglobin can bind with four molecules of oxygen. The haemoglobin is contained within the red blood cells, which are elliptical biconvex discs. They range in size from about 20×14 μm to 10×7 μm, depending upon species, and there may be up to a four- to fivefold difference in the volume of the individual red cells possessed by different fish species. In addition to red cell volume, the abundance of the red cells in the blood also shows considerable variation.

A measure of red cell abundance is given by the *haematocrit* (Hct). This is determined by centrifuging a blood sample and then expressing the volume of the blood cells as a percentage of the total volume of the sample. At one extreme of the haematocrit spectrum lies the icefish, *Chaenocephalus aceratus*, a species with very few, haemoglobin-free, red blood cells. At the other extreme are the scombrids and thunnids, which may have haematocrits of 40–50%. The majority of fish have haematocrits within the range 15–30%, with the more active species tending to be at the top end of the range. It would be expected that the haematocrit and haemoglobin content of the blood would be correlated, and this is found to be the case – the greater the number of red cells, the greater the amount of haemoglobin. Blood volumes (blood volume refers to the proportion of the body occupied by blood) and haemoglobin contents tend, however, to be inversely correlated: fish that have large blood volumes (10% or over) tend to have a low haematocrit and a low haemoglobin content, compared with those species having a more normal blood volume (3–6%).

One reason for this inverse relationship can be seen from an examination of the oxygen transport properties of the blood. The amount of oxygen that could be carried in solution is about 0.5–0.9 ml per 100 ml blood, but with pigment present, fish bloods carry 5–16 ml oxygen per 100 ml. The absence of pigment would mean either that the blood volume would have to be greatly increased or that the blood would have to be pumped much faster.

The importance of this increased capacity to transport oxygen associated with the possession of haemoglobin can be seen by examining the consequences for rates of blood and water flow through the gills. Gas exchange over the gills is optimized when

$$(V_W \times \beta_W)/(V_B \times \beta_B) = 1 \qquad (4.7)$$

where V_W is the ventilation volume, V_B is the blood flow through the gills (usually expressed as cardiac output), and β_W and β_B are the capacitance coefficients of the water and blood, respectively. In the majority of fish species, the presence of haemoglobin in the blood leads to the capacitance coefficient of the blood being much higher than that of the water, so that gas exchange is optimized when gill ventilation volume is about 10–20 times the cardiac output. In other words, far more water is passed over, than blood pumped through, the gills.

In the haemoglobinless icefish, *C. aceratus*, the capacitance coefficient of the blood is much reduced in comparison with fish species possessing haemoglobin. This has important consequences for the relationship between ventilation volume and cardiac output. In this species, optimum conditions for gas exchange at the gills occur when ventilation volume is only 3–4 times cardiac output. Thus, a high tissue oxygen demand would require the fish to both ventilate the gills with large volumes of water and pump large volumes of blood through the gills.

The icefish has relatively low tissue oxygen demands, but the absence of haemoglobin in the blood has resulted in a suite of compensatory adaptations to the cardiovascular system. The blood volume of the icefish is, for example, about two- to fourfold higher than in most other teleosts, and the heart and blood vessels are comparatively large. The icefish heart is about three times the size of that of most teleosts, and makes up a similar proportion of the body mass to the hearts of the highly active tunas, and mammals. In addition the cardiac ouput of the icefish ($150–300$ ml min^{-1} kg^{-1}) exceeds that of even the most active mackerels and tunas ($100–120$ ml min^{-1} kg^{-1}), the high cardiac output of the icefish being achieved through large cardiac stroke volume rather than by a high heart rate (HR $= 25$ bpm). Nevertheless, the absence of haemoglobin in the blood appears to have locked the icefish into a relatively sluggish lifestyle.

Oxygen-binding properties of haemoglobin

The haemoglobin pigment functions by giving up oxygen where oxygen tension is lowest and binding to it at sites where tensions are high. Therefore, oxygenation of the haemoglobin is a function of the oxygen tension with which the pigment is in equilibrium – Hb $+ O_2 =$ HbO$_2$. High values of oxygen tension drive the equilibrium to the right and low to the left. The relationship between the oxygen tension and the percentage saturation of the pigment is not, however, a linear one.

When most of the pigment is in a deoxygenated state, the chance of an oxygen molecule striking one of the four binding sites on the haemoglobin molecule is relatively high. As the percentage saturation increases, the chance of any one oxygen molecule colliding with the diminishing number of binding sites is reduced so that increasingly large numbers of oxygen molecules are required to effect each increment in the saturation of the pigment. Thus, the relationship between percentage saturation and oxygen tension is expected to be an asymptotic curve. In practice the curve tends to be more or less sigmoid. This shows that the simple equilibrium equation is inadequate to explain the system (Fig. 4.10).

The functional haemoglobin molecule consists of four units and the binding of oxygen at one site influences the ease of binding at subsequent sites. These influences are nearly always positive, that is, later bindings are facilitated by earlier ones. In consequence, the binding curve is steeper

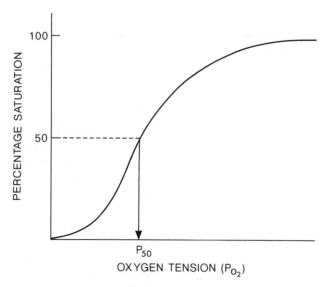

Fig. 4.10 The saturation curve of haemoglobin showing the effects of increasing oxygen tension on the percentage saturation of the haemoglobin. The oxygen tension at which half of the haemoglobin is bound to oxygen, P_{50}, is shown.

than the theoretical and starts with a concave slope. As oxygen binds to progressively more sites the availability of binding sites on the haemoglobin becomes limiting and the shape of the curve becomes sigmoid.

This haemoglobin subunit interaction can be quantified by plotting the oxygen dissociation curve as $\log [S/(100 - S)]$ against $\log P_{O_2}$, where S is the percentage of haemoglobin bound to oxygen (% saturation) and P_{O_2} is the oxygen partial pressure. The slope of the line is denoted by n, and this describes the degree of subunit interaction. If n is 1, there is no interaction. The larger the value of n the more sigmoid is the dissociation curve. It should be clear that the binding site interactions are advantageous in ensuring that large amounts of oxygen are taken, and given, up for relatively small changes in oxygen tension (Fig. 4.10).

Another important property of the haemoglobin that needs to be defined is its ability to bind oxygen – i.e. the oxygen affinity. The oxygen affinity of the haemoglobin is conventionally determined as the partial pressure of oxygen at which half the haemoglobin is saturated with, or bound to, oxygen (the P_{50}) (Fig. 4.10). The P_{50}s of the bloods of fish that live in well-oxygenated waters tend to be higher than those of species that inhabit oxygen-deficient habitats. Thus, if the oxygen dissociation curves are plotted for the haemoglobins of fish from different habitats, it will usually be found that the curves for fish from flowing and well-oxygenated waters will tend to lie to the right of those for fish that inhabit stagnant ponds and oxygen-deficient lakes.

For fish such as salmonids that have a blood with relatively high P_{50}, the haemoglobin can be loaded with oxygen at the gills because of the high oxygen tension of the water flowing over the gill membranes. A high P_{50} gives the added advantage that the tissues do not have to endure a low oxygen pressure before the haemoglobin will begin to offload oxygen.

For fish such as common carp, *Cyprinus carpio*, and tench, *Tinca tinca*, that may experience waters of low oxygen tension, the loading of the haemoglobin at the gills is assisted by the high oxygen affinity (low P_{50}). However, the fact that the haemoglobin has a low P_{50} means that oxygen will not begin to be offloaded at the tissues before the oxygen partial pressure has fallen to quite low levels.

It is important to realize that the P_{50} is not a single fixed value describing the oxygen affinity of a given fish haemoglobin under all possible conditions. The ability of haemoglobin to bind with oxygen will be influenced by a variety of factors, such as pH, temperature, ionic strength and the partial pressure of carbon dioxide (P_{CO_2}) in the water and blood. Thus, the oxygen affinities and dissociation curves of fish haemoglobins will be influenced by the prevailing environmental conditions.

For example, the P_{50} of kawakawa, *Euthynnus affinis*, blood falls from about 42 to 10 mmHg as pH increases from 7.15 to 7.95 (note that a low P_{50} denotes increased affinity for oxygen binding by the haemoglobin). This phenomenon of a decreasing affinity for oxygen binding with reduced pH is known as the *Bohr effect*.

The Bohr effect can be defined as the change in the P_{50} of the blood that occurs for each unit change in pH. Quantitatively, this can be expressed as the Bohr factor – $\delta\log P_{50}/\delta\log$ pH – which allows direct comparisons to be made between the oxygen-binding properties of the haemoglobins of different fish species. As a general rule-of-thumb it can be said that the bloods of fish living in well-oxygenated waters show a large Bohr effect (Bohr factor 0.5–0.6), so that their affinity for binding oxygen is markedly influenced by relatively small changes in pH.

Carbon dioxide will dissolve in water to form carbonic acid, and this will then dissociate to bicarbonate and hydrogen ions – $H_2O + CO_2 = H_2CO_3 = HCO_3^- + H^+$ – so the P_{CO_2} might be expected to exert some influence on pH, and thereby affect the oxygen affinity of the haemoglobin. Increased P_{CO_2} does influence the binding properties of haemoglobin, with an increase in P_{CO_2} resulting in a decreased affinity of the haemoglobin for oxygen. In rainbow trout, *Oncorhynchus mykiss*, for example, the P_{50} of the blood increases from about 16 to 34 mmHg as P_{CO_2} rises from 1 to 6 mmHg.

Gaseous exchange at the tissues

The aerobic metabolic processes of the body tissues result in the production of carbon dioxide. Because the oxygen affinity of the haemoglobin is affected by ambient levels of CO_2 this has a marked effect on the offloading of oxygen from the haemoglobin at the tissues. Since the Bohr effect is pH dependent, effects on offloading will also be induced by acid metabolites, such as lactate, that result from anaerobic metabolism in the body tissues.

Carbon dioxide from the tissues passes to the blood and then enters the red blood cells. In the red blood cells there is a rapid reaction between CO_2 and water, forming first carbonic acid, and then bicarbonate and hydrogen ions. The reaction between carbon dioxide and water normally occurs very slowly, requiring 25–90 s to come to equilibrium, but in the red blood cells the

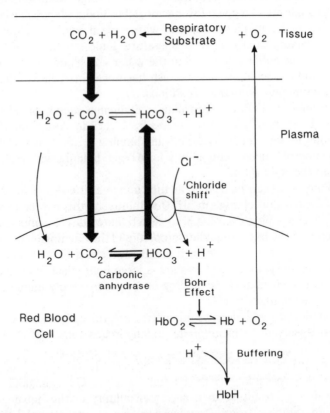

Fig. 4.11 Schematic representation of the offloading of oxygen from the haemoglobin at the body tissues. Hb denotes haemoglobin. Circle indicates ion-exchange mechanisms.

reaction is catalysed by the presence of a zinc-containing enzyme, *carbonic anhydrase.*

The hydration of the carbon dioxide requires that water enter the red blood cell, and this leads to an initial swelling. The hydration of the CO_2 leads to the production of bicarbonate and this is exchanged for chloride over the red cell membrane (chloride or Hamberger shift). This leads to partial restoration of red cell volume. The hydration of the CO_2 also results in the generation of protons (H^+), thereby influencing red blood cell pH. Reduced intracellular pH induces offloading of oxygen from the haemoglobin via the Bohr effect (Fig. 4.11). The oxygen so released diffuses out of the red blood cells and over into the surrounding tissues.

The deoxygenated haemoglobin acts as a buffer and binds to the protons, and will also bind to some of the carbon dioxide. Most of the carbon dioxide (90% or more) is, however, transported in the blood plasma in the form of bicarbonate. Some of the plasma bicarbonate results from the hydration of carbon dioxide that occurs at the uncatalysed rate in the plasma, but the vast majority is derived from the bicarbonate produced in the red blood cells. Thus, as the blood leaves the tissues the haemoglobin is in a deoxygenated state, due to the offloading of the oxygen, and plasma levels of bicarbonate are increasing.

Gaseous exchange at the gills

When the blood reaches the gills it is divided from the surrounding water by the thin membranes of the secondary lamellae (Fig. 4.3). In the majority of circumstances the carbon dioxide content of the water will be very low, so that the water acts as a carbon dioxide sink. However, the short time (1–3 s) that the blood is in the gills is insufficient to allow CO_2 production from bicarbonate at the uncatalysed rate. There is some carbonic anhydrase in the epithelial cells of the secondary lamellae. This could catalyse the reaction from bicarbonate to carbon dioxide, but the contribution made appears to be small compared with that of the red blood cells.

The bicarbonate probably enters the red blood cells from the plasma, and carbon dioxide is produced at the catalysed rate due to the presence of carbonic anhydrase. This intracellular reaction involves the sequestration of protons, leading to a slight increase in the red blood cell pH. The change in pH, in turn, has an influence on the oxygen affinity of the haemoglobin, leading to improved binding and on-loading properties (Fig. 4.12).

The carbon dioxide produced from the bicarbonate diffuses out into the plasma and over the gill epithelium to the surrounding water. Oxygen diffuses in the opposite direction to bind with the intracellular haemoglobin for transport to the tissues. Some bicarbonate can be removed from the plasma via ion-exchange mechanisms, but it is generally con-

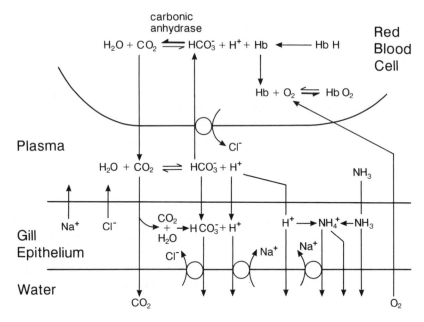

Fig. 4.12 Schematic representation of the binding of oxygen to haemoglobin – on-loading – and the excretion of carbon dioxide and ammonia at the gills. Hb denotes haemoglobin. Circles indicate ion-exchange mechanisms.

sidered that the vast majority of the excretion (about 85%) occurs as CO_2 (Fig. 4.12).

Bohr effect

The significance of the Bohr effect can be summarized by examining oxygen dissociation curves obtained under different conditions of ambient P_{CO_2}. Figure 4.13 represents the oxygen dissociation curves of haemoglobin at two different levels of carbon dioxide tension – those of the arterial (1 mmHg) and the venous (5 mmHg) blood, respectively. Conditions as the blood leaves the gills are defined by point A and those in the tissues are given by V.

As the blood passes into the tissues it encounters a high P_{CO_2} and the haemoglobin would offload some of its oxygen without there necessarily being any need for a fall in the tissue P_{O_2} (the Bohr effect increases the amount of oxygen release at the tissues by an amount indicated by v − V). The conditions prevailing in the tissues are, however, a combination of high P_{CO_2} and low P_{O_2}, which leads to a considerable offloading of oxygen from the haemoglobin (indicated by A − V).

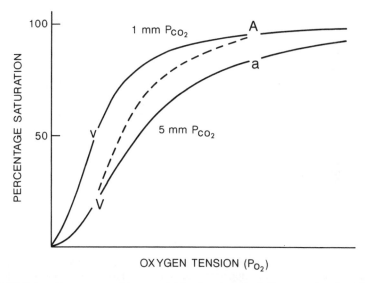

Fig. 4.13 Saturation curves of haemoglobin showing the influence of carbon dioxide (different P_{CO_2}) on the ability of the haemoglobin to bind with oxygen – the Bohr effect. See text for further details.

Conversely, when the blood returns to the gills, the excretion of CO_2 increases the affinity of haemoglobin for oxygen, and assists on-loading (denoted by A − a). Thus, the production of carbon dioxide in the tissues moves the oxygen saturation level to that indicated by the lower curve, whereas the loss of the carbon dioxide across the respiratory epithelium of the gills ensures a return to the upper curve and a greater degree of oxygenation of the haemoglobin. Consequently, the effective dissociation curve is that indicated by the dotted line A–V, with the Bohr effect increasing on- and offloading by amounts equivalent to A − a and v − V, respectively.

Root effect

One consequence of the Bohr effect is that higher oxygen tensions (P_{O_2}) will be required in order for the haemoglobin to reach any given level of saturation. The Bohr effect leads to a displacement of the oxygen dissociation curves but the haemoglobin will eventually become fully saturated provided that the P_{O_2} is high enough. In some fish species, however, the effect of lowering the pH on the affinity of the haemoglobin for oxygen is so great that the blood does not become fully saturated even at very high oxygen tensions (Fig. 4.14).

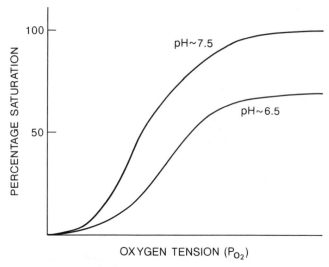

Fig. 4.14 Saturation curves of haemoglobin showing the influence of pH upon the percentage saturation of haemoglobin that can be achieved – the Root effect. See text for further details.

This phenomenon is known as the *Root effect*, and many fish species possess haemoglobins having this peculiar property. Stated simply, the Root effect means that acidification of the blood of the fish will result in a reduced capacity of the haemoglobin to bind oxygen at any P_{O_2}, no matter how high.

Thus, if the pH of the blood is lowered, the Root effect haemoglobins will offload some of their oxygen even if subjected to high P_{O_2}s, enabling extremely high P_{O_2}s to be achieved in certain body tissues. The Root effect is, for example, of great importance for the secretion of gas into the gas bladder (swim bladder) against high hydrostatic pressures (see also Section 8.10).

4.11   CONCLUDING COMMENTS

The basic plan of the circulation in fish is quite simple, with the path of the blood consisting of a single circuit. The heart provides the primary propulsive force for blood flow. The blood leaving the heart flows first through the gills and then through the major vessels to the systemic circulation, before finally returning to the heart. Thus, the entire circulatory system is contained within a closed series of vessels.

The arterial system of fish can be divided into two parts. The first is short, consisting only of the ventral aorta and its branches that lead to the gills. The vessels are thick-walled and elastic; blood flow and pressure within these vessels are generally high. The blood pressure drops by about 30% across the gills, so the pressures in the parts of the arterial system supplying the tissues are lower than in the ventral vessels. Pressures in the venous return system are always low, with blood return towards the heart being assisted by muscular activity and a range of circulatory adaptations.

The interspecific differences in cardiac output may be an order of magnitude or more, with most of these differences between species being explicable on the basis of differences in heart rate. In the majority of species, heart rate is relatively constant and changes in cardiac output are usually the result of changes in stroke volume. The exception to this general rule is seen amongst the thunnids, in which the heart rate of an actively swimming fish may be over double that of a fish at rest.

In the majority of fish species the only oxygen supply to the cardiac muscle is derived from the venous blood within the lumen of the chambers of the heart. Thus, it seems likely that the venous P_{O_2} threshold may limit cardiac performance, and impose a low upper limit on heart rate. The active thunnids have cardiac muscle with a high myoglobin content, and also possess a well-developed coronary blood supply. These adaptations result in increased rates of supply of oxygen to the cardiac muscle, enabling high heart rates to be achieved.

Whilst most fish are unimodal water breathers, using the gills as the respiratory organ, a small number of species are bimodal, being capable of breathing in both water and air. Bimodal breathing usually involves adaptations in the respiratory organs and generally involves the use of tissues other than the gills for aerial gaseous exchange. Bimodal breathing also involves modifications of the circulatory system.

In larval fish the gills are poorly developed. The gills have a small surface area and contribute relatively little to gas exchange. In the larvae, most of the gaseous exchange appears to occur across the body surface, with the importance of the gills increasing as the fish increase in size.

The characteristics of the gill tissues (overall surface area and epithelial thickness), the respiratory mode (bucco–opercular pump or ram ventilation) and the transport properties of the blood (loading and offloading properties of the haemoglobin, blood volume and haematocrit) differ markedly depending upon the lifestyle of the species. Thus, the active species, such as scombrids and salmonids, tend to have gills with a large surface area and secondary lamellae with thin epithelia. When swimming at speed they may ram ventilate the gills and dispense with the bucco–opercular pump. The active species also tend to inhabit flowing, oxygen-rich waters and the haemoglobins of these species generally have higher P_{50}s than the haemoglobins of sedentary fish species found in oxygen-poor habitats such as swamps, and stagnant ponds and lakes.

FURTHER READING

Books

Cameron, J.N. (1989) *The Respiratory Physiology of Animals*, Oxford Univ. Press, Oxford.

Gilles, R. (ed) (1985) *Circulation, Respiration and Metabolism*, Springer-Verlag, Berlin.

Hoar, W.S. and Randall, D.J. (eds) (1970) *Fish Physiology, Vol. IV*, Academic Press, London.

Hoar, W.S. and Randall, D.J. (eds) (1984) *Fish Physiology, Vol. XA*, Academic Press, London.

Hoar, W.S., Randall, D.J. and Farrell, A.P. (eds) (1992) *Fish Physiology, Vols XIIA, B*, Academic Press, London.

Rankin, J.C. and Jensen, F.B. (eds) (1993) *Fish Ecophysiology*, Chapman & Hall, London.

Satchell, G.H. (1991) *Physiology and Form of Fish Circulation*, Cambridge Univ. Press, Cambridge.

Review articles and papers

Burggren, W.W. and Pinder, A.W. (1991) Ontogeny of cardiovascular and respiratory physiology in lower vertebrates. *Ann. Rev. Physiol.*, **53**, 107–35.

Farrell, A.P. (1991) From hagfish to tuna: A perspective on cardiac function in fish. *Physiol. Zool.*, **64**, 1137–64.

Randall, D. (1982) The control of respiration and circulation in fish during exercise and hypoxia. *J. exp. Biol.*, **100**, 275–88.

Rombough, P.J. (1988) Respiratory gas exchange, aerobic metabolism and effects of hypoxia during early life, in *Fish Physiology, Vol. XIA* (eds W.S. Hoar and D.J. Randall) Academic Press, London, pp. 59–161.

Chapter five

Physiological integration – nervous and endocrine systems

5.1 INTRODUCTION

Central integration and coordination of sensory input is achieved by the nervous and endocrine systems, each of which transmits signals to the various effector organs. The nervous system transmits signals via electric–nervous pathways, whereas the pathways of signal transmission of the endocrine system are hormonal–humoral. The signals serve to regulate a variety of body functions, and the regulation is dependent both upon environmental stimuli and upon a series of feedback mechanisms. In other words, the regulation of a particular function requires a two-way exchange of information between the control centre and the effector organ.

The regulating signals are transmitted from the control centre to the effector organ which then responds with a particular set of actions, dependent upon the signals received. The information concerning these actions is then transmitted, or fed back, to the control centre. If the response to the original signal transmission has not been satisfactory, the signals will be adapted and sent back to the effector.

A major difference between the nervous and endocrine systems is concerned with the speed at which they transmit information from the control centre to the effector organs. The nervous system is specialized for the rapid transmission of signals. On the other hand, the endocrine system can be considered to be specialized for the slow transmission of signals, employing the circulatory system for distributing the hormonal messengers throughout the body.

5.2 NERVOUS SYSTEM

The *somatic nervous system* mediates the information from the sensory systems via afferent nervous inputs running from the sense organs to the *central nervous system* (CNS) and *brain*. Thus, much of the processing of sensory information takes place in the central nervous system. There are also efferent nervous connections between the CNS and skeletal muscles via the somatic nervous system. Consequently, much of the somatic nervous activity is under voluntary control.

The nervous supply that serves to regulate the functions of the internal organs (e.g. heart and blood vessels, gut) comprises the bulk of the *autonomic nervous system* (ANS). The efferent motor supply from the brain to the effector organs, apart from the skeletal muscles, is via the ANS. In the periphery of the animal the autonomic and somatic nervous systems are functionally and anatomically almost entirely separate. Within the CNS, the two systems come into intimate contact and connections exist between the two for information exchange and coordination of overall function.

5.3 CENTRAL NERVOUS SYSTEM (CNS)

There are four main components to the central nervous system. These components comprise nerves derived from the brain and spinal cord. The first component comprises the sensory nerves which represent the nervous links between the sense organs, skeletal muscles and joints, and the brain. Nerves conveying motor impulses to the skeletal muscles form the second component of the CNS. These first two components relate to elements of the somatic nervous system. Components three and four, which represent elements of the autonomic nervous system, either convey motor impulses to the internal organs and smooth visceral muscles of the trunk region, or are visceral nerves that carry sensory inputs from the internal organs.

The four components of the CNS can, therefore, be classified according to both their role and to which part of the system they belong. Thus, it is usual to consider the elements as comprising:

1. somatic afferents carrying inputs from receptors in the sense organs and skeletal muscles to the CNS and brain – part of the somatic nervous system;
2. motor, or somatic, efferents carrying impulses from the CNS to the skeletal muscles – part of the somatic nervous system;
3. visceral afferents carrying inputs from the internal organs to the CNS and brain – part of the autonomic nervous system;

4. autonomic efferents carrying impulses from the CNS to the glands, internal organs and smooth muscles – part of the autonomic nervous system.

The four components of the central nervous system are found in the *cranial nerves*, with some of the cranial nerves containing fibres belonging to more than one system, or component. The cranial nerves that contain several components are often termed 'mixed' nerves.

5.4 BRAIN

The brain, which is the main integrative centre of the nervous system, is encased within the skeletal elements of the skull and, as such, is relatively well protected from physical damage. The structure of the brain reveals a series of lobes, or regions (Figs 5.1 and 5.2). Each region, or lobe, tends to have a well-defined functional role, although some of the brain areas are implicated in the integration and coordination of inputs from several sensory systems.

In fish, as in other vertebrates, the efferent fibres running from the olfactory chemoreceptors are intimately associated with the olfactory bulbs, and the nerve fibres from the olfactory system pass from the bulb to the *telencephalon*. Depending upon species, the olfactory bulbs may be either located close to the olfactory organ, or situated in close contact with the telencephalon. In the former case the olfactory nerves are short and the olfactory bulbs are separated from the telencephalon by long olfactory tracts (Fig. 5.2).

The medial and lateral olfactory tracts terminate within different areas of the forebrain. Most of the nervous connections running in the lateral olfactory tract pass to the dorsal part of the telencephalon, but there are some branches that terminate ventrally. The medial olfactory tract divides into two within the forebrain, with one of the branches sending nervous connections exclusively to specific areas of the ventral telencephalon. It is suspected that the ventral telencephalon is the area of the brain that processes the sensory inputs derived from stimulation of the olfactory system by sex pheromones.

Thus, one of the major functions of the telencephalon is the integration of olfactory inputs. Ablation (involving removal of parts of the brain), electrical stimulation and recording experiments have, however, provided evidence that the telencephalon is also involved in other functions.

Electrical stimulation of the telencephalon may lead to the initiation of food search and feeding behaviour, perhaps because the stimulation mimics the sensory input from the olfactory system. The telencephalon

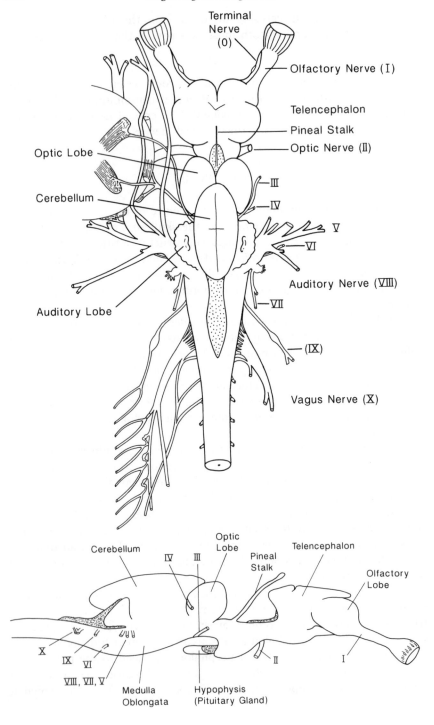

Terminal Nerve (0)

Olfactory Nerve (I)

Telencephalon

Pineal Stalk

Optic Nerve (II)

Optic Lobe

III

Cerebellum

IV

V

VI

Auditory Nerve (VIII)

VII

Auditory Lobe

(IX)

Vagus Nerve (X)

Cerebellum

Optic Lobe

Pineal Stalk

Telencephalon

IV   III

Olfactory Lobe

X

IX   VI

VIII, VII, V

II

I

Medulla Oblongata

Hypophysis (Pituitary Gland)

may also play an important role in reproductive behaviour of fish, and ablation of the telencephalon can result in either a complete loss, or absence of some elements, of reproductive behaviour in some species. These disturbances to the reproductive behaviour may be due to effects of the ablation on the registration and coordination of olfactory inputs derived from pheromonal communication.

The telencephalon may also be involved in some aspects of learning and memory. Telencephalic lesions often result in fish having a reduced ability to learn specific tasks. In some cases there may also be an abolition of the ability to perform previously learned tasks.

The *diencephalon* can be divided into three regions: the epithalamus, the thalamus and the hypothalamus. The major part of the epithalamus is made up of the pineal complex. The pineal organ, which lies on the upper-most surface of the brain, is a light-sensitive organ. This organ is involved in the regulation of the physiological activities that vary in relation to changes in light cycles, either on a daily or on a seasonal basis.

The photoperiod-dependent influences of the pineal organ are mediated via the production and release of an endocrine factor, *melatonin.* Melatonin production and release is linked to the light:dark cycle, with the highest levels of production always being associated with night-time conditions irrespective of the type of activity pattern displayed by the animal. This means that in the nocturnal species, the melatonin levels are highest during the period when the animals are most active, whereas in diurnally active species melatonin levels are highest when the animals are at rest.

The precisely regulated circadian melatonin rhythm imparts important time-of-day information – providing information about both day (low melatonin) and night (high melatonin), and also possibly about the periods around dusk (rising melatonin levels) and dawn (falling melatonin levels). It is also possible that the seasonal changes in melatonin production and release by the pineal, which will be under the influence of the annual cycle of photoperiod, may provide valuable temporal information. In other words, under natural photoperiodic conditions the duration of the elevated melatonin levels at night varies directly with the variations in the light:dark cycle. These variations are most exaggerated in polar regions, at extremes of latitude, but irrespective of latitude there are some annual changes in daylength. These annual changes in daylength may have important influences upon biological systems, with, for example, many species using changes in the light:dark cycle as the major cue for the timing of reproductive development and breeding.

Fig. 5.1 Surface and side views of the brain of an elasmobranch showing general structures and the cranial nerves.

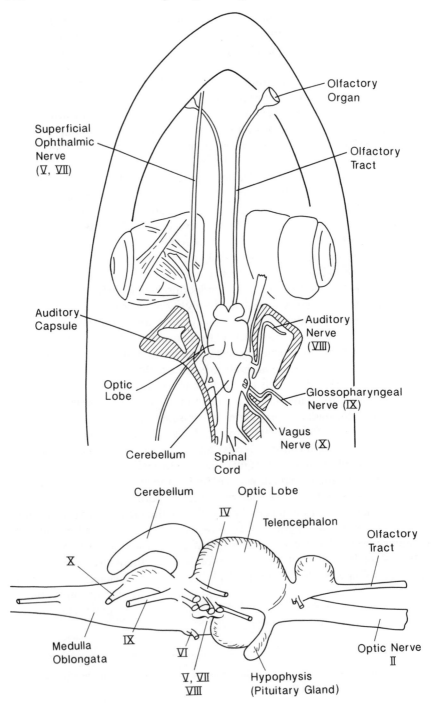

Olfactory Organ

Olfactory Tract

Superficial Ophthalmic Nerve (Ⅴ, Ⅶ)

Auditory Capsule

Auditory Nerve (Ⅷ)

Optic Lobe

Glossopharyngeal Nerve (Ⅸ)

Vagus Nerve (Ⅹ)

Cerebellum

Spinal Cord

Cerebellum

Optic Lobe

Ⅳ

Telencephalon

Olfactory Tract

Ⅹ

Medulla Oblongata

Ⅸ

Ⅵ

Ⅴ, Ⅶ
Ⅷ

Hypophysis (Pituitary Gland)

Optic Nerve Ⅱ

In addition to having nervous connections with the pineal organ, the epithalamus has connections with, and receives sensory inputs from, the olfactory region and telencephalon. Large numbers of nerve fibres run through the thalamus to the hypothalamus. This region of the diencephalon is the major integrative centre of the brain. Afferent fibres from the sensory systems converge in the hypothalamus, and efferent pathways from the hypothalamus lead to other regions of the brain (e.g. telencephalon, medulla, thalamus and cerebellum), the pituitary gland and a variety of internal tissues and organs.

The *mesencephalon*, or midbrain region, is commonly known as the optic lobes. This region comprises two anatomically distinct parts: the optic tectum and the tegumentum. The optic tectum is the major centre for the integration of visual inputs with other sensory information, and there are numerous nervous connections between the optic tectum and other neural centres.

Integration of the sensory inputs that lead to changes in swimming activity and positional changes occurs in the *cerebellum*. Consequently, this region of the brain receives inputs, either directly or indirectly, from all the sensory systems. Anatomically the cerebellum is the most variable structure of the brain. The exact structure, and the number of nervous connections to other regions of the brain, depends upon the relative importance of the different sensory modalities in the life of the given fish species. Ablation of the cerebellum has usually been found to result in a disruption of the ability of the fish to display controlled movements. Electrical stimulation of the cerebellum may lead to increased swimming, increased frequency of turning movements or postural changes.

The *medulla oblongata* is the region of the brain most closely allied to the spinal cord, and it is often difficult to distinguish a clear boundary between the two (Figs 5.1 and 5.2). The spinal cord is well protected by the neural arches of the vertebrae of the spinal column as it runs from the brain towards the caudal region. The cord is made up of nervous tissue and carries both ascending and descending nerve fibres. The inner, grey zone contains the cell bodies, whereas the outer white zone of the cord carries the nerve fibres. The afferent nerve fibres enter the spinal cord dorsally, whilst the efferent fibres that innervate the various internal tissues and organs leave via a ventral route.

The medulla oblongata contains nerve fibres of all four components of the CNS, and, therefore, both receives a range of sensory inputs, and transmits efferent motor impulses. The majority of the sensory nerve fibres

Fig. 5.2 Surface and side views of the brain of a gadoid showing general structures and the cranial nerves.

derive from certain of the *cranial nerves*. There is a close association between the medulla and the nerves carrying impulses to and from the skin and lateral line, gustatory system and viscera (Fig. 5.1).

5.5   CRANIAL NERVES

Sensory inputs from the various sense organs are conveyed to the brain via the cranial nerves and their branches (Figs 5.1 and 5.2; Table 5.1). Some of the nerves carrying somatic sensory inputs (e.g. cranial nerves V, VII, IX and X) also have nerve fibres from other components of the CNS. The cranial nerves with several components are classified as 'mixed' nerves.

Olfactory stimuli detected by the sensory rosettes give rise to impulses which pass along the olfactory nerves (cranial nerve I) to the olfactory bulbs and thence to the olfactory tracts. The terminal nerve (cranial nerve 0) is associated with the olfactory organ, and it has been suggested to have both sensory functions related to the detection of sexual pheromones and special somatic functions. Pheromones may be considered to be chemical messenger substances that are used to convey information between individuals. Whilst the terminal nerve is closely allied with the olfactory nerve, its function or functions are still a matter of debate. The fact that the terminal nerve contains gonadotropin releasing factor (GnRF) and has connections to the brain regions that control reproductive behaviour does, however, strongly implicate this nerve in the transmission of impulses that are important for reproductive activities.

The olfactory tracts run to the forebrain, or telencephalon, and there are neuronal connections to the midbrain, and some directly to the hypothalamus. The sizes of the olfactory bulbs, and their position relative to the forebrain, differ markedly between species. Olfactory rosettes and bulbs tend to be largest in those species, such as ictalurids and certain gadoids, that are highly dependent upon olfaction as a sensory mode. In these species, and in the salmonids, the olfactory bulb is situated close to the olfactory organ and the olfactory nerve is comparatively short. Consequently, the olfactory tract, which connects the olfactory bulb to the forebrain is relatively long (Fig. 5.2). In the thunnids and threespine stickleback, *Gasterosteus aculeatus*, on the other hand, the olfactory bulb is closely associated with the forebrain, and the olfactory nerve makes a comparatively long connection between the olfactory rosette and the olfactory bulb.

The sensory cells of the gustatory system are concentrated in the buccal cavity. There are also components of the gustatory system on the anterior part of the body and its appendages. Sensory inputs from the gustatory system are, therefore, almost certainly collected and conveyed to the brain

Table 5.1 Overview of the functions of the cranial nerves

Number	Name	Fibre type		Function
		Sensory	Motor	
0	Terminal	√		
I	Olfactory	√		Smell
II	Optic	√		Sight
III	Oculomotor	√	√	Supply to and
IV	Trochlear	√	√	receive impulses
VI	Abducens	√	√	from eye muscles
V	Trigemial			
VII	Facial			
	Superficial ophthalmic	√		Reception of impulses from skin
	Deep ophthalmic	√		of the head region
	Maxillary	√		Reception of impulses from upper jaw region
	Mandibular	√	√	Motor fibres to jaw muscles. Sensory fibres from skin overlying lower jaw
	Hyomandibular	√	√	Motor fibres to hyoid muscles. Sensory fibres from roof of the pharynx
VIII	Auditory	√		Hearing
IX	Glossopharyngeal	√	√	Motor fibres to branchial muscles. Sensory fibres from gills and pharynx
X	Vagus	√	√	Motor and sensory fibres from pharynx and viscera. Sensory fibres from lateral line system

in more than one of the cranial nerves. Branches of cranial nerves V, VII, IX and X have been implicated in the transmission of gustatory information.

Sensory impulses from the eye reach the brain via the optic nerves (cranial nerve II). The impulses then pass along the optic tracts to the optic lobes, which are primarily concerned with the processing of visual information.

Impulses from the mechanoreceptors of the inner ear reach the brain via the auditory nerve (cranial nerve VIII), but this nerve does not have branches to the lateral line neuromasts in the head and trunk canals. The lateral line system of the head appears to be innervated by branches of the trigeminal (cranial nerve V) and facial (cranial nerve VII) cranial nerves, whereas the the trunk canal may have connections via the vagus (cranial nerve X) and glossopharyngeal (cranial nerve IX) nerves.

The lateral line canals may, in addition, have a distinct system of innervation with inputs projecting to specific regions of the brain. Thus, the lateral line of many species appears to be innervated by two separate cranial nerves: the anterior lateral line nerve innervating the head, and the posterior lateral line nerve innervating the trunk. These are distinct from all other cranial nerves, including the VIIIth nerve that innervates the otolithic organs and semicircular canals of the inner ear. This functional organization would appear to form a basis for the separation of lateral line and auditory inputs, but there must also be central convergence in order for integration and coordination to be possible.

Anatomically, the system for electroreception is closely allied to the system of lateral line canals, particularly in the region of the head. The sensory epithelium of the ampullae of Lorenzini appears to have afferent innervation via branches of the VIIth cranial nerve.

5.6 AUTONOMIC NERVOUS SYSTEM (ANS)

The autonomic nervous system is essentially an efferent system transmitting impulses to the stomach, intestine, heart, blood vessels and other internal organs. Nevertheless, autonomic nerves do often contain afferent fibres coming from receptors in the inner organs.

Unlike the somatic nervous system the ANS is, for the most part, not subject to voluntary control. The ANS is primarily involved in controlling the various bodily functions via a wide range of reflex arcs. Visceral or somatic afferent fibres convey information from the various receptors, and the ANS provides the efferent fibres that transmit the reflex response based upon this information. Such responses may include contraction or relaxation of the smooth muscles of organs such as the gut and blood vessels. The signals transmitted by the ANS will also influence the functioning of the heart and endocrine organs.

Simple reflexes, such as those controlling a variety of intestinal movements, can be completed within the particular organ concerned. More complex activities, perhaps involving several organ systems, are controlled by higher centres within the spinal cord and brain. The major site of integration of ANS activity is the *hypothalamus*. Within the hypothalamus the functions of the ANS are coordinated with those of the sensory inputs from the somatic nervous system. There is also integration between the nervous and endocrine systems.

The peripheral ANS consists of two subdivisions, termed the *sympathetic* and *parasympathetic* divisions. Most of the internal organs are innervated by fibres from both divisions of the ANS. The two divisions are, for the most part, anatomically and functionally distinct – they employ different chemical messenger substances, or *neurotransmitters*, and the responses produced are often, but not always, antagonistic. Both divisions of the ANS consist of preganglionic nerve fibres that switch to postganglionic fibres in the nerve ganglia.

In fish, parasympathetic preganglionic fibres run from the brain to various regions of the head in some of the cranial nerves (III, V, VII and IX), but the most significant parasympathetic outflow is in the vagus nerve (cranial nerve X), which branches to innervate the gills, heart, gut and some other internal organs. The parasympathetic ganglia are situated close to or even within the organs, so most of the distance between the CNS and the periphery is bridged by preganglionic fibres.

The anatomy of the sympathetic division differs in a number of respects from that of the parasympathetic system. In the sympathetic system, the preganglionic fibres exit from the spinal cord and almost immediately form junctions with the postganglionic fibres in ganglia. These ganglia are linked longitudinally to form a chain, and the postganglionic fibres run from the chain ganglia to the various organs and tissues they innervate. Thus, within the sympathetic system the preganglionic fibres are short, running only from the spinal cord to the chain ganglia, and most of the distance between the CNS and the effector organ is traversed by the postganglionic fibres that arise in the chain ganglia.

Synaptic transmission

Where the preganglionic and postganglionic fibres meet, and where the postganglionic fibres interact with the effector organ there are junctions, or *synapses* (synaptic clefts), that must be bridged by chemical rather than electrical transmission. Thus, for transmission of a signal, the electrical impulse that reaches the presynaptic membrane must cause the release of a chemical transducer, or neurotransmitter substance, into the synaptic cleft.

The neurotransmitter diffuses to the postsynaptic membrane and there initiates a new electrical signal. Since there is no neurotransmitter release from the postsynaptic membrane, the synapse functions as a one-way valve, transmitting only from the presynaptic to the postsynaptic neuron. There are several different types of neurotransmitter substance, but each neuron is generally considered to have only one specific type of neurotransmitter, enabling classifications to be made on the basis of the neurotransmitters employed.

The neurotransmitter substance of the parasympathetic subdivision of the ANS is *acetylcholine.* This neurotransmitter is released both at the synapses between the preganglionic and postganglionic neurons, and at the terminals of the postganglionic neurons with the effector organs (Fig. 5.3). Thus, in the parasympathetic system transmission at both the preganglionic and postganglionic synapses is said to be cholinergic. Acetylcholine is synthesized and stored in the nerve endings, with the amount stored being kept constant by matching the rate of synthesis to the rate of release. The arrival of an electrical impulse at the nerve ending results in the exocytosis of acetylcholine into the synaptic cleft and the neuro-

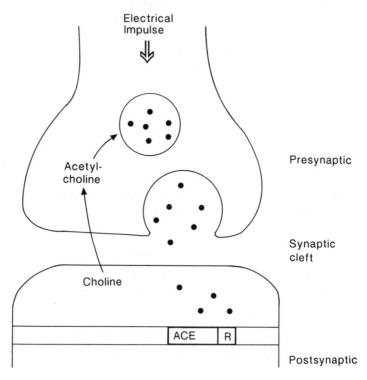

Fig. 5.3 Schematic diagram illustrating the bridging of synaptic clefts by cholinergic transmission. ACE, acetylcholinesterase; R, receptor. See text for further details.

transmitter then diffuses across the cleft to the postsynaptic membrane. The action of the acetylcholine is terminated by hydrolysis of the transmitter by the enzyme acetylcholinesterase.

Within the sympathetic division of the ANS, the synapses occurring within the ganglia, i.e. those between the preganglionic and postganglionic neurons, are cholinergic, but the postganglionic terminals of the sympathetic system employ *catecholamines* (noradrenaline [norepinephrine] and adrenaline [epinephrine]) as neurotransmitter substance. Transmission at this latter type of synapse is said to be adrenergic (Fig. 5.4).

The terminal branches of the sympathetic postganglionic nerve fibres have a series of swellings that both form the synaptic contact with the effector organ and are the sites of synthesis and storage of the neurotransmitter. The amino acid tyrosine is actively taken up over the membranes of the nerve fibre swellings, and this is used as the precursor for catecholamine synthesis. The catecholamines are then stored in granular vesicles until release is stimulated by the arrival of electrical impulses.

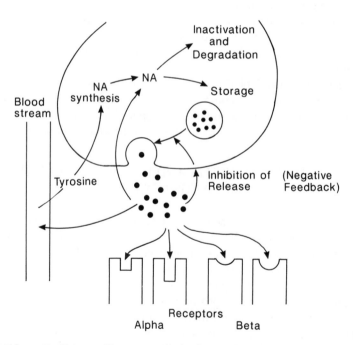

Fig. 5.4 Schematic diagram illustrating the bridging of synaptic clefts by adrenergic transmission. NA, noradrenaline. See text for further details.

Adrenergic receptors

The catecholamines released from the sympathetic postganglionic nerve endings interact directly with receptors in the target tissues (Fig. 5.4). The receptors are of several types, and different receptors may interact specifically with the neurotransmitter substance to produce different types of response (i.e. either stimulation or inhibition).

Within the fish ANS two particular classes of receptors, termed α and β adrenergic receptors, are of importance in providing for alternative responses to catecholamines. Both classes of receptor appear, however, to be heterogeneous. Activation of α-1 adrenoceptors is stimulatory. Thus, at effector sites where there is vascular smooth muscle, activation of α-1 adrenoceptors will lead to muscular contraction. On the other hand, activation of α-2 adrenoceptors tends to be inhibitory. The β adrenoceptors are also of two types. Activation of the β-1 adrenoceptors usually has a stimulatory effect, whereas activation of β-2 adrenoceptors results in inhibition. Inhibition via β-2 adrenoceptor activation would, for example, result in a relaxation of smooth muscle.

Termination of catecholamine action is generally via resorption of the catecholamines into the nerve endings. Some of the catecholamines may, however, diffuse into blood capillaries and be carried to other tissues for inactivation. In addition to the direct inactivation mechanisms, the presence of catecholamines in the synaptic cleft has a negative feedback effect on the nerve endings, inhibiting further release of the neurotransmitter (Fig. 5.4).

The nervous control of the functioning of the various internal organs via the ANS is extremely complex. It involves both antagonism between the parasympathetic and sympathetic systems, and interactions between different groups of adrenoceptors in the effector tissues and organs.

5.7 ENDOCRINE SYSTEM AND HORMONES – CHEMICAL MESSENGER SYSTEM

The hormones are synthesized in specific hormone-producing cells, and following release into the bloodstream act on target organs or cells. These may either be subordinate hormone-producing glands or non-endocrine effector organs. Thus, hormones may be released into the circulatory system and have target organs at distant sites in the body – *endocrine* effect. Some hormones may also have local effects caused by the diffusion of the hormone from the hormone-producing cells to cells in the near vicinity – *paracrine* effect. Others may be secreted or excreted into the environment, where they can act as chemical messengers between animals – *pheromonal* effect.

Hormone receptors

Since hormones are released into the blood it is clear that there will be a wide range of hormones circulating simultaneously. Consequently, there must be some way in which the hormone and its specific target organ(s) recognize each other. The cells in the target organ possess receptors for the hormone, and recognition is achieved when the hormone binds to these specific sites. The receptors are specific for a given hormone, and the affinities of the receptors are extremely high because the plasma concentrations of circulating hormones are usually very low (a few ng ml^{-1} blood plasma).

Three main groups of hormones are recognized on the basis of chemical structure. These are hormones having a peptide or glycopeptide structure, a series of hormones derived from the amino acid tyrosine, and the steroid and related hormones. The peptide hormones comprise chains of amino

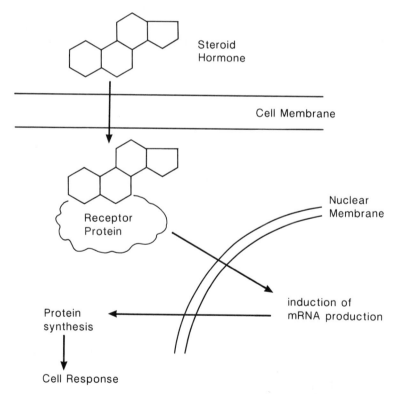

Fig. 5.5 Transduction of endocrine messages across cell membranes and the intracellular events that occur in response to receptor activation by steroid hormones. mRNA, messenger ribonucleic acid.

acids linked together in specific sequences, whereas the glycopeptides are formed by the binding of a peptide to a carbohydrate residue. The steroid hormones are usually bound to specific proteins when they are being transported via the circulatory system, e.g. the sex steroid hormones bind to globulin, and cortisol is transported bound to transcortin.

The steroid hormones differ from the peptide hormones in that they must enter the target cells in order to exert their effect. The steroid hormones are lipid soluble and can transverse the cell membrane relatively easily. Once they have entered the target cell, the hormones bind with specific cytoplasmic receptor proteins. The hormone–receptor complex then attaches to the cell nucleus where it induces processes that lead to changes in protein synthesis (Fig. 5.5). The thyroid hormones – tyrosine derivatives – also have intracellular receptors, located on the nucleus of the cell.

The receptors for the peptide and glycopeptide hormones are situated on the outer surface of the target cell. When the hormone binds to the receptor site this leads to the release of an intracellular transmitter, or 'second messsenger'. The binding of the hormone to its receptor site will initiate a series of reactions that leads to the synthesis and release of cyclic

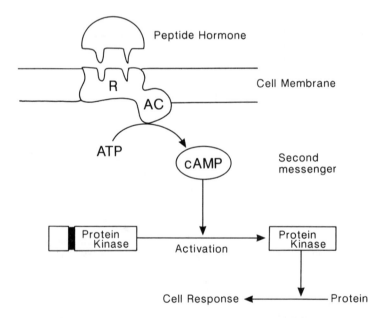

Fig. 5.6 Transduction of endocrine messages across cell membranes and the intracellular events that occur in response to receptor activation by peptide hormones. R, receptor; AC, adenyl cyclase; ATP, adenosine triphosphate; cAMP = cyclic adenosine monophosphate.

adenosine monophosphate (cAMP), the most commonly encountered second (intracellular) messenger. Whilst a large number of hormones are able to activate the production of cAMP, the specificity of hormone action is ensured by the fact that the cell membrane receptor sites are highly specific (Fig. 5.6).

The release of the cAMP leads to the intracellular activation of protein kinases, which in turn effect the response. The specific cellular response to hormonal stimulation will, therefore, ultimately depend upon the nature of the protein kinase present in the target cell. This means that if different 'families' of target cells have receptor sites for a specific hormone, but contain different protein kinases, then the cellular responses to stimulation by one and the same hormone can differ markedly between these different families of target cells. Thus, it is possible that a given hormone can exert different effects in different target cells or organs, and this obviously increases the variety and flexibility of responses.

5.8 HORMONAL HIERARCHY AND FEEDBACK CONTROL

The release of hormones will often occur as a result of input signals received from the peripheral sense organs and transmitted to centres in the brain via the central nervous system. Within the brain there are connections to the *hypothalamus*, which is the major integrative centre controlling the release of a range of hormones. Stimulation of the hypothalamus initiates the synthesis and release of a series of hormones and hypothalamic messenger substances. These messengers are low-molecular-weight peptide molecules composed of a small number of amino acids. Synthesis occurs in certain nerve cells within the hypothalamus, and the peptides are released following stimulation of these cells. These hypothalamic neuropeptides, in turn, have effects upon the production and release of hormones from the *pituitary gland* (or hypophysis).

The pituitary gland is an endocrine gland located directly below the hypothalamic region of the brain to which it is connected by a 'hypophysial stalk'. The gland consists of two lobes, the anterior, or *adenohypophysis*, and the posterior, or *neurohypophysis* (Fig. 5.7). In the teleosts the tissues of the neurohypophysis and adenohypophysis are interdigitated.

The hypophysial stalk contains the fibres that run from the cell bodies of the neurosecretory cells, located in the hypothalamus, to the neurohypophysis and adenohypophysis. In many fish the stalk is virtually absent, and the pituitary is then located very close to the hypothalamus, but in other species the stalk may be quite long.

The two lobes of the pituitary gland are derived from different embryonic tissues and may differ in the form of their connections with the

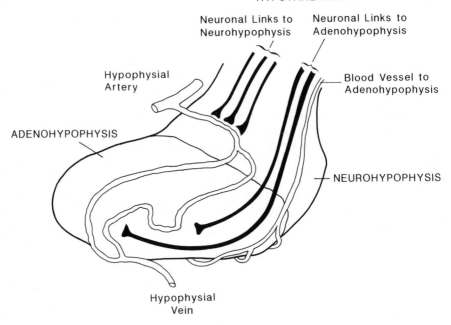

Fig. 5.7 General structure of the pituitary gland (hypophysis) showing the possible neuronal and humoral links between the hypothalamus and different regions of the pituitary.

hypothalamus. The connection between the hypothalamus and the neurohypophysis is a neuronal one, whereas the hypothalamic neuropeptides may reach their target cells in the adenohypophysis either via a neuronal route or via a series of blood vessels that form localized portal systems similar to those seen in higher vertebrates (Fig. 5.7). Neuronal connections running from the hypothalamus could also secrete hormones into the capillary system within the neurohypophysis, and these secretions could then reach, and stimulate, the cells of the adenohypophysis. There is also the possibility of a third blood-borne route of stimulation, since a small number of blood vessels appear to run directly from the hypothalamus to the adenohypophysis, but whether or not this pathway has importance for the transport of neurosecretory products has not been particularly well investigated (Fig. 5.7).

Whilst a capillary network of blood vessels seems to be present in some species, its occurrence is not universal. The teleosts, for example, lack a hypothalamic–pituitary portal system, and the neurons from the hypothalamus project directly into the adenohypophysis. Thus, whilst it has been suggested that the portal vessel route may be an important drainage system from the hypothalamus and neurohypophysis to the adenohypophysis in some species, the functional significance remains in doubt. In the teleosts, the cells of the adenohypophysis seem to be stimulated by direct neuronal connections running from the hypothalamus, and there is consensus that this is a major, if not the sole, route of information transfer between the hypothalamus and pituitary in teleost fish.

The adenohypophysis is the site of synthesis, storage and release of several different peptide hormones (Fig. 5.8). The adenohypophysis is generally considered to consist of the *pars distalis*, the site of secretion of most of the hormones, and the *pars intermedia*. The pars distalis may be further subdivided into two regions. This subdivision is undertaken on the basis of anatomy and the locations of the different secretory cell types. In species having a capillary drainage system, both the pars distalis and pars intermedia will receive the blood that has passed through the network of blood vessels surrounding the neurohypophysis, but it is the cells of the pars distalis that appear to have the greatest innervation of neurons directly from the hypothalamus.

The hormones that are released into the circulatory system from the neurohypophysis are not produced there, but are transported from the site

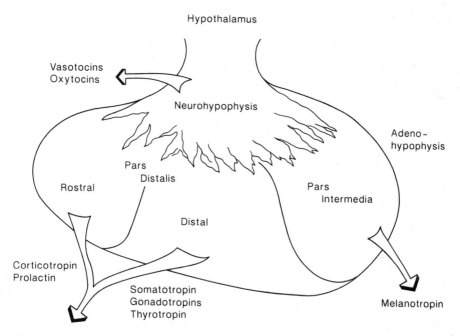

Fig. 5.8 Structure of the pituitary gland of a teleost fish showing the probable sites of synthesis and release of the different pituitary hormones.

of production in the hypothalamus to the site of release in the form of membrane-encapsulated granules (about 200 nm in diameter). The transport is neuronal, and the granules release their hormones from the nerve endings in response to specific stimulation of these neurons.

The hypothalamic neuropeptides that control the release of hormones from the adenohypophysis may have either a releasing (releasing hormone [RH] or factor [RF]) or an inhibiting (inhibiting hormone [IH] or factor [IF]) function. The interplay between these factors plays an important role in the feedback control of the release of several of the pituitary hormones. As their name suggests, the releasing factors stimulate the synthesis and liberation of hormones from the adenohypophysis, and the pituitary hormones are then transported to their target organs in the circulatory system.

The pituitary hormones may not only act upon the target organs, but can also have a direct feedback effect upon the hypothalamus, resulting in the inhibition of the secretion of releasing factor. This, in turn, will eventually result in a cessation of the liberation of the pituitary hormone. This is a *negative feedback system* since the stimulation of the hypothalamus by the pituitary hormone results in actions that lead to an inhibition or decrease in response.

Feedback control does not only occur via direct effects of the pituitary hormone on the hypothalamus, and there are several possible feedback loops and mechanisms (Fig. 5.9). For example, if pituitary hormone stimulates release of a hormone from a peripheral target organ, this hormone can provide the feedback signal to the hypothalamus and/or to the secretory cells of the pituitary itself. Furthermore, the hormones released from

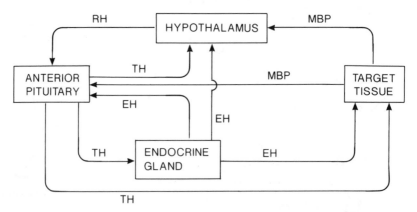

Fig. 5.9 Schematic diagram showing some possible modes of negative feedback action within the endocrine system. EH, end hormone; MBP, metabolite or breakdown product; RH, releasing hormone; TH, tropic hormone.

the peripheral endocrine target organs will produce metabolic changes in their own target cells and organs, and the end products of these metabolic reactions can also function as feedback signals to the hypothalamus and pituitary.

In addition to links between the hypothalamus and the pituitary gland, there are also neuronal connections from the hypothalamus that innervate some endocrine organs, or cells, in the periphery. Connections of this type include those associated with the gastrointestinal tract and the chromaffin tissue located close to the kidney.

The *chromaffin tissue* is a neuroendocrine transducer which converts the nervous signals from the hypothalamus into chemical signals that can be disseminated via the circulatory system. The chromaffin tissue is localized close to the kidney adjacent to the large blood vessels, the posterior cardinal veins, that drain the venous blood returning from the body directly into the heart. The *catecholamines*, which are the chemical messengers released from the chromaffin tissue, can therefore be dispersed to various locations throughout the body in the general circulation.

The catecholamines are generally present in the circulation in only extremely low concentrations, but plasma concentrations increase markedly in response to emergency conditions. The neuronal links between the hypothalamus and the chromaffin tissue ensure a rapid release of catecholamines following detection of an acute emergency. The fact that the catecholamines are liberated into blood that drains into the heart ensures that these hormones are rapidly distributed to all parts of the body. Thus, the time between the detection of an emergency situation and that taken for the catecholamines to come into contact with the various effectors (target tissues and organs) that are sensitive to them is comparatively short. Consequently, this system of combined neuronal and hormonal transmission is well suited to bringing about rapid responses in widely dispersed target tissues.

5.9 PITUITARY HORMONES

The hormones of the pituitary gland consist of a range of peptides that have a variety of effects upon other endocrine organs and body tissues. The pituitary hormones can be broadly grouped into the neurohypophyseal and adenohypophyseal hormones on the basis of their sites of synthesis and release (Fig. 5.8). The neurohypophyseal hormones are synthesized in the hypothalamus and are transported to the neurohypophysis for release, whereas the adenohypophyseal hormones are both synthesized in, and released from, the cells of the adenohypophysis.

Neurohypophyseal hormones

As a group the vertebrates are known to produce a total of about ten active neurohypophyseal peptides, and most of these have been identified in different fish species. These small peptides can be divided into two groups based upon their structure. One group forms the vasotocin family and the other, the oxytocin family. The exact roles of the neurohypophyseal hormones have not been identified with any degree of certainty, but they appear to be involved in both salt and water balance, and in the coordination of certain reproductive functions.

Adenohypophyseal hormones

The adenohypophyseal peptide hormones include the gonadotropins (GTH), growth hormone (GH or somatotropin), adrenocorticotropic hormone (ACTH or corticotropin), thyroid stimulating hormone (TSH or thyrotropin), prolactin and melanocyte stimulating hormone (MSH or melanotropin). The majority of these hormones are synthesized and secreted by specialized cell types in different locations within the pars distalis, but MSH appears to derive from cells located within the pars intermedia.

The differentiation and development of the different hormone-producing cells is not synchronous. In the rainbow trout, *Oncorhynchus mykiss*, for example, groups of cells capable of producing prolactin and MSH first appear in the pituitary of the embryonic fish a few days after fertilization of the egg. Cells that synthesize and secrete GH and ACTH are the next to appear, and those that secrete TSH appear just before hatching. In the rainbow trout, cells capable of producing GTH are not found in the pituitary until several days after the fish have hatched, and just prior to the time at which sexual differentiation takes place. The order in which the different endocrine cells appear, and the rates at which they differentiate, seem, however, to vary from species to species.

MSH is involved in the processes leading to colour changes in fish. The contraction and dispersal of the pigment granules in the chromatophores of the skin may, however, also be under direct nervous control.

ACTH has structural similarities to MSH, but synthesis and release seem to be largely confined to the pars distalis of the adenohypophysis. The release of ACTH is controlled both via hypothalamic factors and by negative feedback effects resulting from the secretion of corticosteroid hormones from the interrenal tissue of the head kidney.

It has been known for many years that the pituitary gland of fish produces gonadotropic hormones, but there has been considerable uncertainty as to how many gonadotropins (GTHs) are synthesized and secreted. The gonadotropins originate from cells in the pars distalis, and there is now convincing evidence that two gonadotropins – GTH-I and GTH-II – are produced by different cell types.

The first form of gonadotropin may be secreted early in the life of the fish and be involved in the control of sexual differentiation. GTH-I is also secreted during the early stages of the reproductive cycle, and is involved in the physiological processes related to oocyte formation and vitellogenesis (yolk production) in female fish. The second form of gonadotropin appears to have functions related to the final maturation of the gametes, and the events related to gamete release and spawning.

In mammals and birds the adenohypophyseal hormone prolactin has a role in the control of reproductive processes, but evidence for such a role for prolactin in fish is equivocal. The pars distalis of fish does, however, contain cells that produce and secrete prolactin. Prolactin may have some metabolic functions related to the storage and metabolism of lipids, especially via interactions with the thyroid hormones. The major role of prolactin in fish appears, however, to be in connection with salt and water balance.

Prolactin has effects on the movements of water and sodium across the membranes of the gills, kidneys and urinary bladder, the skin and the gut. The hormone is usually considered to promote the mechanisms required for the fish to achieve osmoregulatory homeostasis when in fresh water. It is known, for example, that prolactin-producing cells appear very early in the embryonic development of salmonid fish, which lay their eggs in fresh water. In the sea bass, *Dicentrarchus labrax*, on the other hand, the pituitary does not seem to contain cells capable of producing prolactin until several days after the fish have hatched. Thus, the early appearance of prolactin cells in the freshwater teleosts, and the delayed appearance of similar cells in the marine teleosts, suggests that prolactin may play a role in water and salt balance of fish embryos that develop in freshwater environments.

Growth hormone (GH), as its name suggests, plays a central role in the regulation of growth processes. Thus, GH, usually with the mediation of growth factors (produced in the liver) and other endocrine factors (e.g. thyroid hormones, insulin), governs both skeletal growth and the rates of synthesis of tissue proteins. Growth hormone may have effects on appetite, and is also known to influence the patterns of metabolism of stored and ingested nutrients. There is also convincing evidence that GH plays an important role in salt and water balance, via mediating changes in the cell types and enzyme systems involved in ion exchange and osmoregulation. This role of GH appears to be most pronounced when fish move between media of different osmotic strengths, and particularly when they migrate from freshwater to marine environments.

Whilst the adenohypophyseal hormones prolactin and growth hormone can influence the metabolic effects of the thyroid hormones, the release of the thyroid hormones appears to be largely under the control of a third endocrine factor – thyrotropic hormone (TSH). This hormone is also produced in the adenohypophysis, by cells of the pars distalis.

5.10   NON-PITUITARY ENDOCRINE FACTORS

Together the hypothalamus and pituitary gland form a system allowing for the regulation of physiological processes in the peripheral organs and tissues. The majority of the pituitary hormones have effects upon the production and release of endocrine factors by other glands and organs (e.g. the gonads, thyroid gland, interrenal tissue), but some of the pituitary hormones also exert direct influences upon the physiological functions of a range of tissues.

Whilst many of the endocrine organs of the periphery (Fig. 5.10) are influenced by the pituitary hormones, other forms of stimulation and inhibition are also widespread. Influences on the endocrine glands and organs are possible, for example, via neuronal connections from the hypothalamus, or via chemical messengers from sources other than the pituitary, e.g. nutrients or a variety of metabolites.

Steroid hormones

Steroid hormones are stored in very limited quantities at their sites of production (i.e. the gonads and interrenal tissue), and are synthesized as needed. Cholesterol, a 27-carbon (27-C) compound, is the basic precursor for all the steroid hormones.

The first step in the production of steroids involves the conversion of cholesterol to pregnenolone, a 21-C compound. All the steroid hormones can be derived from pregnenolone via a series of chemical steps (Fig. 5.11). The different steroid hormones can be broadly divided into:

1. progesterone, a 21-C compound, that both acts as a potent female sex hormone, and can be converted into other steroids;
2. the 21-C steroids of the interrenal, e.g. cortisol;
3. the male sex hormones (androgens), which are 19-C compounds, e.g. testosterone;
4. the female sex hormones (oestrogens), which are 18-C compounds, e.g. oestradiol.

Thus, cholesterol and pregnenolone are the starting points for the production of the steroids, but which hormone will be produced in specific endocrine cells depends upon the types of enzymes that are present within

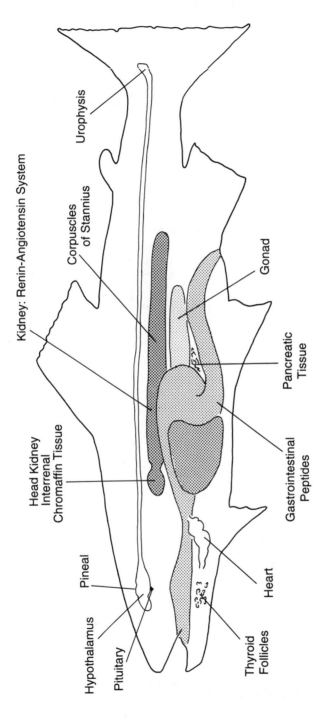

Fig. 5.10 Schematic diagram illustrating localization of the different endocrine organs in fish.

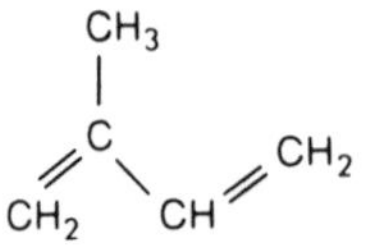

'Isoprenoid Unit'

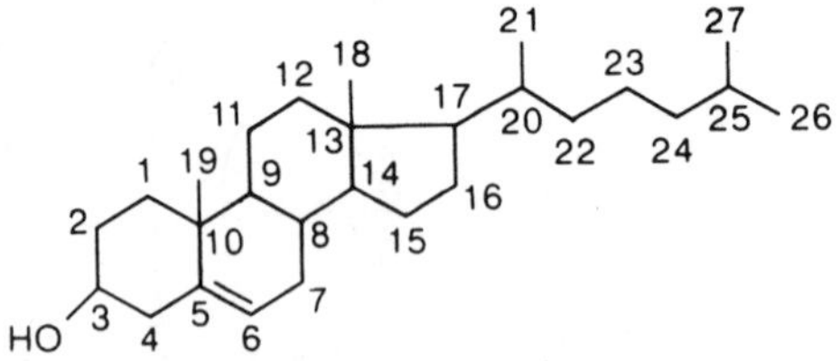

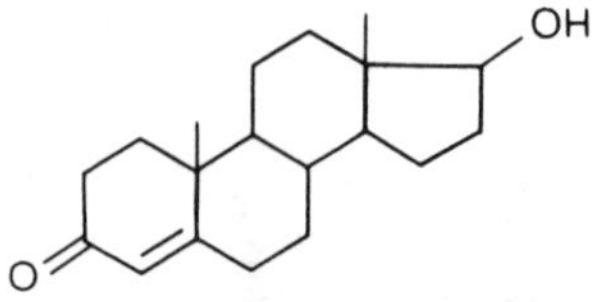

Testosterone

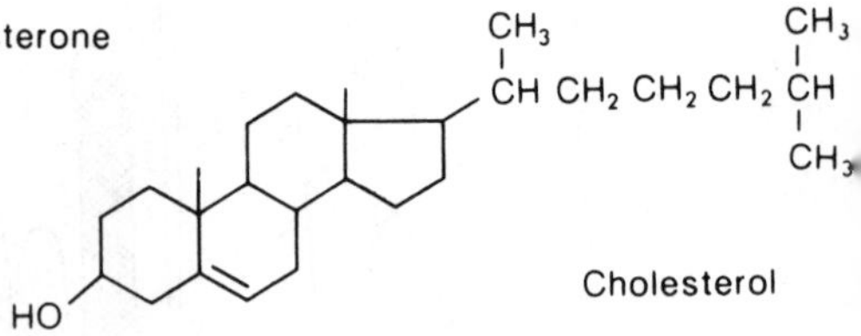

Cholesterol

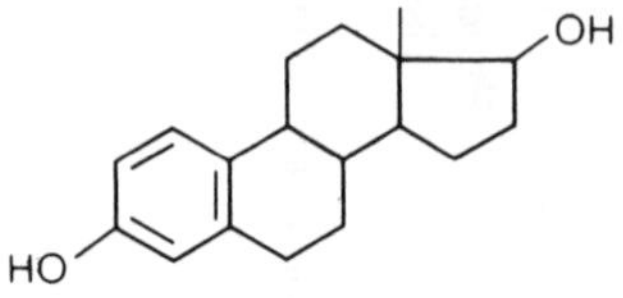

Oestradiol

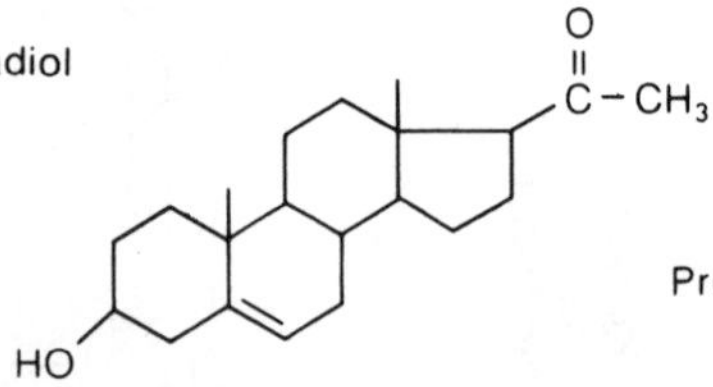

Pregnenolone

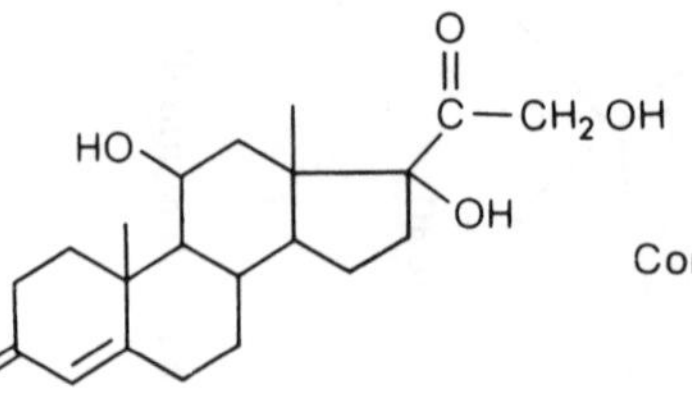

Cortisol

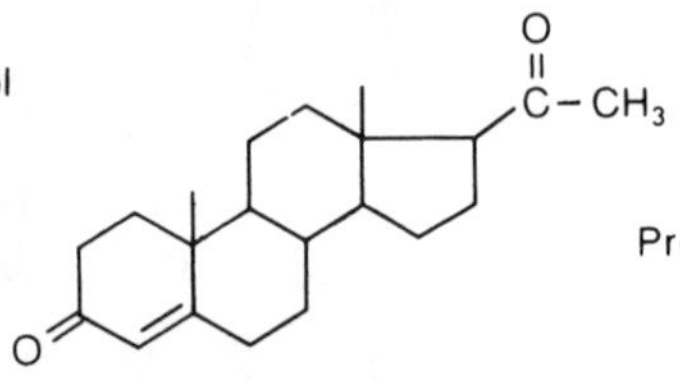

Progesterone

Fig. 5.11 The basic chemical structure of cholesterol and the structures of some of the steroid hormones derived from it.

the cell. Different types of enzymes will modify the molecular structure of the steroid at different points, leading to the possibility of a range of end products from a single precursor. For example, the cells may contain hydroxylase enzymes which introduce a hydroxyl group (−OH) into the steroid at some specific carbon atom of the molecule − a 17-hydroxylase would introduce a hydroxyl group onto C atom 17, and an 11-hydroxylase leads to the introduction of a hydroxyl group onto C atom 11. Cortisol, for example, has hydroxyl groups (−OH) on both carbons 11 and 17, whereas both testosterone and oestradiol have hydroxyl groups on carbon 17, but not carbon 11.

The corticosteroid hormones of the *interrenal tissues* play a part in both the regulation of salt and water balance, and in the metabolism of carbohydrates and proteins. In the majority of teleost fish species, cortisol is the major corticosteroid hormone, although other corticosteroids, e.g. corticosterone, 11-deoxycorticosterone and 11-deoxycortisol, are also produced in the interrenal tissue. The corticosteroids, and particularly cortisol, have been implicated in the 'stress responses' shown by fish following exposure to unfavourable conditions.

The *gonadal steroids*, the androgens and oestrogens, have a variety of reproductive functions. These include the induction of the production of vitellogenin in the liver, the development of secondary sexual characteristics and the initiation of a number of behavioural changes related to spawning activity.

The steroid hormones are removed from the blood as they circulate through the liver, and breakdown of the hormones takes place there. Usually the steroids are coupled to either a sulphate ion or glucuronic acid, and are then excreted in the bile or urine. The metabolism of the steroids to sulphates or glucuronides leads to increased solubility of the compounds in water, and therefore aids in increasing the ease with which they are excreted.

Tyrosine derivatives

The endocrine factors derived from the amino acid tyrosine are the catecholamines, produced in the chromaffin tissue of the kidney, and the thyroid hormones.

The *thyroid* consists of a series of spherical, or almost spherical, follicles (50−500 μm in diameter) that have their walls composed of a single layer of cuboidal or columnar cells. In fish the follicles are not generally aggregated to form a compact gland, but are scattered in small groups in the tissues of the pharyngeal region. The cells of the thyroid tissue are capable of trapping inorganic iodine and then using this for the synthesis of the thyroid hormones thyroxine (T_4) and triiodothyronine (T_3).

The thyroid hormones are stored within the central cavity of the follicle bound to a glycoprotein, thyroglobulin. Both the synthesis and release of the thyroid hormones are promoted by thyroid stimulating hormone (TSH) which is released from the adenohypophysis of the pituitary. The thyroid hormones are released into the blood via enzymatic action on the thyroglobulin, and the breakdown products that result are recycled. The thyroid hormones bind to specific plasma proteins and, in this form, are transported to the various tissues of the body.

The thyroid hormones have a number of effects upon growth, cellular and organ differentiation and metabolism, but many of these effects are mediated in concert with other endocrine factors. Of the two thyroid hormones, triiodothyronine appears both to be much more potent than thyroxine, and exerts its effects much more rapidly. Much of the triiodothyronine found in the blood is, however, not directly secreted by the thyroid follicles, but results from the action of deiodination enzymes on thyroxine. Thus, thyroxine is often considered to be the storage form, or prohormone, of the thyroid hormones, whereas triiodothyronine is the metabolically active form. The enzymes responsible for the production of triiodothyronine from thyroxine occur in a number of tissues, e.g. in the liver and kidney, and the ratio of the circulating levels of the two thyroid hormones is usually controlled within relatively restricted limits.

Peptide hormones

In addition to the peptide endocrine factors of the hypothalamus and pituitary, peptides having a hormonal function are produced in a number of other tissues including the heart, the gut and its associated organs, the corpuscles of Stannius and the urophysis.

Gastrointestinal hormones

The gastrointestinal tract is the site of the production and release of many peptide hormones, most of which act on the gut and its associated organs. Thus, the majority of these peptides have a role to play in the control and regulation of the digestive processes, via influences on either gastrointestinal motility, enzyme secretion or the mechanisms of nutrient absorption. For example, both the secretion of enzymes from the pancreas and the release of bile from the gall bladder are known to be regulated by peptide hormones produced in the gut tissue.

In addition to acting upon the gut itself, some of the gastrointestinal peptides may have roles to play in the regulation of metabolism. Some of the hormones of the gastrointestinal tract have also been implicated in appetite control.

The release of the peptide hormones from the endocrine cells of the gut wall is usually stimulated by the ingestion of food. Thus, the stimulation for hormone release will usually be the presence of specific nutrients in the lumen of the gastrointestinal tract, rather than signals emanating from the hypothalamus and pituitary.

Pancreatic hormones

Although the pancreas may be considered to be part of the gastrointestinal tract, the secretions from the endocrine tissues of the pancreas are not directly involved in digestive functions. The pancreas may be divided into the exocrine pancreas, which produces enzymes and other digestive fluids, and the endocrine pancreas, which produces and secretes a range of peptide hormones.

All fish have an endocrine pancreas, but there is much variability in anatomy between species. The endocrine tissue is not embedded within the exocrine pancreas, but the two tissues are usually closely associated with each other. The endocrine pancreas may contain a principal islet, an accumulation of endocrine cells visible to the naked eye. Alternatively, the cells may form the Brockmann body. The Brockmann body is formed when the endocrine cells become surrounded by a rim of exocrine tissue. The cells of the endocrine pancreas are generally located close to the gall bladder, or may occur in scattered locations around the gall bladder and between the pyloric caeca of the foregut.

The pancreatic hormones have important regulatory functions related to nutrient metabolism and growth. The principal hormones produced by the endocrine pancreas are *insulin* and *glucagon* (or more correctly a range of peptides of the glucagon family). These hormones may both stimulate metabolism and have a range of anabolic actions. Other peptide hormones may also be produced and secreted by the endocrine pancreas, e.g. *somatostatins*. The somatostatins seem to have inhibitory or suppressive, rather than stimulatory, effects on a range of processes linked to metabolism and growth.

Glucagon appears to have an important role in the regulation of carbohydrate metabolism in the liver. Insulin also seems to play a role in carbohydrate metabolism in fish. In addition, there are clear indications that insulin influences both protein metabolism and protein synthesis. The major stimulus for the release of insulin from the endocrine cells of the pancreas is the ingestion of food. Both carbohydrates and amino acids evoke insulin release, but some amino acids (e.g. arginine, leucine, lysine) appear to be more potent stimulators of secretion than carbohydrates (e.g. glucose).

Insulin may stimulate glucose uptake by the cells of the liver and skeletal muscle, but the effects upon the uptake of amino acids into the cells are usually more pronounced. In addition, the hormone increases the incorporation of amino acids into proteins in various organs, particularly those of the liver and skeletal muscle. Thus, the hormone appears to have a role to play in the regulation of growth processes, but the exact mechanisms by which the anabolic effects are mediated are incompletely understood.

Insulin-like growth factors

A series of peptides similar in structure to insulin are produced in the liver. The synthesis and release of these peptides occur in response to stimulation by growth hormone (GH). These insulin-like growth factors (IGFs) have, as their name suggests, important functions in the regulation of growth and tissue differentiation.

Atrial natriuretic peptide (ANP)

A peptide hormone, atrial natriuretic peptide, is released from the tissues of the heart in response to a stretching of the atrial myocytes. The peptide is involved in blood volume regulation, and in salt and water balance. ANP may, for example, stimulate fluid and solute secretion by the rectal gland of elasmobranchs, and lead to increased ion efflux over the gills of fish held in sea water. Whilst the peptide was initially isolated from the atrium of the heart, ANP is now known to have a wider distribution, being found in many tissues, including those of the brain.

Corpuscles of Stannius

The corpuscles of Stannius are small spherical bodies that lie on, or embedded within, the kidney. The corpuscles synthesize and secrete glyco-peptide hormones that are intimately involved in the control of the ionic balance of the body fluids. The corpuscles appear to be more active when fish are in sea water than when they are in fresh water, and it is now known that the corpuscles of Stannius are the prime organs exerting an effect over body loadings of calcium.

Urotensins

At the caudal tip of the spinal cord in elasmobranchs and teleosts there is a neurosecretory and neurohaemal area known as the *urophysis*. The urophysis is a neurosecretory organ that releases at least two biologically active peptides – urotensin I and urotensin II – into the blood vessels in its close vicinity. The urotensins are vasoactive compounds that appear to be released from the urophysis in response to excessive salt loading in the body tissues. The urotensins released into the blood are carried directly to the kidney and bladder in the vessels draining the caudal region, and the urotensins may exert their greatest effects on these tissues. There are clear indications that the urotensins have marked effects upon ion transport, and there is good evidence that they play a major role in osmoregulatory control.

'Tissue hormones'

All of the hormones considered hitherto are synthesized and secreted by distinct endocrine tissues or organs. The release of these hormones usually leads to the induction of physiological changes at sites distant from the site of release. Some hormones may, however, exert effects on tissues both distant from and close to the site of release. The latter type of action is known as a paracrine effect. There are, in addition to the classical hormones, a number of compounds that seem to exert their major effects via local, or paracrine, actions within tissues. These hormones may be considered to be 'tissue hormones'. The distinction between the 'classical' hormones and the tissue hormones is not, however, a clear one, since the tissue hormones may also be released into the blood and carried to distant sites before they exert their regulatory actions.

The tissue hormones include a number of vasoactive compounds which may have either vasodilator – bradykinin, histamine – or vasoconstrictor – angiotensin, serotonin – effects. In addition, the tissue hormones can be considered to encompass the eicosanoids, a heterogeneous group of compounds derived from the 20-C unsaturated fatty acids.

Histamine

Histamine, which is a derivative of the biotransformation of the amino acid histidine, acts as a vasodilator and leads to increased capillary permeability. These actions of histamine lead to local mucosal swelling and a drop in blood pressure as part of the inflammation response.

Histamine released by cells in the wall of the stomach evokes secretion of gastric acid.

Angiotensin

In mammals, the maintenance of blood pressure and the regulation of salt and water balance are very much under the influence of the *renin–angiotensin* system. Renin, formed in the kidney, is released into the bloodstream where it acts on the polypeptide angiotensinogen to form angiotensin. The effects of angiotensin lead to elevated blood pressure and increased sodium retention by the kidney.

The renin–angiotensin system is present in teleosts, but its exact physiological roles remain almost unknown. Nevertheless, in view of the fact that the release of renin tends to increase following the transfer of fish from sea water to fresh water, it is likely that the renin–angiotensin system plays a role in the control of salt and water balance.

Serotonin

Serotonin, or 5-hydroxytryptamine, is a derivative of the amino acid tryptophan, and it has a wide distribution in nervous tissue. It is, for example, found in many regions of the vertebrate brain, and one of its major roles is as a transmitter substance.

Serotonin is also a powerful vasoconstrictor, and it is released from the blood platelets in areas of blood vessel damage. The local vasoconstriction caused by the serotonin serves to provide a temporary sealing of the vessel prior to the formation of the blood clot that will effectively seal the damaged area.

Eicosanoids

The eicosanoids are derived from 20-C polyunsaturated fatty acids, and those eicosanoids produced by enzymatic transformation of the $(n-6)$ series fatty acids have homologues derived from biotransformation of 20-C fatty acids of the $(n-3)$ series. Most, if not all, tissues respond to eicosanoids, and in the majority of cases the eicosanoids will act as paracrine mediators. Under certain circumstances, however, such as where there is extensive tissue damage, large amounts of these biologically active compounds will be released into the general circulation.

There are two major enzymatic pathways for the production of eicosanoids from fatty acids – the cyclooxygenase pathway and the lipoxygenase pathway. Metabolism of the fatty acids via the cyclooxygenase pathway leads to the production of prostanoids – prostaglandins, thromboxanes and prostacyclins. Leukotrienes and lipoxins are formed by the action of the lipoxygenase enzymes.

Seemingly small changes in eicosanoid structure can result in dramatic changes in bioactivity, since the receptors for the eicosanoids are highly selective. As a generalization it can be said that the eicosanoids derived from the $(n-6)$ series fatty acids are much more biologically active than their homologues in the $(n-3)$ series. In addition, compounds that are apparently very similar can exert counteracting effects, so a balance between the different eicosanoids is required for the maintenance of homeostasis. Consequently, an imbalance in relative production of one class of eicosanoid compared with that of another could lead to dramatic effects that result in pathological changes in the affected tissues.

The earliest observed effects of the eicosanoids were related to their influences on smooth muscle contractions. Smooth muscles within the cardiovascular, reproductive, respiratory and gastrointestinal systems are all affected, with some of the eicosanoids having a stimulatory, and others an inhibitory, effect.

The most dramatic effects of the eicosanoids appear as inflammatory and immune reactions. Some of the prostaglandins promote the inflammatory response by increasing vascular permeability and vasodilation. Leukotrienes also operate as intermediaries of inflammatory reactions, whereas lipoxins promote arteriolar dilation and have pronounced effects upon vascular smooth muscle. Some thromboxanes are highly potent proaggregatory agents, that is, they cause the blood platelets to aggregate, stick together and form clumps. In addition, these eicosanoids cause strong constriction of blood vessels. Thus, overproduction of leukotrienes and thromboxanes is, potentially, deleterious.

In fish, it is the roles that the *prostaglandins* play in the reproductive processes that have received most attention. The ovaries and ovarian follicles of fish produce prostaglandins during the final stages of maturation, just prior to the time of ovulation. The increase of prostaglandin synthesis by the ovarian tissue is progestogen dependent, and this increase, in turn, triggers ovulation. As the amounts of prostaglandins produced by the ovarian tissue increase, there is a corresponding rise in plasma levels of these eicosanoids. The circulating prostaglandins seem to act upon the brain to stimulate female sexual behaviour. Following ovulation some of the prostaglandins may be released into the environment, and water-borne prostaglandins are known to be potent olfactory stimulants. Thus, the prostaglandins may also act as sex pheromones, inducing the males to display courtship, and other sexual, behaviours.

5.11   CONCLUDING COMMENTS

In complex multicellular animals such as fish, coordination and control can be achieved either by the delivery of regulatory substances to the bloodstream (hormonal control) or via nervous mechanisms. The two types of control system have often been viewed as being distinct, but it is clear that the systems are closely integrated within the hypothalamus (Fig. 5.12).

The neurons innervating the neurohypophysis of the pituitary gland release biologically active compounds (peptide hormones) into the bloodstream. This points to the possibility of further links between the endocrine

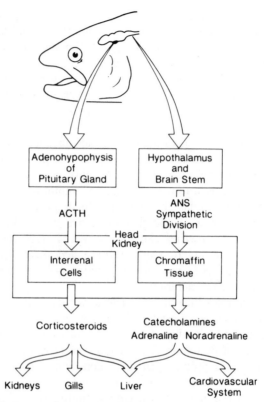

Fig. 5.12 Central linking and integration between the nervous and endocrine systems illustrated by the involvement of the hypothalamic–pituitary–interrenal axis in the stress response in fish.

and nervous systems with respect to the production and secretion of chemical messenger substances.

It is now clear that the same, or similar, compounds are able to function as regulatory substances in both the nervous and endocrine systems. Furthermore, there may be many similarities in the mechanisms of production, release and actions of hormonal and neurotransmitter substances. In no case is this better demonstrated than with the various groups of regulatory peptides.

There are at least four different modes by which peptides can act as regulatory substances: endocrine, paracrine, neuronal and neuroendocrine (Fig. 5.13).

1. Endocrine action – the peptides are produced in specialized cells and are released into the bloodstream, thereby ensuring their distribution to the

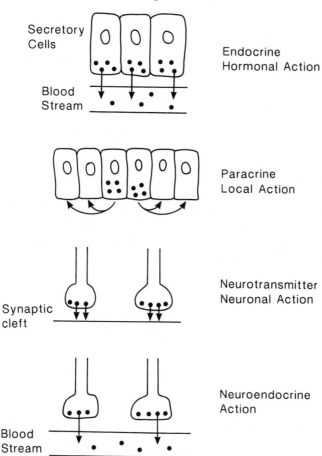

Fig. 5.13 The possible roles of peptides as regulatory substances.

target cells and organs. Since the hormone is circulated in the blood supply it reaches all tissues, and the specificity of the response to the hormone depends upon the target cells possessing the appropriate receptors.

2. Paracrine action – secretions from scattered endocrine cells result in responses being shown by neighbouring cells, giving the possibility of a form of local control.

3. Neuronal action – the chemical transmission of signals over the synapse has been known for many years, with the catecholamines and acetylcholine being considered to be the neurotransmitters. It is now evident that a number of amino acids, amino acid derivatives and peptides can also have a neurotransmitter function. The neuropeptides often function

over greater distances and for longer time periods than the classical neurotransmitters.

4. Neuroendocrine action – when a peptide is secreted from a nerve ending into the bloodstream it may reach distant target tissues and have a hormonal effect. For example, neuroendocrine actions of this type are properties displayed by the peptides of the neurohypophysis.

One important feature of regulatory peptides is that they can occur in both endocrine cells and neurons. This has the consequence that, in different parts of the body, the same substance can act as a hormone, a paracrine agent or a neurotransmitter. The picture becomes even more complex when it is realized that not only may neurons produce and release both peptide and classical neurotransmitter substances, but cases are known where individual neurons may contain four or more such substances. The differential release of such transmitter substances would obviously allow for the chemical encoding of a wide range of different messages, but it does not ease the task of allocating specific functions and roles to the different forms of chemical transmitter substances.

FURTHER READING

Books

Hoar, W.S. and Randall, D.J. (eds)(1969) *Fish Physiology*, Vol. II, Academic Press, London.

Hoar, W.S. and Randall, D.J. (eds)(1970) *Fish Physiology*, Vol. IV, Academic Press, London.

Holmgren, S. (ed)(1989) *The Comparative Physiology of Regulatory Peptides*, Chapman & Hall, London.

Matty, A.J. (1985) *Fish Endocrinology*, Croom Helm, London.

Nilsson, S. (1983) *Autonomic Nerve Function in the Vertebrates*, Springer-Verlag, Berlin.

Review articles and papers

Bern, H.A. (1967) Hormones and endocrine glands of fishes. *Science*, **158**, 455–62.

Dockray, G.J. (1988) Regulatory peptides. *Sci. Prog.*, **72**, 21–35.

Dulka, J.G. (1993) Sex pheromone systems in goldfish: Comparisons to vomeronasal systems in tetrapods. *Brain Behav. Evol.*, **42**, 265–80.

Eales, J.G. and Brown, S.B. (1993) Measurement and regulation of thyroidal status in teleost fish. *Rev. Fish Biol. Fisheries*, **3**, 299–347.

Holst, J.J. and Schmidt, P. (1994) Gut hormones and intestinal function. *Baillière's Clin. Endocrinol. Metab.*, **8**, 137–64.

Kaltenbach, J.C. (1988) Endocrine aspects of homeostasis. *Amer. Zool.*, **28**, 761–73.

Leatherland, J.F. (1982) Environmental physiology of the teleostean thyroid gland: a review. *Env. Biol. Fishes*, **7**, 83-110.

Mommsen, T.P. and Plisetskaya, E.M. (1991) Insulin in fishes and agnathans: History, structure and metabolic regulation. *Rev. Aquat. Sci.*, **4**, 225–59.

Murdoch, W.J., Hansen, T.R., and McPherson, L.A. (1993) A review – Role of eicosanoids in vertebrate ovulation. *Prostaglandins*, **46**, 85–115.

Nilsson, S. (1984) Adrenergic control systems in fish. *Mar. Biol. Lett.*, **5**, 127–46.

Nishioka, R.S., Kelley, K.M. and Bern, H.A. (1988) Control of prolactin and growth hormone secretion in teleost fishes. *Zool. Sci.*, **5**, 267–80.

Peter, R.E. (1973) Neuroendocrinology of teleosts. *Amer. Zool.*, **13**, 743–55.

Peter, R.E., Yu K.-L., Marchant, T.A. and Rosenblum, P.M. (1990) Direct neural regulation of the teleost adenohypophysis. *J. exp. Zool. Suppl.*, **4**, 84–9.

Plisetskaya, E.M. (1989) Physiology of fish endocrine pancreas. *Fish Physiol. Biochem.*, **7**, 39–8.

Plisetskaya, E., Woo, N.Y.S. and Murat, J.-C. (1983) Thyroid hormones in cyclostomes and fish and their role in regulation of intermediary metabolism. *Comp. Biochem. Physiol.*, **74A**, 179–87.

Popper, A.N. and Fay, R.R. (1993) Sound detection and processing by fish: Critical review and major research questions. *Brain Behav. Evol.*, **41**, 14–38.

Reiter, R.J. (1993) The melatonin rhythm: both a clock and a calendar. *Experientia*, **49**, 654–64.

Sardesai, V.M. (1992) Biochemical and nutritional aspects of eicosanoids. *J. Nutr. Biochem.*, **3**, 562–79.

Schreibman, M.P., Leatherland, J.F. and McKeown, B.A. (1973) Functional morphology of the teleost pituitary gland. *Amer. Zool.*, **13**, 719–42.

Underwood, H. (1990) The pineal and melatonin: Regulators of circadian function in lower vertebrates. *Experientia*, **46**, 120–28.

Chapter six

Digestion and absorption

6.1 INTRODUCTION

Foodstuffs differ markedly in physical and chemical properties and in nutrient composition. Fish may be herbivorous, omnivorous or carnivorous. Specialization upon given food types requires particular sets of morphological and physiological adaptations for ingestion, digestion and absorption of nutrients. Thus, there may be marked differences in the morphology of the alimentary tract between fish species and there may be differences in the profiles of the digestive enzymes present. Finally, there may be marked interspecific differences in absorptive capacity for nutrients within the three major classes – proteins, lipids and carbohydrates.

Food consists of nutrients that are either large molecules, e.g. proteins, are relatively insoluble in water, e.g. lipids, or both, e.g. chitin. Before these nutrients can be absorbed they must be broken down into smaller units and transformed into a more soluble form. It is the role of the digestive enzymes to perform this function. The digestive enzymes that catalyse the breakdown of nutrients in the gut are hydrolases. As the name suggests, the reactions leading to nutrient breakdown involve the introduction of water molecules, and this leads to hydrolysis of the nutrient. Hydrolysis of the large molecules can proceed in the absence of hydrolase enzymes, but the reaction would take a very long time to be complete at temperatures within the normal biological range. Thus, the enzymes act as biological catalysts, speeding the reaction.

All enzymes are themselves proteins, and the hydrolases are broadly grouped according to the type of molecule they hydrolyse, i.e. proteases hydrolyse proteins, and lipases hydrolyse lipids. Within this framework there are digestive hydrolases with different catalytic properties, both with respect to the points in the specific molecular structure that are attacked, and with respect to the conditions, e.g. pH, temperature, under which they display maximum catalytic activity.

It is possible to make a division between intracellular and extracellular digestion of nutrients. Extracellular digestion is that which occurs outside the cell boundaries, such as within the lumen of the gut, whereas intracellular digestion is particularly important in connection with the turnover and renewal of body tissues. Intracellular digestion also plays a role in the breakdown of nutrients that are absorbed into the cells as large molecules. Thus, nutrients may undergo further digestion, transformation and metabolism following absorption from the lumen into the cells of the gut wall. Consequently, the nutrients do not always reach the bloodstream in exactly the same form as that in which they were absorbed from the lumen of the gut.

6.2 ALIMENTARY CANAL – GENERAL MORPHOLOGY

The major divisions of the alimentary canal are the mouth and buccal cavity, the pharynx, the oesophagus, the stomach, the intestine with the pyloric caeca and related organs (i.e. the liver, gall bladder and pancreas), and the rectum and anus. In some fish the digestive tract is a more or less straight tube leading from the mouth to the anus. More often the alimentary canal is looped and folded and can be divided into structurally functional parts. Thus, it is usually possible to distinguish oesophagus, stomach, intestine, rectum and anus, with muscular valves or sphincters often separating the different regions of the alimentary tract. There are a variety of modifications in gut structure and enzyme production superimposed upon the basic gut plan (Fig. 6.1), these differences being related both to evolutionary position of the species and to dietary habits.

The main layers of the wall of the alimentary tract remain similar from region to region, although the relative thicknesses of the different layers and the proportions of the various cell types vary in different parts of the tract. The main layers are the mucosa (inner epithelium and related tissues), submucosa, muscularis (containing both a circular and a longitudinal muscle layer) and the serosa.

The digestive processes often begin in the mouth and pharynx, with some mechanical breakdown of the ingested food. Depending upon the fish species, the teeth within the buccal cavity may act to either crush or lacerate the prey, whereas the pharyngeal teeth generally have a grinding and crushing function. Breakdown of prey in this portion of the alimentary tract is mechanical rather than chemical, since there is no enzymatic secretion by the tissues of the buccal cavity and pharynx.

In most fish the oesophagus is short, but greatly distensible to allow the passage of large food items to the stomach. Cells in the wall of the oesophagus produce mucus which acts both as a lubricant and prevents the oesophagus wall being damaged during the swallowing of food. As far as is known there are no digestive enzymes produced in the oesophagus.

6.3 STOMACH

The stomachs of fish show many variations in size and shape but can be considered as organs for short-term storage, mixing and primary digestion of food. In many piscivores the stomach is an elongated or J-shaped tube, but in most carnivorous species taking a mixed diet of fish, crustaceans and worms the stomach is generally more sac- or bag-like. A common feature of the stomachs of predatory fish is that they have highly elastic muscular walls which can expand to hold relatively large amounts of food.

Some detritivorous and microphagous species, such as mullets (Mugilidae), have a stomach with thick muscular walls. The circular muscles in the distal region close to the pylorus are particularly well developed. Other fish having thick-walled stomachs of this type include those species of surgeonfish (*Acanthurus* and *Ctenochaetus* spp.) and girellids that may ingest particles of sediment along with the algae and micro-organisms upon which they feed. The function of the stomach is thought to be that of fragmenting the food mass, and grinding is aided by the presence of grit and sand in the stomach. Thus, the stomach in these species appears to be analogous to the gizzard found in birds. In other words, the thick-walled, gizzard-like stomach serves to break down the cell walls of bacteria, bluegreen algae, diatoms and filamentous red and green macroalgae that are ingested along with some quantities of sand and other sedimentary material.

A second type of adaptation to a microphagous diet may consist of a loss

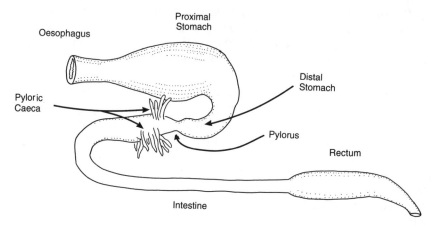

Fig. 6.1 General structure of the alimentary canal.

of the stomach or a marked decrease in its size. The stomach will also tend to be lacking any form of morphological specialization. In some microphagous and herbivorous species, such as the parrotfish (scarids), the herbivorous odacids and the garfish, *Hyporhamphus melanochir*, the stomach is absent and the grinding of the food is accomplished by the pharyngeal teeth. Thus, in these species, the finely ground ingested plant material passes directly from the pharyngeal mill to the intestine in which the pH may be neutral or slightly alkaline.

By contrast, other herbivorous species of fish rely on acid lysis in order to gain access to the nutrients contained within the cells of the green and red algae upon which they feed. The low pH (range 1.2–4.3) within the thin-walled stomach results in lysis of the cell walls of the ingested plant material, thereby releasing their contents and enabling enzymatic digestion to take place. Herbivorous fish relying upon this type of digestive adaptation may include the acanthurids, some sparids, the milkfish, *Chanos chanos*, the monkeyface prickleback, *Cebidichthys violaceus*, and some of the rabbitfishes (Siganidae).

Stomach motility

The stomach of many fish species is morphologically and functionally divisible into a proximal region with a thinner elastic wall and a thick-walled muscular distal region. In the distal stomach the highly muscular antrum is the area closest to the pyloric sphincter which opens to the upper intestine.

The proximal stomach has the short-term storage function, and the first response to the ingestion of food is a receptive reflex. This reflex response results in a relaxation of the muscles of the proximal stomach wall and enables the stomach to accommodate further food. This is followed by local stimulation of the stomach wall, leading to muscular contractions in both the proximal and distal stomach. The tonic contractions of the stomach ensure that the food passes to the distal stomach where peristaltic contractions result in more vigorous activity.

The peristaltic contractions are thought to be initiated in pacemaker tissue that lies between the proximal and distal regions of the stomach (Fig. 6.2). Waves of peristaltic contraction spread from the pacemaker tissue towards the pylorus, with the contractions being particularly powerful in the antrum. In the muscular antrum the gastric contents are pressed towards the pylorus, followed by compression and retropulsion as the pylorus closes. This leads to further mechanical breakdown of the food and mixes the particles with the secretions of the stomach. Thus the antrum has a grinding and mixing function (Fig. 6.3), aiding in breaking down large food particles to a semiliquid chyme.

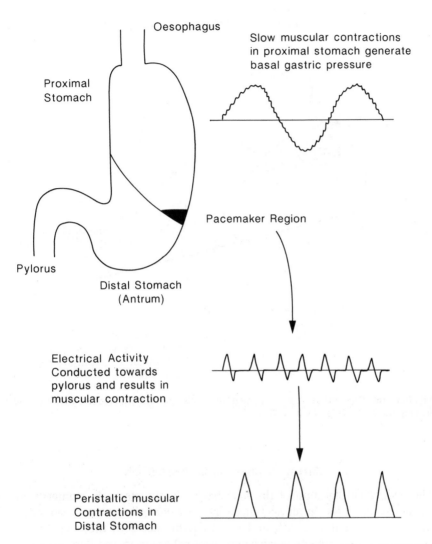

Fig. 6.2 Schematic diagram showing the electrical activity and muscular contractions in different regions of the stomach.

The emptying of the chyme from the stomach is dependent upon there being a pressure difference between the stomach and upper intestine. This pressure difference will, in turn, be dependent upon both gastric and upper intestinal muscular activity, and upon pyloric opening. Gastric emptying will be influenced by any factors that effect changes in motility leading to alterations in the pressure gradient.

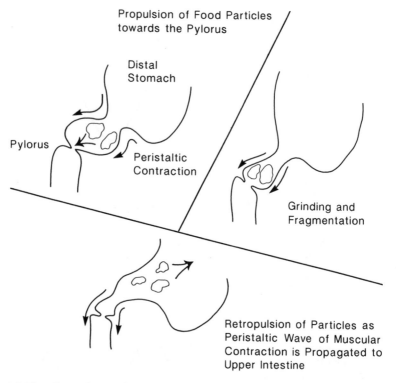

Fig. 6.3 The effects of muscular contractions of the stomach wall on the grinding and fragmentation of large food particles.

Enzymatic digestion in the stomach

The enzymatic digestion of the food begins in the stomach, primarily by the action of *pepsins* in an acid medium. Only one cell type has been found in the gastric glands of fish and it is thought that both the acid (hydrochloric acid) and proteolytic enzymes (pepsins) are produced here.

Mucus-producing cells are also found in the wall of the stomach. It is thought that the major function of the mucus is to prevent damage to the epithelial tissues of the stomach wall by the acid and the proteolytic enzymes, i.e. to reduce the danger of self-digestion.

The enzymes (pepsins) are secreted from the cells of the gastric glands in the form of inactive zymogens (pepsinogens) which are activated by acidic conditions below pH 6. Enzyme extracts from the stomachs of some species of fish have maximum peptic activity in the pH range 4–5, but maxima at pH 2–3 have been reported for other species. Measurements of stomach pH have shown that the stomach contents rarely fall as low as 2–3, except during the latter stages of gastric digestion.

Pepsins are endopeptidases which hydrolyse peptide bonds between aromatic (e.g. tryptophan, phenylalanine, tyrosine) and dicarboxylic amino acids (e.g. glutamic acid, aspartic acid) (Fig. 6.4). Therefore, pepsins attack only a few of the bonds in a large protein molecule, but since they are

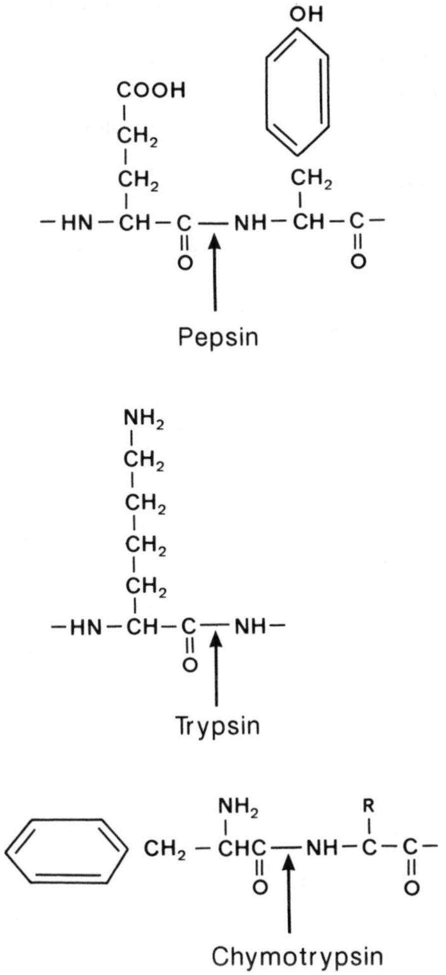

Pepsin

Trypsin

Chymotrypsin

Fig. 6.4 Sites of action of the endopeptidase enzymes – pepsin, trypsin and chymotrypsin. Pepsin hydrolyses peptide bonds between aromatic and dicarboxylic acids, trypsin requires residues from either lysine or arginine to be present before it can act on the peptide bond, and chymotrypsin attacks peptide linkages in which aromatic amino acid residues occur.

endopeptidases they will attack internal peptide linkages and break down the large protein molecule into smaller polypeptides and peptides.

The pH in the stomach of an unfed fish may range from 3 to 7; it may increase slightly, shortly after feeding. This rise in pH may be due to the ingestion of food with a high buffer capacity or due to intake of water along with the food. Towards the end of the gastric phase of digestion the pH within the stomach is lower (e.g. European flounder, *Platichthys flesus*, pH 5.5; Atlantic cod, *Gadus morhua*, pH 2–2.5; plaice, *Pleuronectes platessa*, pH 2.5–4; elasmobranchs, pH 1.3–2.2) and the reduction in pH with time is due to the secretion of hydrochloric acid from the gastric glands.

Increased production and secretion of hydrochloric acid may occur in response to both mechanical (gastric distension) and chemical (peptides, amino acids etc.) stimulation of receptors in the stomach wall. The regulation of acid secretion in fish appears to differ from that reported for mammalian species, in that the hormone gastrin does not appear to be active in fish species, although the tissue hormone histamine does stimulate the secretion of gastric acid in species such as the Atlantic cod. Production

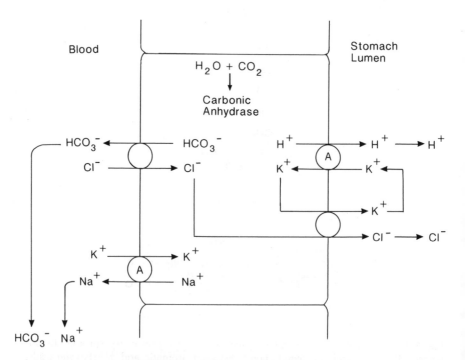

Fig. 6.5 Schematic diagram illustrating the secretion of gastric acid. Open circles indicate ion transport mechanisms, and a circled A indicates an active transport mechanism.

and secretion of the hydrochloric acid is dependent upon the catalytic activity of the enzyme carbonic anhydrase, and ATPase enzymes that function in ion transport. In the presence of carbonic anhydrase, the reaction between carbon dioxide and water proceeds at the catalysed rate to result in the production of protons (H^+) and bicarbonate ions (HCO_3^-). The protons are actively accumulated in the gastric lumen by active transport, and for each proton secreted into the lumen one bicarbonate ion leaves the cell on the blood side in exchange for a chloride ion (Fig. 6.5).

Proteolytic enzymes other than pepsins have been reported from the stomachs of a number of fish species and there may also be a variety of non-proteolytic enzymes present. Amylase (a carbohydrase), lipase and chitinase have been extracted from fish stomachs but their effectiveness has been little investigated.

Thus, in the stomach the food is broken down both mechanically and chemically, and the acidic mixture that leaves the stomach via the pylorus contains a wide range of partially digested products: peptides and amino acids, acylglycerols and fatty acids, poly-, oligo- and monosaccharides.

Control of gastric emptying

As portions of the partially digested mixture begin to enter the upper intestine from the stomach the nature of both the gastric secretions and muscular contractions may be affected by a series of nervous and hormonal feedback mechanisms. These feedback mechanisms are induced, partially, in response to distension of the intestine. There are also receptor cells located in the wall of the uppermost part of the intestine that appear to be sensitive to chemical stimuli such as acid, certain amino acids and free fatty acids. Stimulation of these receptors leads to both the initiation of nervous feedback reflexes, and the liberation of a range of peptide hormones. For example, low pH and the presence of fatty acids in the upper intestine may lead to the liberation of various hormones (e.g. secretin, gastric inhibitory peptide [GIP], vasoactive intestinal polypeptide [VIP]) that inhibit gastric secretion, but these peptide hormones have multiple roles. GIP, for example, in addition to inhibiting secretion also has an inhibitory effect on gastric motility.

On the other hand, the hormone cholecystokinin (CCK), which is secreted in response to the presence of certain amino acids (particularly phenylalanine and tryptophan) and fatty acids in the upper intestine, appears to inhibit the tonus of the proximal stomach, whereas it increases the frequency and strength of the contractions in the distal stomach, leading to increased antral activity. CCK also seems to stimulate contraction of the pyloric musculature, leading to a closing of the pylorus. Increased peristaltic activity in the antrum leads to more efficient grinding and mixing of the food, but closure of the pylorus prevents the muscular contractions of the distal stomach forcing particles into the upper intestine. Thus, the overall effect is to increase the rate at which the food is broken down into smaller particles, but the secretion of CCK also results in a slowing in the rate at which food is emptied from the stomach.

The length of time the food remains in the stomach varies considerably, depending upon both physical characteristics and chemical composition. Large pieces of food must, for example, be broken down into smaller particles before they are allowed to pass through the pylorus. The pylorus and antrum seem to have a sieving function leading to the preferential retention of large food particles in the stomach. Small particles pass readily through the pylorus, but the retention times for larger particles appear to be related to their size, density and the ease with which they can be broken down.

One consequence of the sieving action of the pylorus and antrum is that there is a preferential retention of large indigestible particles (e.g. bones, exoskeletons of insects and crustaceans) in the stomach. These indigestible particles tend to congregate in the antral region during the course of the gastric digestion of a meal. It is possible that this material may contribute to a more rapid grinding of the digestible, more friable components of the meal. Thus, towards the end of the digestive phase of gastric digestion the stomach content may comprise a large proportion of indigestible material, and these components are, generally, the last to be evacuated from the stomach. These indigestible components of prey items are, however, eventually evacuated, and do not seem to accumulate in the stomach with time.

The evacuation of the indigestible material takes place during the interdigestive phase, during which the pattern of muscular contractions in the stomach differs from that seen during the digestive phase. During the digestive phase the sieving action of the pylorus and antrum is maintained by the feedback signals arising as the result of stimulation of the chemoreceptors in the wall of the upper intestine. Once the digestible material has been evacuated from the stomach these chemoreceptors will no longer receive stimulation. There will then be either a cessation or a change in the form of the feedback from the upper intestine to the stomach and pylorus. In the absence of feedback the muscles of the pylorus will remain relaxed and the pyloric opening will be comparatively large. Under these circumstances, muscular contractions of the stomach can drive the indigestible material out of the stomach, through the pylorus and into the upper intestine.

If the gastric contractions of the interdigestive phase are responsible for emptying indigestible solids from the stomach, any manipulations that delayed the onset of these contractions (e.g. feeding small amounts of food at regular intervals so the stomach never becomes completely empty of digestible material) would delay the emptying of the indigestible material. In a series of experiments conducted on the Atlantic cod, it was shown that the regular feeding of small portions of food delayed the emptying of indigestible solids, and that substantial emptying of this material did not commence until almost all traces of digestible nutrients had been evacuated from the stomach. Thus, in this species at least, there appears to be a clear division between the digestive and interdigestive phases of gastric motor activity.

Gastric evacuation times

The time required for food to be evacuated from the stomach may vary from a few hours to several days, with emptying times being markedly influenced by factors such as water temperature, the size of the meal and the characteristics of the food. In general, an increase in water temperature reduces the time required for the stomach to empty. On the other hand, an increase in meal size will increase gastric emptying time (Fig. 6.6).

Small prey organisms are usually broken down and evacuated from the stomach more rapidly than large, so a meal consisting of zooplankton will usually be processed more rapidly than one consisting of larger crustaceans or fish. However, within the size range spanned by a single prey

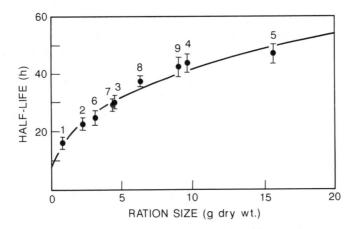

Fig. 6.6 The effects of different meal sizes (ration sizes) of herring, *Clupea harengus*, on the time required by Atlantic cod, *Gadus morhua*, to evacuate 50% of the meal from the stomach (half-life). Meals comprised different numbers of herring prey of different wet weights: 1, 1 × 4 g = 4 g; 2, 1 × 8 g = 8 g; 3, 1 × 16 g = 16 g; 4, 1 × 32 g = 32 g; 5, 1 × 48 g = 48 g; 6, 4 × 4 g = 16 g; 7, 2 × 8 g = 16 g; 8, 8 × 4 g = 32 g; 9, 2 × 16 g = 32 g.

species the size of individual prey items does not appear to have any marked influence upon the time required to empty the stomach, i.e. a meal of a given size will be evacuated at more or less the same rate irrespective of whether it is made up of a single large prey item or a number of smaller items. This is illustrated in Fig. 6.6, which shows that whilst the overall size of the meal influenced the gastric emptying time of Atlantic cod, the size of the individual prey items, i.e. small versus large herring, *Clupea harengus*, did not.

The physical form of the food can have a marked effect upon the time required for it to be emptied from the stomach. For example, when cod were fed on herring, approximately 50% of the meal had been emptied within the space of 24 h, but when minced herring was fed the rate of emptying was greater, and over 80% of the meal had been evacuated from the stomach within the space of 24 h.

The physical properties and the chemical compositions of different prey types have significant effects on the rates at which they are digested and emptied from the stomach. For example, brown trout, *Salmo trutta*, evacuate meals of gammarids and oligochaete worms much more rapidly from the stomach than similar-sized meals of caddis-larvae. When gadoids are fed meals of brittle stars or large crustaceans, some hours may be required before the prey begins to be broken down and evacuation from the stomach commences, whereas small crustaceans are digested and evacuated rapidly. Thus, meals of small crustaceans, such as euphausiids (krill), are emptied from the cod stomach more rapidly than are meals consisting of large crustaceans, such as prawn, *Pandalus borealis*, and fish, e.g. herring or capelin, *Mallotus villosus*. Most of these differences in rates of gastric digestion and emptying can be attributed to differences in toughness or thickness of the exoskeleton of the various prey organisms, but the chemical composition of the food may also affect the rate at which it is evacuated from the stomach. Foods with a high lipid content tend to be emptied from the stomach more slowly than other food types, but the influences of chemical composition on rates of evacuation can be masked by the factors that determine prey friability.

Small prey species are usually more easily broken down than large prey items, and there is clear evidence that fish species that eat dissimilar prey types display differences in the rates at which they digest and evacuate food from the stomach. Microphagous fish, which feed on detritus or small planktonic organisms, seem to show more rapid rates of gastric evacuation than either mesophagous (eating molluscs, insects, crustaceans and worms) or macrophagous (feeding on large crustaceans or fish) fish species. Whilst the greater part of these differences can be explained on the basis of prey characteristics, the trends may be maintained when fish are fed upon pellet feeds of defined chemical composition. Thus, the rates of gastric digestion and evacuation of microphagous species seem to be higher than those of meso- and macrophagous species when the fish are fed identical diets, so there may be inherent differences in gastric physiology linked to the natural prey types of individual fish species.

Stomachless fish

Whilst the vast majority of fish species possess a stomach, this part of the alimentary tract is absent in some taxonomic groups, such as the cyprinids and catostomids. Scarids, labrids and blennids are also stomachless. All of these stomachless fish lack both acid secretion and the enzymes responsible for gastric digestion. In addition, the storage function of the stomach has been lost. A reduction in size, or loss, of the stomach is often considered to be a secondary feature amongst fish and several hypotheses have been put forward to explain why there should have been a loss of the stomach in certain taxonomic groups.

Many of the groups of stomachless fish contain a predominance of species that are either microphagous, detritivorous or herbivorous. In these species the ingestion of water along with the small food particles could lead to increased alkalinity in the stomach, thereby reducing the value of acid secretion and inhibiting the action of digestive enzymes that require an acidic environment. Some herbivores, however, have an acid-secreting stomach and the pH of the stomach contents may be reduced to 1–1.5, leading to lysis of the ingested plant cells.

The ingestion of food particles that are small means that they do not require extensive working in order to be broken down to increase the surface area available for attack by the enzymes of the intestine. Any mechanical grinding and mastication of the refractory food particles is usually carried out by the well-developed pharyngeal mill. Microphagous and herbivorous diets contain considerable quantities of indigestible ballast in the form of sediment particles or cellulose, and the fish must process large amounts of such material to extract sufficient nutrients from these food types. The stomach is a receptacle with a short-term storage function, but this storage and restraining function is obsolete in fish consuming microphagous diets containing large amounts of indigestible material.

All representatives of the stomachless groups of fish are not microphagous or herbivorous, and many stomachless fish are either omnivores, planktivores or benthophages. There are even a few piscivorous species. In the carnivorous representatives of the stomachless fish the foremost part of the intestine is usually enlarged to form an intestinal bulb. The intestinal bulb can be viewed as being an anatomical analogue of the stomach, and it has a short-term food-storage function. It must be stressed, however, that the digestive processes of these carnivorous species are limited by taxonomic constraints. For example, the northern and Colorado squawfish, *Ptychocheilus oregonensis* and *P. lucius*, are piscivorous cyprinids that feed on juvenile salmonids and other small fish. In common with all cyprinids they have reduced jaw teeth, are stomachless and therefore lack the ability to secrete acid and there is no peptic digestion of the prey.

6.4   INTESTINAL STRUCTURE AND FUNCTION

The intestine is usually a relatively long coiled tube which functions to complete the digestion and absorption of food. It also has important roles in the absorption of water and electrolytes. In a number of species the intestine comprises a single tube, but many species possess a greater or lesser number of diverticula located in the upper intestine close to the pyloric opening. These pyloric caeca consist of finger-like projections opening from the intestine, and small particles of food may enter the caeca from the main body of the intestine. The numbers and forms of the caeca vary markedly between species, ranging from very few (two to four) as in the pleuronectids, to several tens or more, as in some salmonids and gadoids. The pyloric caeca probably have a digestive function, and also provide an increased surface area over which absorption of nutrients can occur.

The gross morphology of the intestinal tract appears to be linked to the feeding habits of individual fish species. For example, the length of the intestine tends to be short in carnivores and elongated in herbivorous species, leading to the following generalization. The ratio of intestinal length to body length is usually less than unity in carnivorous species, being lower for piscivores than for more generalized carnivores that feed on molluscs, worms, crustaceans and insects. In omnivores the ratio may increase to around 2–3, and the ratio is even higher in herbivores and those fish species which consume diets with a high roughage content (Table 6.1). This generalization refers to interspecific differences, but there is also evidence that the same trends apply on an intraspecific basis. For example, individual carp, *Cyprinus carpio*, and roach, *Rutilus rutilus*, that have consumed diets containing large quantities of indigestible matter tend to have longer intestines than those fed on animal prey. Similarly, grass carp, *Ctenopharyngodon idella*, that have been fed on animal diets tend to have shorter intestines than individuals fed predominantly on plant material.

Whilst the generalization that herbivorous and detritivorous fish have long intestines usually holds true, there may be some exceptions. Consequently, relative gut length should not be viewed in isolation, but should be considered along with other morphological and physiological adaptations to particular types of diet. For example, the digestive tracts of herbi-

Table 6.1 The intestinal lengths of a range of fish species in relation to their feeding habits

Species	Feeding habits*	Ratio IL:BL†
Atlantic salmon, *Salmo salar*	C	0.75–0.85
Largemouth bass, *Micropterus salmoides*	C	0.75–0.9
Northern squawfish,*Ptychocheilus oregonensis*	C	0.7–0.9
Atlantic cod, *Gadus morhua*	C	1.0–1.50
Flagfish, *Jordanella floridae*	H	2.5–3.0
Silver carp, *Hypophthalmichthys molitrix*	H	4.5–7.0
Calbasu, *Labeo calbasu*	H/D	4.5–10.0
Mud carp, *Cirrhina molitorella*	H/D	6.0–13.0

*Habits: C, carnivorous; D, Detritivorous; H, herbivorous.
†Ratio of intestinal length (IL) to body length (BL).

vorous marine fish show a wide range of specializations for handling diets of algae and sea-grasses. Feeding may range from browsing or cropping, to scraping and the concomitant ingestion of inorganic sediment.

Within this broad framework there appear to be four main digestive mechanisms by which the herbivorous species release the nutrients locked within the plant cells. These mechanisms are the possession of a thin-walled, highly acidic stomach in which plant cell walls are lysed due to the low pH, the use of a thick-walled, gizzard-like stomach, or a pharyngeal mill to grind and rupture the plant cell walls by mechanical means, or the possession of a hindgut caecum in which microbial fermentation serves to break down the plant material. Certain combinations of these mechanisms, such as a highly acidic stomach and hindgut fermentation, may occur in some species. The type of digestive mechanism employed, together with the relative proportions of nutritive and inert material ingested, will determine the type of intestinal morphology required to ensure the absorption of adequate quantities of nutrients.

For example, the striped mullet, *Mugil cephalus*, feeds by sucking detritus from bottom muds or by grazing small algae from the surfaces of rocks and submerged plants. Some selection of particles is possible using the fine comb-like teeth and the long gill rakers, but sand or other small sediment particles may often make up 50% or more of the stomach content. Thus, the nutrient concentration in the diet is relatively low. The mullet has a muscular gizzard-like stomach and relies on the muscular activity of the stomach coupled to the abrasive action of the sand grains to lyse the algal cells present in the ingested material. Nutrient supply is maintained by the fish feeding almost continuously and by swift passage through the long gut. Under normal feeding conditions the time required for the food to be passed through the gut, which may be 5.5–6.0 times the body length, may be no more than 2–6 h.

Like the mullet, the diets of most parrotfish (scarids) contain a high proportion of inorganic material. The inorganic material is ingested along with the algae which the fish grazes from rock surfaces and coral rubble using the fused beak-like teeth. The algal cell walls are ruptured by the grinding action of the pharyngeal mill, and the scarids lack a stomach. The relative intestinal length of the parrotfish is shorter than that of the mullet, being 2–2.5 times the body length. Nutrient supply is maintained by the fish consuming a large volume of food each day and by rapid transit of food through the digestive tract (4–6 h).

In contrast to the mullets and parrotfish, the herbivorous monkeyface prickleback does not ingest very much inorganic material along with its food, and the nutrient concentration is, therefore, somewhat higher. The fish grazes on macroalgae which are attacked by acid secreted from the walls of the thin-walled stomach. The pH of the stomach contents is generally within the range 2.2–3.0, and the fish feeds intermittently to allow the breakdown of the plant material within the acidic environment of the stomach. The relative gut length, about 1.5 times the body length, is similar to that found in a number of omnivorous fish species, and food transit through the gut may take up to 2 days or more.

The silver drummer, *Kyphosus sydneyanus*, is a herbivorous marine fish found in the coastal waters of Oceania. The fish browses on macroalgae and ingests little inorganic material along with the food. Breakdown of the plant cell walls probably begins in the acidic environment (pH 2.5–3.0) of the thin-walled stomach, but the majority of the digestion of the plant material may take place in the hindgut caecum by means of microbial fermentation. The intestine of the fish is relatively long, and 20–25 h may be required for the food to pass through the gut and be evacuated.

The generally long intestines of the herbivorous, microphagous and detrivorous fish would be expected to provide a large surface area for absorption of nutrients, but the external dimensions of the intestine give few indications of the absorptive area presented to the lumen. The mucosal and submucosal layers of the intestinal wall may be extensively folded, and the mucosal epithelium also possesses numerous finger-like projections, the *villi*, which lead to the surface area being increased several-fold over that seen in a smooth cylinder. The inner epithelial surface facing towards the intestinal lumen is formed by a layer of various types of cells, including *enterocytes*, which are the cells with absorptive function. The brush border of the enterocyte is decked with numerous microvilli, which increase the absorptive surface. The epithelium also has numerous mucous cells, and the mucus produced and secreted by these cells serves a protective and lubricant function.

At the base of the villi, there may be groups of undifferentiated cells which divide to give rise to the epithelial cells of the villi. The cells at the tips of the villi are continuously being sloughed off, and they are replaced by the cells produced at the base. Thus, the entire epithelium of the intestine is renewed at intervals of a few days. The intestinal epithelium also contains endocrine cells, secretory cells and cells with immunological protective functions. The endocrine cells, which probably possess receptors on the luminal side, release their peptide hormones into the blood, whereas the secretory cells release enzymes and other secretions into the lumen.

Intestinal motility

The intestine receives a nervous supply from the autonomic nervous system, and a plexus within the submucosa contains the sensory neurons from the chemo- and mechanoreceptors of the mucosal layer of the intestinal wall. Signals from these nerves, and from stretch receptors located within the muscle, elicit peripheral and central reflexes that influence both intestinal motility and secretory mechanisms. Some of the intestinal movements may occur independently of external innervation, including the to-and-fro movements of individual villi, and the pendular and segmentation movements that mix the intestinal contents.

The peristaltic movements that propel the contents along the intestine are initiated by stimulation of stretch receptors. The presence of the intestinal contents stimulates stretch receptors, and this leads to a reflex response causing a narrowing of the lumen in the wake of the material and an increase in the cross section of the lumen in the section in advance of the intestinal content. These changes result from a series of linked contractions and relaxations of the circular and longitudinal muscles of the intestinal wall as the contents pass along the intestinal lumen.

Enzymatic digestion in the intestine

Most of the enzymatic digestion of the foodstuffs occurs in the intestine. The enzymes responsible for digestion within the intestine may come from one of two endogenous sources – the pancreas or the secretory cells of the intestinal wall itself, with the pancreas probably secreting the greater variety and quantities of enzymes (Table 6.2). Digestive enzymes may also be produced and secreted by the gut microflora. These exogenous enzymes will aid in the breakdown of food particles and nutrients into a form that can be absorbed.

In the majority of fish species the pancreas is not a discrete organ, but the pancreatic tissue is dispersed in the regions between the pyloric caeca. The secretions produced and released by the pancreas contain both diges-

Table 6.2 Localization of the digestive enzymes of the fish gut, their substrates and the products resulting from their digestive actions

Source site of secretion	Enzyme	Site of action	Substrate	Product
Stomach	Pepsins	Stomach	Protein	Peptides
Pancreas	Trypsin	Intestine	Protein/peptides	Peptides
Pancreas	Chymotrypsin	Intestine	Protein/peptides	Peptides
Pancreas	Carboxypeptidase	Intestine	Protein/peptides	Amino acids, Peptides
Intestine	Aminopeptidase	Intestine	Protein/peptides	Amino acids, Peptides
Intestine	Di-/tripeptidases	Intestine	Di-/tripeptides	Amino acids
Pancreas	Lipase	Intestine	Triacylglycerols	Fatty acids, Monoacylglycerols
	Esterases	Intestine	Esters	Alcohols, Fatty acids
Pancreas	Amylase	Intestine	Starches	Disaccharides
Intestine	Disaccharidases	Intestine	Disaccharides	Monosaccharides
Pancreas and gut microflora	Chitinases	Intestine	Chitin	N-acetyl-glucosamine
Gut microflora	Cellulase	Intestine	Cellulose	Saccharides

tive enzymes and bicarbonate, the latter acting to neutralize the gastric acid. Neutralization of the acidic gastric chyme is also aided by the secretion of *bile* from the gall bladder.

Bile is produced in the liver, but it does not necessarily flow immediately to the intestine, and the bile is usually stored in the gall bladder. The bile is produced more or less continuously, but is only secreted in large amounts in response to the presence of material in the intestine. Bile accumulates in the gall bladder, and the bladder will gradually increase in size if a fish does not feed for a prolonged period of time. During storage in the gall bladder the bile may become more concentrated due to the absorption of water, so the resulting concentrate may contain relatively large quantities of specific bile components in a small volume. The presence of food in the intestine results in the bile being emptied from the gall bladder into the uppermost region of the intestine along the bile duct.

Intestinal proteases

The predominant protein-digesting enzyme within the intestine appears to be *trypsin*, but in many cases the enzyme has not been definitely isolated and identified. Most researchers have just tested for proteolytic activity within the pH range 7–11 and have assumed that any activity was due to the presence of trypsin. Specifically, trypsin is an endopeptidase that hydrolyses peptide linkages adjacent to amino acids possessing two amino groups (Fig. 6.4). Thus, trypsin requires residues from either lysine or arginine to be present before it can act on the protein molecule. Trypsin is produced in the pancreas as the inactive zymogen trypsinogen, and this is activated by enterokinase produced in the intestinal wall. Trypsinogen is also activated by active trypsin so trypsinogen is activated more and more rapidly as more trypsin is formed (*autocatalytic activation*). Activation occurs when a small peptide (four to ten amino acids in length, dependent upon species) is removed from the trypsinogen molecule, the peptide bond attacked being the first one from the amino-end that has a lysine residue.

The proteolytic enzyme *chymotrypsin* is produced and secreted from the pancreas as the inactive precursor chymotrypsinogen. Activation occurs once a small peptide molecule has been split from the chymotrypsinogen, and activation occurs due to the action of trypsin. Chymotrypsin, like trypsin, is an endopeptidase, but chymotrypsin attacks peptide linkages in which there is an aromatic amino acid (tyrosine, phenylalanine, tryptophan) or a large hydrophobic residue, such as that found in methionine (Fig. 6.4).

The pancreas also produces and secretes an exopeptidase, a *carboxypeptidase*, that splits off single amino acids from the carboxyl ends of peptide chains (Fig. 6.7). Carboxypeptidase acts best on a carboxy-terminal

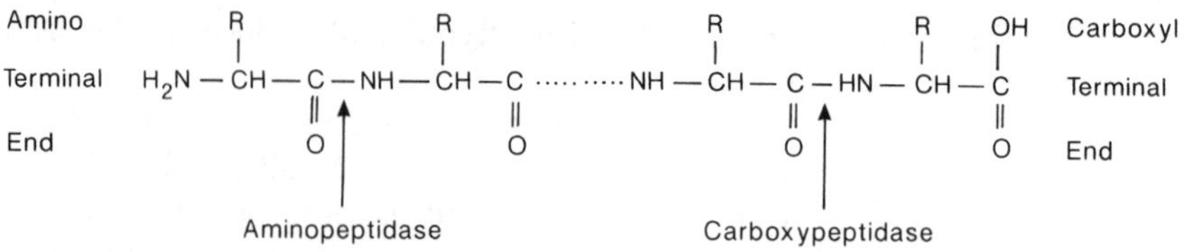

Fig. 6.7 Sites of action of exopeptidases.

residue with an aromatic or large aliphatic side chain (i.e. the products resulting from breakdown of proteins due to the action of pepsin and chymotrypsin). The enzyme appears to be inhibited when there is an amino group close to the carboxyl end of the peptide that is being attacked.

The epithelial cells in the intestinal wall produce a number of proteolytic enzymes, including an *aminopeptidase*. Aminopeptidase is an exopeptidase that splits off single amino acids from the amino ends of peptide chains (Fig. 6.7). The aminopeptidase may be inhibited by the presence of a carboxyl group close to the amino-terminal end of the peptide molecule being attacked. The epithelial cells may also have dipeptidases and tripeptidases associated with the brush border, and these proteolytic enzymes are probably responsible for the final stages of protein digestion immediately prior to absorption.

The proteolytic enzymes secreted by the pancreas and intestinal epithelium act best in slightly alkaline conditions (pH 7–9) but the food entering the intestine from the stomach will be mixed with acidic secretions and the overall pH will be lower than that required for optimal function of these proteases. On entering the intestine from the stomach the food is rapidly mixed with the pancreatic secretions and with bile from the gall bladder. Both of these secretions have a pH within the range 5.5–8.0, and they are responsible for increasing the pH of the intestinal contents (usual range 6.4–9.0). Fluids secreted from the intestinal wall may also play some role in neutralizing the gastric acid.

Pancreatic secretion

The release of the pancreatic secretions and bile into the intestine occurs in response to the presence of food in the intestine, but the nature of the nutrients present has some influence upon the composition of the secretions. Secretion by the pancreas is under both nervous and endocrine control, with the intestinal peptide hormones *secretin* and *cholecystokinin* (CCK) appearing to play major roles.

The stimulus for release of secretin is provided by the presence of lipids and acidic conditions in the upper intestine. Secretin reaches the pancreatic tissue via the bloodstream, and both increases the flow of secretions from the pancreas and raises their content of bicarbonate ions.

Amino acids and lipids in the upper intestine stimulate the release of CCK which, in turn, increases the enzymatic content of the pancreatic secretions. In addition to its influence on the pancreatic tissue, CCK also causes contraction of the gall bladder, resulting in the release of bile into the upper intestine. The production of bile, however, appears to be more influenced by secretin, with release of this peptide hormone from the endocrine cells of the gut resulting in increased production of bile in the liver.

Lipid digestion

In addition to its role in maintaining intestinal pH, bile also has a very important function in assisting in the digestion and absorption of lipids, particularly the triacylglycerols. Most storage lipids of plant or animal origin are esters formed between glycerol and long-chain fatty acids. These acylglycerols are highly insoluble and are not easily hydrolysed. *Lipases* are the enzymes responsible for the hydrolysis of neutral lipids, particularly the triacylglycerols. The lipases are secreted largely by the pancreas, but they are also found in the gastric and intestinal secretions of some species.

Lipases are active mainly at the lipid–water interface formed by the emulsification of the lipids due to mechanical mixing. Relatively small drops of lipid emulsion (1–2 μm in diameter) present a large surface area for the lipases to attack, but because of the insoluble nature of the lipids, the digestive process may be aided by the presence of a detergent that can bring the lipases and lipids into closer contact. Bile acts in this detergent capacity. Bile is a mixture of water, electrolytes, bile salts (e.g. taurocholates), bilirubin (resulting from the breakdown of red blood cells), detoxified xenobiotics and residues of steroid hormones. It is the bile salts which have the greatest role in lipid digestion.

The pancreatic lipase becomes active in the presence of calcium ions and a colipase. The colipase arises from the action of trypsin on a pro-colipase produced in the pancreas. Enzymatic activity of the lipase appears to be maximized when the lipase-to-colipase ratio is about one. Once activated the lipase hydrolyses the triacylglcerols to free fatty acids and monoglycerides, with the ester linkages in the 1 and 3 positions being attacked. Lipases from some fish species appear to preferentially attack triacylglycerols containing long-chain unsaturated fatty acids. The fact that the lipases have a higher affinity for lipids with unsaturated fatty acids than those with saturated fatty acids means that these enzymes will have little effect on wax esters, which generally have relatively short-chained, saturated fatty acids (e.g. 14:0 or 16:0).

The phospholipids consumed in the food are broken down in the presence of calcium and bile acids by *phospholipase*. Phospholipase is produced in the pancreas as an inactive pro-enzyme and is activated by trypsin in the intestinal lumen. Wax esters, the storage products of some zooplanktonic species, are less easily digested than the acylglycerols and phospholipids, and they are hydrolysed much more slowly by lipase. Esters of this type and the ester linkage in the 2 position of the triacylglycerols are attacked by pancreatic *esterase*, usually referred to as non-specific lipase.

Fish that feed on zooplankton seem to show three adaptations to overcome the problem of the digestion of wax esters: an increase in the output of pancreatic lipase, elevated levels of non-specific lipases and esterases, and increased retention time of food due to the development of special morphological features such as pyloric caeca. The pyloric caeca are 'blind alleys' in which digestion of food occurs slowly but digestion tends to be complete. In other words, there will be few, if any, undigested remains to be returned to the intestine.

Intestinal carbohydrases

Of the carbohydrates present in fish foods the most important are chitin, cellulose, glycogen and starch. Chitin and cellulose are the structural carbohydrates of animals and plants, respectively. Glycogen is the main storage carbohydrate of animals, whereas that of most plants is starch.

Chitin-digesting enzymes appear to be found particularly in fish species that eat crustaceans. The enzymes may be secreted by the stomach and pancreas, although an additional source of the chitinolytic activity may be bacteria living within the intestine. The pyloric caeca appear to be amongst the most important sites for chitin digestion. *Chitinase* activity results in the production of di- and trimers of N-acetyl-glucosamine, and these are attacked and broken down into the monomer by the enzyme N-acetyl-glucosaminidase.

Very few, if any, fish produce *cellulases* for the chemical digestion of plant cell walls, and any cellulase activity present in the digestive tract of fish is probably derived from intestinal bacteria. Since the cellulase activity in the gut of fish is usually bacterial in origin, there may not be close links between the cellulytic activity recorded in the gut and the normal diet of the fish. For example, carnivorous species, which consume little or no cellulose, sometimes have high levels of intestinal cellulase activity. In herbivorous fish species the plant cell walls may be largely broken down by mechanical, rather than chemical means, either by the action of the pharyngeal teeth or by grinding and mixing in a highly modified and muscular stomach.

The storage carbohydrates are digested by a range of enzymes, the most widespread of which is *amylase,* produced by the pancreas. The amylase will attack the α 1,4 glycosidic linkages found in starch but will have little effect on other forms of carbohydrate. Digestion of starch by amylase results in the production of a range of oligosaccharides, but carbohydrates are absorbed in the form of monosaccharides. Thus, the oligosaccharides must be further digested prior to absorption. This is done by carbohydrases in the pancreatic secretions, but more especially by enzymes associated with the brush border of the intestinal epithelium.

The algal storage carbohydrate laminarin consists of glucose residues bound via β 1,3 and 1,6 glycosidic linkages, so digestion of this poly-saccharide requires specific *laminarinase* enzymes to be present. Enzymes of this type have been isolated from micro-organisms, fungi, and several species of invertebrates. In vertebrates, laminarinase activity has been recorded in the guts of phytophagous and microphagous fish species. For example, the nase, *Chondrostoma nasus,* a cyprinid which includes a high proportion of freshwater algae in its diet, has a laminarinase that is capable of hydrolysing the 1,3 glycosidic linkages, but lacks a β 1,6 hydrolase. Thus, the nase is incapable of complete enzymatic digestion of laminarin. By contrast, some species of tilapia, *Oreochromis* spp., possess both 1,3 and 1,6 hydrolases and are, therefore, capable of almost complete enzymatic degradation of the laminarin molecules.

Diet and enzyme production

The secretion of both the digestive enzymes and fluids (gastric acid, pancreatic bicarbonate, bile) in response to the ingestion of a meal is influenced by the nutrient composition of the meal, but there appear to be interspecific differences in the abilities of fish to produce and secrete digestive enzymes. This ability is probably linked to the general feeding habits of the fish species. For example, the amylase activities in the guts of omnivorous and herbivorous species, e.g. some cyprinids and tilapias, tend to be higher than those in the guts of carnivores, e.g. salmonids and percids. Proteolytic enzyme activity also appears to be correlated with feeding habits and relative intestinal length. Carnivorous species, with the shortest guts, tend to have high pepsin levels, but lower activities of pancreatic proteases. Herbivorous, detritivorous and microphagous species have relatively long guts, and relatively high levels of trypsin and chymotrypsin, but peptic enzyme activity may be low or absent. Omnivorous species take an intermediate position. Fish that consume insects and other animals with chitinous exoskeletons have high levels of intestinal chitinases, and those that consume planktonic copepods tend to produce relatively large quantities of lipases, esterases and non-specific lipase. Cellulase activity does not, however, appear to be closely linked with the dietary habits of the fish.

In general, there is a link between the production and secretion of digestive enzymes and the feeding habits of the species. Intraspecific variations, related to the dietary habits of individual fish, are also known to exist. For example, cyprinids, such as roach and rudd, *Scardinius erythrophthalmus*, show high proteolytic activity when fed animal prey, but amylase activity may be unaffected by dietary composition. Dietary manipulation also resulted in the maltase, amylase and protease activities of carp, *Cyprinus carpio*, being altered within the space of 1 week, with these changes in enzyme activity being correlated with the type of dietary change. In addition, tryptic and amylase activities have been shown to be correlated with diet in certain species of tilapias. Thus, there is good experimental evidence that a high dietary protein content can lead to an adaptive increase in the production and secretion of proteolytic enzymes. On the other hand, diets high in carbohydrates may result in a gradual increase in amylase activity over the course of a few days.

Enzymatic adaptation to diet

The mechanisms by which the enzymatic adaptations to specific diets are brought about have been little studied in fish species. Work carried out on the adaptive responses of the pancreatic enzymes of mammalian species to dietary change indicates that there are a number of possible pathways involved (Fig. 6.8). For example, the stimuli for increased enzyme production could be either nutrients within the lumen of the gut, i.e. pre-absorptive effects, or the nutrients may exert their effects following absorption from the gut, i.e. post-absorptive effects. There are also several possible

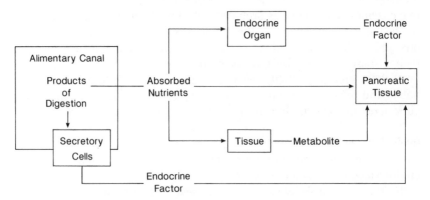

Fig. 6.8 Possible routes by which dietary manipulations can lead to adaptive changes in the synthesis and storage of pancreatic enzymes.

routes of mediation of the effects of dietary change on enzyme production, and the chemical messengers involved could be:

1. gastrointestinal hormones released in response to the presence of nutrients or digestive hydrolysis products in the gut lumen;
2. hormones released in response to absorption of a specific nutrient or digestive product;
3. a direct response to the absorbed nutrient;
4. a response to a metabolite resulting from the metabolism of a specific nutrient.

The stimulus for increased production of pancreatic proteases appears to be the presence of proteins or peptides in the gut lumen, and there is little or no response to intravenous amino acids. Thus, the mediator for the increased protease production resulting from elevated levels of dietary protein appears to be a gastrointestinal hormone released in response to the presence of proteins and peptides in the gut lumen. The hormone CCK has been suggested as being the most probable mediator of the effect.

Increases in pancreatic amylase production occur both in response to the presence of carbohydrates in the gut lumen and following intravenous injections of hydrolysis products. The effects of carbohydrates on pancreatic amylase production appear to be both direct and indirect. Glucose may directly influence enzyme production by pancreatic tissues. In mammals, glucose is a potent stimulant for release of the hormone insulin, and insulin may also have effects upon pancreatic enzyme production. Thus, indirect effects resulting from insulin stimulation of the pancreas also appear to be important for the increased amylase production resulting from increased levels of dietary carbohydrate.

Increased pancreatic lipase activity may be stimulated both by luminal and intravenous fatty acids. The presence of fatty acids in the gut lumen leads to the release of the hormone secretin, and intravenous infusion of secretin has been shown to lead to increased rates of pancreatic lipase synthesis in some mammalian species. Thus, the effects of luminal fatty acids appear to be mediated via a hormonal route. Hormonal mediation cannot, however, explain the observed effects of intravenous fatty acids, and it is suggested that the increase in pancreatic lipase activity that results from intravenous fatty acid infusion occurs as a response to an increase in circulating levels of certain products of fatty acid metabolism (ketones).

Intestinal absorptive mechanisms

One of the major functions of the intestine is the absorption of the products of digestion. The enterocytes are the absorptive cells of the intestinal epithelium. There are several possible routes by which the products of

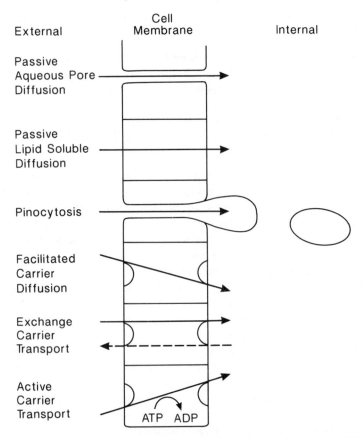

Fig. 6.9 Cell membrane transport systems and absorptive mechanisms. ATP, adenosine triphosphate; ADP, adenosine diphosphate.

digestion can be taken up from the gut lumen and into the enterocytes. Uptake may be either passive (not requiring the expenditure of energy) or active, and may or may not require a special 'carrier' molecule for the transport of the nutrient molecule over the cell membrane (Fig. 6.9).

Absorption of lipids

Triacylglycerols are hydrolysed to free fatty acids and monoacylglycerols by pancreatic lipase. In the presence of the bile salts that enter the intestine following contraction of the gall bladder, there is a spontaneous formation of micelles. The micelles are formed from small groups of molecules, and they comprise bile salts, monoacylglycerols and long-chained fatty acids, the latter being almost insoluble. The fatty acids with

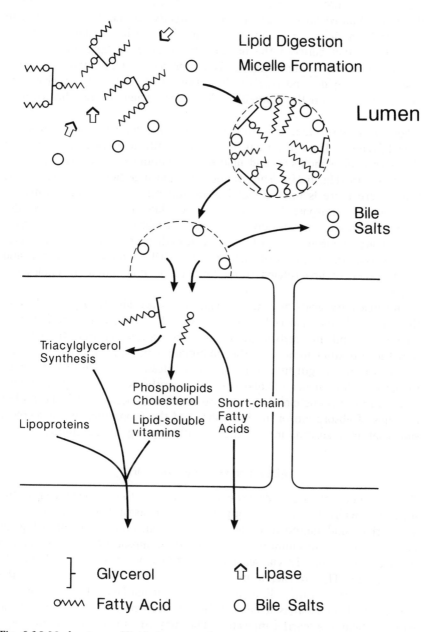

Fig. 6.10 Mechanisms of lipid absorption from the intestine.

short carbon chains are relatively soluble in water and the absorption of the short-chained fatty acids can occur directly, i.e. does not require bile salts. With a diameter of only 3–6 nm the micelles make possible an intimate contact between the breakdown products of lipid digestion and the epithelial wall of the intestine. This represents an essential step in the absorption of lipids, and passive absorption of the monoacylglycerols and free fatty acids into the epithelial cells occurs when the micelles contact the brush border.

Following absorption into the epithelial cells there is a resynthesis of triacylglycerols. These are then incorporated into lipoproteins, along with cholesterol, phospholipids, and lipid-soluble vitamins, for transport in the blood system (Fig. 6.10). The products of lipid digestion then pass to the liver where there is further metabolism and incorporation into transport lipoproteins. Transport lipoproteins exist in different sizes; there are also differences in the composition of the types of lipids transported. The very-low-density lipoproteins (VLDL) contain a high proportion of triacylglycerols, whereas the high-density lipoproteins (HDL) contain cholesterol, and the low-density lipoproteins (LDL) tend to be dominated by cholesterol esters.

The free fatty acids and monoacylglycerols are absorbed passively into the intestinal epithelial cells when the micelles contact the brush border membrane, and this leads to the bile salts being released from the micelles into the intestinal lumen. In the mammalian intestine the bile salts are resorbed from the gut lumen in the lower intestine and are then returned to the liver for further metabolism and resynthesis (*entero–hepatic circulation*). Entero–hepatic circulation is also thought to occur in fish species but the sites of absorption of the bile salts, and the proportion of the secreted salts that are resorbed, are not known with any degree of certainty.

Absorption via active transport

The absorption of carbohydrates and proteins into the epithelial cells of the intestinal walls is usually thought to occur as a transport of monosaccharides and amino acids against a concentration gradient using an active transport mechanism. Transport of monosaccharides and amino acids occurs primarily via secondary co-transport mechanisms linked to sodium ions. The transfer from the absorptive cell to the blood, over the basal membrane of the cell, is thought to occur via facilitated diffusion.

Facilitated diffusion is a passive form of transport across a membrane with the help of a 'carrier molecule'. The substances to be transported bind reversibly to special carrier proteins located in the membrane wall, and the transported substances then separate from the carrier on the other side of the membrane. Transport of this type is saturable. Each type of carrier molecule will bind to and transport a limited number of specific substances, each having a related chemical structure. Thus, transport of different classes of substances requires that there be a range of different carrier molecules on the cell membrane. Facilitated diffusion can only transport substances down an electrochemical gradient, whereas active transport mechanisms can transport substances against a concentration gradient.

Transport of substances against a concentration gradient by active mechanisms requires energy expenditure. In primary active transport, the hydrolysis of ATP provides the energy for transport via the ATPase 'pumps'. The 'sodium pump', or Na^+-K^+ ATPase enzyme, is ubiquitous in cells, and other ATPases, such as Ca^{2+}-ATPase and H^+-ATPase, occur in certain tissues. These pumps act in the direct transport of sodium, potassium, calcium and hydrogen ions. Secondary active transport means that the active transport of a substance (e.g. monosaccharide, amino acids) is coupled by a carrier molecule to the passive transport of an ion (e.g. Na^+, H^+). In this case, the Na^+ gradient is the driving force for transport, but this gradient is maintained by a primary active transport of Na^+, in the opposite direction, at another site on the membrane. The active transport mechanisms are characterized by having the following properties.

1. Transport is specific, i.e. only certain, usually chemically-similar, substances can be transported by a given carrier system.
2. Chemically similar substances may be transported to differing extents, i.e. carrier systems have different affinities for related substances.
3. The transport mechanisms are saturable, i.e. above a certain rate of transport the carriers are unable to transport any further load.

The transport rate (J) of a saturable transport process can usually be described by the Michaelis–Menten equation:

$$J = J_{max} \times C/(K_m + C) \qquad (6.1)$$

where C is the concentration of the substance to be transported, J_{max} is the maximum rate of transport by the system, and K_m is the concentration at half saturation, i.e. at $0.5\,J_{max}$.

An important characteristic of the carriers in the various transport systems is that they are specific for molecules having a particular structure. Thus, in the transport of carbohydrates there are carrier systems specific for aldohexoses (e.g. glucose) and ketohexoses (e.g. fructose), and, since there are groups of amino acids having different molecular structures, there are several carriers that participate in the transport of free amino acids over the cell membrane from the gut lumen into the enterocyte.

Amino acid and protein absorption

The typical structure of an amino acid (Fig. 3.5) is a central carbon atom to which there is attached a carboxyl group (–COOH), an amino group (–NH$_2$), a hydrogen atom (–H) and a side chain (designated the R-group). The side chain takes a number of forms, and it is the structure of the side chain that is used in classification of, and determines the properties of, the amino acids. All four of the groups attached to the central carbon atom are involved in the binding of the amino acid to the transport site of the carrier molecule, so amino acid carriers are specific for molecules having both a carboxyl and an amino group in their structure. The nature of the side chain differs between the amino acids.

1. The neutral (monoaminomonocarboxylic) amino acids have many types of side chain – they may have an aliphatic chain, straight or branched, may have an aromatic ring or the side chain may be in the form of a heterocyclic ring.
2. The side chain may carry a second amino group, giving rise to the basic (diamino) amino acid series.
3. A side chain with a carboxyl group gives rise to the acidic (dicarboxylic) series amino acids.

The structure of the side chain is the major factor determining which transport system, or systems, are utilized by an amino acid, and it also influences the degree of affinity for transport. There appear to be four major transport systems for amino acids in the intestine, these being classified as:

1. a system transporting most neutral acids;
2. a system specific for basic amino acids and cystine;
3. a system transporting acidic amino acids;
4. a system transporting imino acids (proline and hydroxyproline), glycine and related molecules (e.g. betaine).

Thus, several amino acids may be transported by the same system, leading to the possibility of competition between amino acids for binding to the transport sites on the carrier molecule. For example, methionine, valine and threonine have transport systems in common, and absorption of methionine can occur at the cost of valine and threonine transport. Similarly, leucine and lysine appear to compete for binding to carrier molecules, and high concentrations of leucine in the intestinal lumen may inhibit absorption of lysine. In addition, some amino acids may be transported by more than one system, and not all interactions between amino acids result in competition on binding sites with reduced absorption as a result. For example, uptake of basic amino acids (e.g. arginine, lysine) can be stimulated by some neutral amino acids, with absorption of the basic amino acids apparently being stimulated by alanine, leucine, methionine and phenylalanine.

Small peptides may be hydrolysed to amino acids by di- and tripeptidases in the brush border of the intestinal epithelial cells. These free amino acids will then be absorbed over the cell membrane via the amino acid transport systems. The di- and tripeptides need not, however, be completely

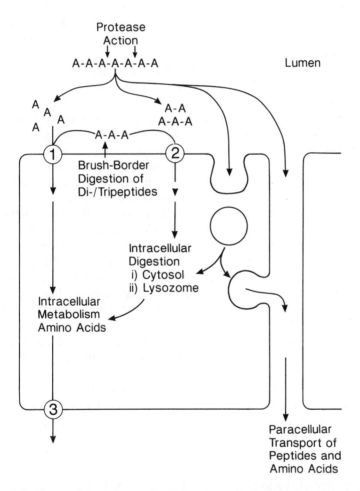

Fig. 6.11 Absorption of proteins, peptides and amino acids from the intestine. A-A-A-A indicates chains of amino acids linked by peptide bonds. (1) Transport of amino acids – this is mostly active sodium-linked transport. (2) The transport of dipeptides and tripeptides is mostly proton-linked active transport. (3) Transport over the basal membrane is thought to occur largely by facilitated diffusion.

hydrolysed to free amino acids before they are taken up into the enterocytes, and there are transport systems for these small peptides (Fig. 6.11). The transport systems, and carrier molecules, for the di- and tripeptides are distinct from those transporting the free amino acids. The uptake of amino acids in the small peptides may be much more rapid than the absorption of the same amino acids in the free form. The absorption of the small peptides occurs via active transport, but unlike the active transport of the free amino acids, the uptake of peptides is not via a sodium-mediated co-transport system. The uptake of the di- and tripeptides into the enterocyte appears to be dependent upon the active transport of protons (H^+) as the driving force for absorption, and this form of uptake is probably of much greater importance than previously realized.

From this it follows that a specific amino acid may be transported by a number of different systems, both in the free form and when bound to other amino acids in a small peptide. This flexibility is of clear advantage under conditions where there may be competition between amino acids for a particular carrier, or where there are genetic errors leading to defects in the development of particular carrier systems.

The facts that some amino acids may compete for transport sites on the carrier molecules and that absorption of di- and tripeptides appears to be more rapid than that of free amino acids, tend to suggest that dietary supplementation with free amino acids may not be the best method for improving the quality of proteins that are known to be deficient in certain amino acids. These findings would also call into question the validity of using 'free amino acid' mixtures for the determination of the quantitative amino acid requirements of animals, but, fortunately for the researcher, such studies will lead to an overestimation, rather than an underestimation of the quantitative requirement. Reliance upon a range of transport systems and absorptive mechanisms also provides a partial explanation for why the rates of growth of animals are almost invariably poorer when the amino acids are provided in the free form rather than being fed as whole proteins.

Whilst active transport via saturable carrier systems is thought to be the major pathway for absorption of the products of carbohydrate and protein digestion, other absorptive mechanisms are known to exist (Fig. 6.11). There may be some uptake of intact proteins, either directly into the blood system via a paracellular route, or proteins and larger peptides may be taken up into the enterocytes via pinocytosis.

Irrespective of the form in which the amino acids are absorbed into the enterocyte – free amino acids, small peptides, larger peptides or protein – they will usually enter into some form of intracellular digestive or metabolic processes rather than being transported directly through the enterocyte. Within the enterocyte there may be transamination and deamination of amino acids, and synthesis of peptides and proteins. Thus, the composition of the mixture of amino acids that eventually reaches the blood system need not necessarily give an exact reflection of the mixture absorbed from the gut lumen.

Nutrient transporters and diet

The nutrient transport systems and carriers are not rigidly fixed, but are, to some extent, both related to the natural diet of the species and, for given individuals, may be influenced by changes in dietary composition. Not all dietary constituents, or nutrients, affect the transport systems in the same manner, and there appear to be fundamental differences between the effects of the macronutrients (e.g. carbohydrates, amino acids) and micronutrients (e.g. vitamins, minerals) on the active transport mechanisms.

A dietary switch leading to an increase in a particular macronutrient class will generally lead to an induction of transporters for that nutrient class, whereas increased dietary concentrations of micronutrients may lead to a repression of the transport system. The macronutrients are required as respiratory substrates, and some are also essential nutrients necessary for normal growth and development. A common characteristic for all these macronutrients is that they can be transformed and stored within the body if they are consumed and absorbed in excess of immediate requirements.

By contrast, large doses of some vitamins and minerals may be toxic, and if there were induction of transport systems with increasing dietary concentrations of micronutrients, the risk of toxicity would be correspondingly increased. The micronutrients are, however, essential dietary constituents that are required in very small quantities, and very low dietary concentrations of these nutrients tend to lead to an induction of the transport systems needed for absorption. Thus, the inverse relationships between dietary micronutrient concentrations and the levels of intestinal transporters appear to be adaptations that ensure effective absorption of these nutrients when dietary concentrations are low, whilst, at the same time, reducing the risk of toxicity as dietary levels of the micronutrients are increased.

The effects of dietary composition on the intestinal transporters of fish have been examined for a small number of species, with the transport systems for glucose and the imino acids (e.g. proline) being those that have been most widely studied. When tested at saturating concentrations the overall capacity for the maximal uptake of the imino acids by the intestine appears to be similar for carnivorous, omnivorous and herbivorous species of fish. On the other hand, the capacity for glucose uptake differs markedly between species, with rates of glucose uptake tending to be higher in herbivores than in carnivores. Thus, when the capacities of glucose to proline uptake are expressed in the form of ratios, the ratios for carnivorous species are within the range 0.1–0.35 (i.e. the capacity for glucose uptake is much lower than that of proline), whereas those of herbivorous fish may be 0.7–1.5 (i.e. capacities for glucose and proline uptake are approximately equal). Omnivorous species tend to take an intermediate position. These results refer to experiments carried out on fish that had been feeding on their natural diets, but even when fish representing different feeding categories are fed on the same type of diet for a protracted period the general trends are still observed.

When the diet of a fish is changed, there are some changes in the intestinal transporters with time, so there is a certain degree of phenotypic flexibility. The changes that occur are, however, not sufficiently large to mask the inherent differences that exist between species having different feeding habits.

Ontogenetic changes in nutrient transport

Many species of fish display ontogenetic changes in diet, with the fish feeding on small planktonic organisms as larvae and juveniles, and, perhaps, changing to a piscivorous or herbivorous diet when they become adult. The changes in intestinal transporters that accompany ontogenetic development have been examined for the monkeyface prickleback, a species that is carnivorous in the juvenile stage, and almost exclusively herbivorous as an adult. When fish were raised in the laboratory and fed on the same type of diet throughout the rearing period, there were nevertheless found to be changes in the intestinal transporters with time. In other words, the juvenile fish had glucose:proline uptake ratios of 0.1–0.4, typical of a carnivorous species, whereas the uptake ratios were found to be within the range 0.5–2.5 in larger fish. Thus, the ontogenetic changes in the development of the intestinal transport systems seem to follow a predetermined genetic programme, rather than being a phenotypic response induced by dietary changes as the fish increase in size.

6.5   CONCLUDING COMMENTS

There are considerable differences in the anatomy, morphology and physiology of the alimentary tracts of fish species that feed upon different prey types. Whilst there may be some phenotypic plasticity in both the anatomical and physiological adaptability of the gut to changes in diet, the changes that can be induced are of relatively limited proportions. Thus, the overall

Table 6.3 The efficiencies with which a range of prey types are digested and absorbed by different families of fish

Food type	Fish species	Absorption efficiency (%)	
		Protein	Energy
Detritus	Cichlids	~45	~40
Macroalgae	Gobiids, pomacentrids, cichlids, stichaeids	45–80	30–70
Aquatic macrophytes	Cichlids, cyprinids	60–80	30–60
Oligochaetes	Salmonids, percids, cyprinids, pleuronectids	80–95	70–85
Crustaceans	Salmonids, percids	~90	80–88
Fish	Percids, gadoids	85–98	80–97

morphology and enzyme complement of the alimentary tract appears to be, to a considerable extent, 'hard-wired' to the normal diet of the fish.

Nevertheless, despite the fact that there appears to be an overall morphological and enzymatic adaptation of the alimentary tract to dietary type, the impression is gained that fish are better able to digest and absorb animal foods than plant foods. This seems to apply for the efficiency with which both the protein content and the total nutrient content (energy content) are digested and absorbed (Table 6.3). One reason for this may be the lack of cellulase activity in the guts of most fish.

Carbohydrates are absorbed as monosaccharides and the lack of cellulase activity means that the glucose units making up the cellulose molecule are virtually unavailable to the fish. Similarly, other plant material, such as hemicelluloses and lignin, is very refractory and is poorly digested. In the absence of enzymatic breakdown, the plant cells are attacked by mechanical means.

The grinding actions leading to the mechanical breakdown of plant cell walls may, however, be incomplete so that the cell contents may not be fully exposed to the digestive enzymes. Thus, if a proportion of the ingested plant cells remains complete this would lead to some nutrients not being available for further digestion and absorption. Other factors that could contribute to the poor digestion and absorption of plant materials include the formation of complex aggregates between plant fibres, phytates and nutrients, and reductions in enzyme activity caused by the presence of enzyme inhibitors.

FURTHER READING

Books

Cowey, C.B., Mackie, A.M. and Bell, J.G. (eds)(1985) *Nutrition and Feeding in Fish,* Academic Press, London.

Halver, J.E. (ed)(1989) *Fish Nutrition,* Academic Press, London.

Hoar, W.S., Randall, D.J. and Brett, J.R. (eds)(1979) *Fish Physiology, Vol. VIII,* Academic Press, London.

Matthews, D.M. (1991) *Protein Absorption: Development and Present State of the Subject,* Wiley-Liss, New York.

Review articles and papers

Brannon, P.L. (1990) Adaptation of the exocrine pancreas to diet. *Ann. Rev. Nutr,* **10,** 85–105.

Bromley, P.J. (1994) The role of gastric evacuation experiments in quantifying the feeding rates of predatory fish. *Rev. Fish. Biol. Fisheries,* **4,** 36–66.

Ferraris, R.P. and Diamond, J.M. (1989) Specific regulation of intestinal nutrient transporters by their dietary substrates. *Ann. Rev. Physiol.,* **51,** 125–41.

Gardner, M.L.G. (1988) Gastrointestinal absorption of intact proteins. *Ann. Rev. Nutr.,* **8,** 329–50.

Gildberg, A. (1988) Aspartic proteinases in fishes and aquatic invertebrates. *Comp. Biochem. Physiol.,* **91B,** 425–35.

Hirst, B.H. (1993) Dietary regulation of intestinal nutrient carriers. *Proc. Nutr. Soc.,* **52,** 315–24.

Holst, J.J. and Schmidt, P. (1994) Gut hormones and intestinal function. *Baillière's Clin. Endocrinol. Metab.,* **8,** 137–64.

Horn, M.H. (1989) Biology of marine herbivorous fishes. *Oceanogr. mar. Biol. ann. Rev.,* **27,** 167–253.

Horn, M.H. (1992) Herbivorous fishes: feeding and digestive mechanisms, in *Plant–Animal Interactions in the Marine Benthos* (eds D.M. John, S.J. Hawkins and J.H. Price) Clarendon Press, Oxford, pp. 339–62.

Kapoor, B.G., Smit, H. and Verighina, I.A. (1975) The alimentary canal and digestion in teleosts. *Adv. mar. Biol.,* **13,** 109–239.

Karasov, W.H. and Diamond, J.M. (1988) Interplay between physiology and ecology in digestion. *BioScience,* **38,** 602–11.

dos Santos, J. and Jobling, M. (1991) Factors affecting gastric evacuation in cod, *Gadus morhua* L., fed single-meals of natural prey. *J. Fish Biol.,* **38,** 697–713.

Sire, M.F. and Vernier, J.-M. (1992) Intestinal absorption of protein in teleost fish. *Comp. Biochem. Physiol.,* **103A,** 771–81.

Spiller, R.C. (1994) Intestinal absorptive function. *Gut; supplement,* **1,** S5–9.

Chapter seven

Osmotic and ionic regulation – water and salt balance

7.1 INTRODUCTION

The internal body fluids of the vast majority of fish species differ markedly both in osmotic strength and in ionic composition from the surrounding medium. Since the outer surface of the fish can be considered to be a semipermeable membrane this means that there may be substantial movements of water and ions from the medium and into the fish, or vice versa.

It is usual to describe the concentrations of substances in solution either in terms of molality (amount of a substance in a given weight of solution) or molarity (amount of a substance in a given volume of solution). Osmolality and osmolarity are terms that describe osmotic concentrations of dissolved substances, and these terms take into account the fact that there may be dissociation of substances once they go into solution. For example, a solution of 1 mol NaCl (58.45 g) dissolved in 1 l water has an osmolarity of 2 Osm since NaCl dissociates to Na^+ and Cl^- when it goes into solution in water.

The osmolality of sea water falls within the range 800–1200 mOsm kg^{-1} (1 gram molecule of an undissociated substance dissolved in 1 kg of water is said to have an osmolality of 1 Osm kg^{-1} and exert an osmotic pressure of 22.4 atm), and the majority of the ionic strength is derived from sodium chloride. On the other hand, the osmolality of the blood of the majority of marine teleosts is within the range 370–480 mOsm kg^{-1}, so there will be a tendency for water to pass from the fish into the surrounding medium and for salts, particularly sodium and chloride ions, to pass in the opposite direction. Thus, active regulation is required if the

Osmotic and ionic regulation

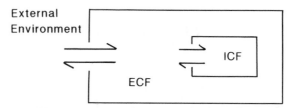

Fig. 7.1 Schematic diagram indicating possible sites of ionic exchange and water flux. ECF, extracellular fluid; ICF, intracellular fluid.

fish is to maintain its body fluids more dilute than the surrounding sea water.

Because the marine teleosts maintain the osmotic and ionic strengths of their body fluids at different levels from those of the surrounding medium they are said to be capable of both osmotic and ionic regulation. Some animals are incapable of precise regulation, and allow the osmotic and ionic strengths of their body fluids to fluctuate in step with changes in the surrounding medium. Animals showing these types of responses are known as conformers. It is important to realize that these changes refer to alterations in the extracellular body fluids (ECF). Thus, in the case of regulators, reference is being made to control mechanisms situated at the barrier between the ECF and the external medium (Fig. 7.1).

Within the body, additional regulation will occur at the cell wall interface between the ECF and intracellular fluid (ICF). In the case of conforming animals there may be quite wide fluctuations in the ionic composition of the ECF, but regulatory mechanisms are invoked in order to maintain the ionic strengths of the ICF within strictly defined limits.

With respect to ICF, a number of generalizations can be made. Firstly, the most abundant intracellular solute is potassium (K^+), and the intracellular concentration (usually 120–150 mM in fish species) is much higher than the extracellular concentration (2–10 mM). In contrast, the intracellular concentrations of sodium (Na^+) and chloride (Cl^-) are low (usually around 50 mM). Since the major extracellular solutes are sodium and chloride, the concentrations of these ions in the ICF are much lower than in the ECF. The cell membranes are permeable to these three ions and the maintenance of the differences in concentrations between the ICF and ECF requires some form of active regulation. Maintenance of low intracellular Na^+ requires the presence of an active extrusion mechanism in the membrane, and the active transports of Na^+ (out of the cell) and K^+ (into the cell) are coupled together.

The ICF is always isosmotic with, or slightly hyperosmotic to, the ECF. The osmotic effectors making up the osmolarity inside and outside the cell are, however, very different. Outside of the cells much of the osmolarity will usually be due to inorganic ions (predominantly Na^+ and Cl^-). Much of the intracellular osmotic pressure, on the other hand, will be created by the presence of organic molecules that are too large to pass through the cell membrane but are small enough to create a substantial osmotic pressure. Should the ECF become diluted there would be a tendency for water to flow into the cell because of the presence of the organic molecules. This would lead to cell swelling. In the absence of a rapid restoration of intracellular and extracellular osmotic pressures by active means, the continued osmotic flow of water into the cell could eventually lead to rupturing of the cell membrane.

The intracellular concentrations of inorganic ions are held within tightly constrained limits. Consequently, the active restoration of ICF osmolarity to balance that of the ECF is achieved by manipulation of the intracellular concentrations of organic molecules, usually via controlling the levels of free amino acids within the cell. Not all intracellular amino acids are involved in this mechanism, and it is only the non-essential amino acids that are implicated. Amongst the non-essential amino acids there is considerable interspecific variation in the osmotic significance of the various amino acids, although proline, glycine and alanine often dominate. Organic osmolytes other than amino acids may also be implicated in the mechanisms required for cell volume regulation. In teleost fish, taurine and sarcosine, which are metabolically inert, are of considerable importance as osmotic effectors. The elasmobranchs, on the other hand, employ urea and trimethylamine oxide (TMAO) as the major osmotic effectors.

7.2 MARINE TELEOSTS

The osmolality of sea water is approximately 1000 mOsm kg^{-1} and that of the body fluids of most marine teleosts is about 400 mOsm kg^{-1}. This requires that the fish actively regulate water and ionic flows in order to maintain the body fluids more dilute than the surrounding medium.

Water loss through the skin is hindered by the scales and mucus. On the other hand, water loss over the gills can be quite substantial. This is the case because the thin-walled secondary lamellae of the gills bring the blood into near contact with the external medium to facilitate gaseous exchange.

The loss of water over the gills is replaced by the fish drinking sea water. The drinking rate varies both among species and with environmental salinity, the higher the salinity the greater the rate of drinking. Most species can drink up to 35–40% of their body weight per day, although 10–20% is more usual. At first sight it is difficult to see how the drinking of sea water can assist the fish in maintaining water balance since it would be expected that water would flow from the body fluids of the fish across the gut wall and act to dilute the sea water. Although the exact mechanisms involved in the regulation of water intake are still unclear, it appears that the various parts of the gut differ in their permeabilities to water and ions.

The foremost part of the gut, the oesophagus, is impermeable to water but is permeable to sodium and chloride ions. These ions diffuse down their concentration gradients, over the gut wall and into the body fluids. There is also an active uptake of these ions in this region of the gut. The driving force for the active absorption of ions is provided by Na^+-K^+ ATPase activity located in the basal membrane of the absorptive cells. Transport of the ions from the gut lumen and into the cells appears to be mediated via a Na^+-K^+-$2Cl^-$ co-transport system located in the apical brush border membrane.

As a consequence of the absorption of monovalent ions in the foregut the water entering the intestine is less concentrated than sea water. Active uptake of ions continues in the intestine and, since the wall of the intestine is permeable to water, water can be taken up from this region of the gut. In other words, the active uptake of ions leads to the absorption of water via passive means, a phenomenon known as *solute-linked water transport*. About 60–80% of the water drunk appears to be absorbed from the gut in this way and hence the process of drinking can assist in maintaining water balance.

However, the absorption of the monovalent ions (Na^+, K^+, Cl^-) from the sea water imposes a salt load on the fish and these ions must be excreted. These ions are transported in the blood to the gills where they are excreted via the 'chloride cells'. Although these cells are largely concentrated in the gills they also appear in the skin of the head and operculum of a number of species. The exact mechanisms by which these cells excrete the various ions are not known, although the excretion appears to involve a sodium–potassium exchange mechanism. There is convincing evidence that operation of the 'sodium pump' involves the enzyme Na^+-K^+ ATPase, and the fact that levels of this enzyme increase in the gills of saltwater-adapted fish indicates the importance of this enzyme in the ionoregulatory processes. The excretion of the excess sodium and chloride ions appears to be linked, and the main active energy-requiring component for excretion appears to be the sodium pump of the chloride cells.

Sea water contains several other ions in addition to sodium and chloride, but only small amounts of the divalent ions such as magnesium (Mg^{2+}) and sulphate (SO_4^{2-}) are absorbed from the sea water during its passage through the gut. Approximately 80% of the divalent ions drunk along with the sea water appear to be expelled in the faeces and only 20% or so are absorbed. The divalent ions are transported to the kidney and are excreted in the urine. Water losses in the urine of marine teleosts are

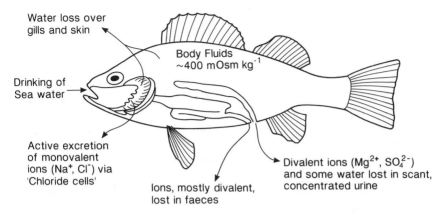

Fig. 7.2 Summary of ionic exchanges and osmoregulation in a marine teleost.

reduced to a minimum, and these fish have a urine flow rate that amounts to about 1–2% of the body weight per day (Fig. 7.2).

It is commonly believed that the glucocorticoid hormone cortisol plays a key role in the regulation of ion and water balance in marine teleosts. The actions of this hormone have been reported to include enhanced excretion of sodium ions, improved water absorptive capacity in the intestine and urinary bladder, and increased Na^+-K^+ ATPase activity in gill tissue. Further evidence for the involvement of cortisol in osmoregulation by marine fish has been gained in studies in which larvae have been exposed to low doses of water-borne cortisol and then exposed to a salinity shock. Exposure to cortisol improved the ability of Asian seabass, *Lates calcarifer*, larvae to withstand a hypersaline shock (i.e. 60‰ salinity), but did not have any effects upon the ability of the larvae to survive the challenge of exposure to fresh water. Thus, it appears that exposure to cortisol enhanced the ability of the fish to hypo-osmoregulate in the hypersaline environment, but was ineffective in promoting hyperosmoregulatory ability in the larvae exposed to fresh water.

7.3 CHLORIDE CELLS AND THE SODIUM PUMP

Around 1930 it was reported that there was an extrarenal site for the secretion of sodium chloride in seawater teleosts, and it was demonstrated that the anterior part of the body was capable of secreting salts. Using a gill perfusion technique it was shown that chloride ions (chloride ions were easily measured using titration against silver nitrate, but there were no convenient analytical methods for other ions, such as sodium) could be transported against a steep chemical gradient across the gills from the perfusion medium to a bathing solution of sea water. Histological examination of the gill tissue revealed that there were certain cells showing a number of unique morphological features, and these were thought to be the cells involved in ion extrusion – they were given the name chloride-secreting cells or 'chloride cells'.

As analytical methods for other ions improved, especially following the invention of the flame photometer, it became clear that other ions, in addition to chloride, were also transported against their concentration gradients. Attention became increasingly directed towards the study of sodium transport. This change in direction was sparked by the discovery that the enzyme Na^+-K^+ ATPase was of central importance in the transport of ions over epithelial tissues.

A number of findings soon pointed to the involvement of Na^+-K^+ ATPase in ion transport over the fish gill, and the idea of active sodium transport became established (with the assumption then being that chloride followed passively).

1. The greater the external salinity, the greater was the sodium efflux over the gill tissue.
2. Na^+-K^+ ATPase activity in the tissues of euryhaline species was found to be greater in the tissues of animals adapted to sea water than in those held in fresh water.
3. Sodium ion efflux appeared to be dependent upon the presence of potassium ions in the bathing solution. Since Na^+-K^+ ATPase involves the exchange of $3Na^+$ for $2K^+$, the dependence of the sodium efflux on the presence of K^+ seemed to provide convincing evidence for the involvement of the enzyme in ion transport over the fish gill.

Thus, the idea was launched that Na^+-K^+ ATPase concentrated on the apical membrane of the chloride cell served as the site for the active extrusion of sodium ions over the gill epithelium. The chloride ions were thought to diffuse passively in order to restore the electrochemical balance. It soon became clear that this was an inadequate model for the secretion of ions over the gills, there being several lines of evidence that called this hypothesis into question.

Firstly, experimental evidence became available indicating that sodium pump activity was located on the basal, rather than the apical, membrane of the chloride cells. On the basis of the hypothesis, inhibition of Na^+-K^+ ATPase activity would be expected to lead to a cessation, or at least a marked reduction, of ion efflux. The hypothesis would also predict that the greatest effects should be observed by application of the inhibitory substance to the apical membrane of the chloride cell.

Ouabain is a specific inhibitor of Na^+-K^+ ATPase activity and can, therefore, be used in experiments designed to test for both the presence and the location of the enzyme. The gills have a very complex structure and are not well suited to experiments of this type. In some species of fish, such as the mummichog, *Fundulus heteroclitus*, chloride cells are present in large numbers in the epithelial tissues lining the operculum. This epithelium can be dissected off as a thin sheet and mounted between two chambers, allowing the examination of the effects of different bathing solutions on ion transport.

Results of experiments with ouabain provided evidence that ion transport was inhibited more strongly when the ouabain was in the solution bathing the basal membrane of the epithelium than when ouabain had been added to the medium bathing the apical membrane. Thus, the Na^+-K^+ ATPase, representing the putative sites of ion transport, appeared to be located on the basal, rather than the apical membrane.

The greater Na^+-K^+ ATPase activity seen in seawater-adapted fish than in those held in fresh water, might imply a larger number of 'transport sites' in the seawater-adapted individuals. Larger numbers of transport sites would be expected to result in a hypertrophy of the regions of the cell membrane containing the Na^+-K^+ ATPase. Histochemical studies have demonstrated that the Na^+-K^+ ATPase activity is located on the basolateral membranes of the chloride cells (Fig. 7.3), and it is these membranes, and not the apical membrane, that are hypertrophied in seawater-adapted animals.

In addition, the membranes of the gill and opercular epithelial tissues, in common with other biological membranes, appear to be more permeable to cations (positively charged ions, such as Na^+) than anions (e.g. Cl^-). This would not be expected if sodium ions were being actively secreted from the chloride cells and the chloride ions were following passively.

Finally, electrical potential differences recorded across the epithelial tissues appear to be opposite to those required for the active extrusion of sodium ions from the apical membrane followed by passive diffusion of chloride. In other words, the transepithelial potential difference is such that the external environment is negatively charged relative to the interior.

Thus, the evidence indicates that Na^+-K^+ ATPase is present in very large amounts in the chloride cells, suggesting that the sodium pump is involved in ion transport and salt extrusion. The chloride cells of fish adapted to sea water have a characteristic morphology, with only a small area of the apical membrane – the apical pit – being exposed to the external environment. The chloride cells are metabolically very active, have high succinic dehydrogenase enzyme activity, and contain many mitochondria, but the most characteristic feature is the extensively

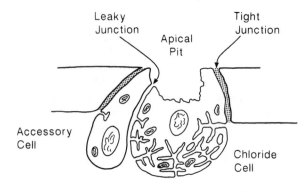

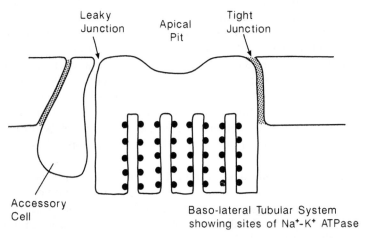

Fig. 7.3 Schematic diagram showing the structure of the chloride cell, and the association with the accessory cells.

branched tubular system of the baso-lateral membrane (Fig. 7.3). The Na^+-K^+ ATPase is concentrated in the baso-lateral membrane, and would appear, therefore, to pump sodium from the chloride cells and into the lumen of the tubular system. It is not immediately obvious how this could contribute to the transport of ions over the gill epithelium from the interior of the fish to the external environment.

The sodium pumps remove sodium from the chloride cells and concentrate it in the tubular system. This will generate, and maintain, a large gradient for passive sodium entry into the cell. Chloride appears to enter the cell in co-transport with sodium, as the sodium ion moves down its electrochemical gradient (Fig. 7.4). The sodium pumps extrude the sodium

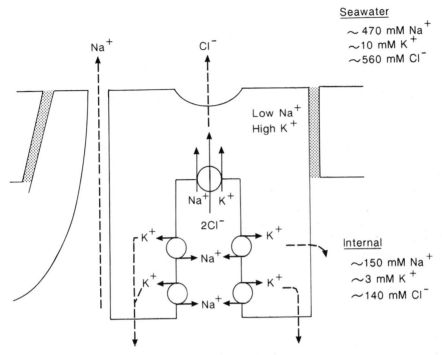

Fig. 7.4 Schematic diagram illustrating the involvement of the chloride cell in ionic exchange in marine fish. Circles indicate ion transport mechanisms.

ions back into the tubular system, and a negative intracellular potential is generated. Chloride ions that entered with the sodium do not accumulate in the chloride cell, but some pass out across the apical membrane. Thus, the extrusion of the chloride ions appears to be driven by the active transport of sodium across the baso-lateral membrane, and this can explain why the inhibition of Na^+-K^+ ATPase activity with ouabain leads to a cessation of chloride extrusion.

Consequently, it is currently believed that the chloride cells actively secrete chloride ions. The passage of sodium across the gill membrane from the interior to the exterior is thought to occur passively. It is thought that the sodium ions pass between the cells and leave via thin 'leaky' junctions (Fig. 7.4).

The chloride cells of the gill epithelium of fish adapted to sea water tend to be located on the primary lamella (or filament) close to the points at which the filaments form junctions with the secondary lamellae. Thus, these chloride cells are closely associated with the vessels that carry the blood flow through the gills. The chloride cells are grouped together and form interdigitations with a series of accessory cells. The junctions between the chloride cells and the accessory cells are much shallower than those between other cell types. It is thought that these junctions are comparatively leaky, allowing sodium ions to diffuse relatively easily from the interior to the external environment (Figs 7.3 and 7.4). Whilst this is a generally accepted view of the way in which the chloride cells participate in the secretion of ions, it must be emphasized that the mechanisms by which ion transport occurs have not been fully elucidated.

There also remains some debate as to whether there is a single type of chloride cell that undergoes morphological and functional changes when fish are transferred from fresh water to salt water, or whether clearly distinct types of cells are involved in ion transport in the different media. Based both upon the morphology of the apical membrane and upon the localization of the cells, two distinct types of chloride cells have been distinguished on the filament (primary lamella) epithelium of certain teleost fish (Fig. 7.5).

The first type – the α chloride cell – is located on the filament in the angle formed between the filament and the secondary lamellae. This type of chloride cell seems to be present on the branchial epithelia of both

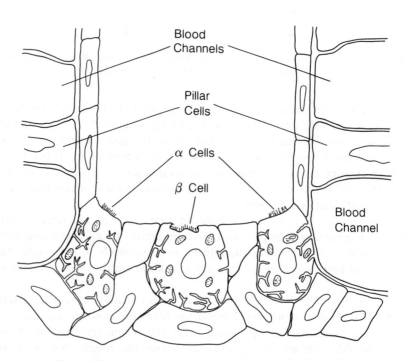

Fig. 7.5 Locations of the two types – α and β – of chloride cells on the gills of a teleost fish.

freshwater- and saltwater-adapted teleosts, although the cells appear to be morphologically different in fish that have been held in the different media. It has been suggested that this type of chloride cell undergoes differentiation when freshwater-adapted fish are exposed to sea water. These cells may gradually develop the extensive baso-lateral channel system and interdigitations with accessory cells typical of the chloride cells seen in the seawater teleosts.

The second type of chloride cell – the β chloride cell – is located on the filament in the open area between the secondary lamellae. This type of chloride cell seems to be most frequent on the branchial epithelia of freshwater-adapted fish. It has been suggested that these cells may degenerate when fish are transferred to sea water.

Chloride cells, probably of the β type, have also been reported to occur on the secondary lamellae of the gills in some species. These chloride cells are probably more related to ion absorption in fresh water than to the excretion of excess ions in saltwater-adapted fish. The chloride cells on the secondary lamellae of the freshwater species tend to become more prominent when the fish are exposed to water of low ionic content. For example, the area of the secondary lamellae occupied by chloride cells has been found to be greater in fish held in water deficient in calcium or sodium than in fish held in water having greater ionic content.

Exposure to acidic water may also lead to an increase in the prominence of the chloride cells in freshwater fish. To what extent this is the result of cell proliferation or an 'unmasking' of deep-lying, chloride cells due to the retraction of the superficial epithelial cell layer remains to be investigated.

7.4   KIDNEY STRUCTURE AND FUNCTION

The kidneys are seen as long reddish-brown structures running dorsally in the body cavity above the gas bladder and beneath the vertebral column. The teleost kidney, which comprises a multitude of nephrons (Fig. 7.6) interspersed with a capillary network, is not uniformly structured along its entire length. There are also differences between freshwater and marine teleosts in the structural units forming the nephron.

The glomeruli of the nephrons are often concentrated in the posterior part of the kidney, whilst the more distal tubular portions of the nephrons tend to run within the anterior lobes of the kidney. Blood from the dorsal aorta passes along the renal artery to the kidney and then passes through the capillary beds of the glomeruli. Capillaries leaving the glomeruli may then pass alongside the proximal kidney tubules (proximal segment) before eventually draining into the posterior cardinal veins, via which the blood is returned to the heart.

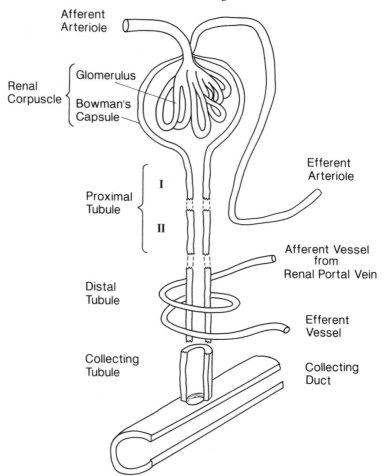

Fig. 7.6 Schematic diagram of a glomerular nephron in the kidney of a fish. Note that there are independent blood supplies to the renal corpuscle and proximal tubule, and the more distal parts of the nephron.

The glomerulus is a filter that allows blood plasma containing dissolved materials to pass into the kidney tubule – blood cells and large molecules such as proteins cannot pass through the filter. As the filtered fluid passes down the kidney tubule various substances are resorbed and secreted; glucose and amino acids are resorbed in the proximal tubule, or segment, whereas ions may be resorbed and secreted in both the proximal and distal segments.

The distal segments, or tubules, which lie in the anterior lobes of the kidney, receive their blood supply, not from the renal arteries, but from the renal portal veins that branch from the caudal vein returning blood from the posterior part of the body to the heart. There is a distinct distal segment in the nephrons of freshwater teleost species, but this is absent in the marine teleosts. In the marine teleosts the more distal portion of the nephron consists of the collecting tubule, which opens into the collecting duct.

The renal portal veins subdivide to form a network of capillaries surrounding the distal parts of the nephron, and it is often in this region of the nephron that much of the ionic exchange occurs. Branches of the caudal vein also run to the urinary bladder, and it is known that the bladder not only serves as a urinary reservoir, but is actively involved in ion and water exchange. The blood draining from both the vessels supplying the urinary bladder and the capillaries surrounding the distal tubular network eventually passes into the posterior cardinal veins before being returned to the heart.

Thus, the glomerulus and proximal tubule of the nephron are generally supplied with blood from a different source than the more distal portions of the nephron and the urinary bladder. This may have important consequences for the endocrine control of water and ion exchange in the kidney and urinary bladder. For example, blood returning from the caudal region may have been in close contact with the urophysis, a neurosecretory organ that lies at the tip of the spinal cord. The urophysis secretes peptide hormones, the urotensins, that are known to influence the transport of monovalent ions and water over the epithelium of the urinary bladder. Thus, it is possible that the urotensins secreted from the urophysis could directly stimulate the mechanisms responsible for the resorption of monovalent ions (Na^+ and Cl^-) in the distal parts of the nephron and in the urinary bladder. This would, in turn, lead to increased resorption of water due to solute-linked water transport, and the net result would be reduced water loss in the urine. Since the glomeruli and proximal tubules do not receive any blood supply via the caudal vein, the effects of the urotensins on the distal parts of the nephron and urinary bladder would occur without there being any influences on the remaining parts of the nephrons.

The glomeruli and proximal tubules receive their blood supply via the renal arteries. Any arterial blood-borne factors that influence capillary dilation, ion exchange mechanisms or epithelial permeability would have effects on the glomeruli and proximal tubule, without necessarily exerting any influences on the remainder of the nephron or on the urinary bladder. One endocrine factor that may reach the kidney via the arterial route is atrial natriuretic peptide (ANP), a hormone first isolated from the heart but now known to occur in a range of tissues. ANP may cause relaxation of renal vascular smooth muscle, may induce glomerular hyperfiltration, and can give rise to increased salt excretion. Circulating levels of ANP have been reported to increase following the exposure of a number of euryhaline and marine fish species to full-strength (35‰) sea water, suggesting a role for this peptide in salt excretion.

The way in which the resorption and secretion of ions, other solutes and water are regulated by the kidney is not well understood. It is, however, known that the different parts of the nephron have different roles to play.

The amount of urine produced by a fish will, to a large extent, be determined by the amount of blood filtered by the glomeruli, i.e. the *glomerular filtration rate* (GFR). Glomerular filtration rate is defined as the volume filtered per unit time by the glomeruli, and will be dependent upon the capillary pressure driving filtration, the permeability of the glomerular filter and the number of filtering glomeruli. Thus, GFR can be altered by changes in blood pressure, in epithelial permeability or in glomerular perfusion patterns.

Experimentally, the GFR may be calculated indirectly from the urine flow rate (U_f) using a marker technique. The marker introduced into the blood must, however, have a number of special properties: it must be inert, and not influence renal function; it must also enter the tubule only by filtration; it must pass down the tubule without being either resorbed or metabolized. This means that all the marker filtered must appear in the urine, so that the rate of filtration equals the rate of excretion of the marker. In other words, GFR $\times$ $[M_P]$ $=$ $U_f \times [M_U]$, or GFR $=$ $U_f \times [M_U]/[M_P]$, where $[M_P]$ and $[M_U]$ are the concentrations of the marker in the plasma and the urine, respectively. In experimental studies designed to measure GFR, the most commonly used marker substance is the carbohydrate inulin.

Kidney function in marine teleosts

In the majority of marine teleosts there is relatively little blood filtered by the glomeruli and the GFR amounts to approximately 0.5 ml kg^{-1} h^{-1}. The composition of the glomerular filtrate will be similar to the blood plasma, apart from the fact that it will not contain the large plasma proteins, such as globulins, lipoproteins and albumin-like proteins.

The composition of the filtrate will begin to be altered by resorption and secretion in the proximal tubule. The major role of the proximal tubule appears to be the resorption of organic compounds such as amino acids and sugars. There is, however, also some secretion of organic acids and nitrogenous waste products in this region of the nephron. There is some exchange of ions between the filtrate and the blood in the first part of the proximal tubule, but the majority of the ion exchange occurs more distally.

The major ion exchanges can be summarized as resulting in a net secretion of divalent ions (particularly Mg^{2+} and SO_4^{2-}), whereas there is a resorption of monovalent ions, accompanied by solute-linked water transport. Further resorption of monovalent ions and water may occur in the urinary bladder. This is assisted by the fact that the residence time of the

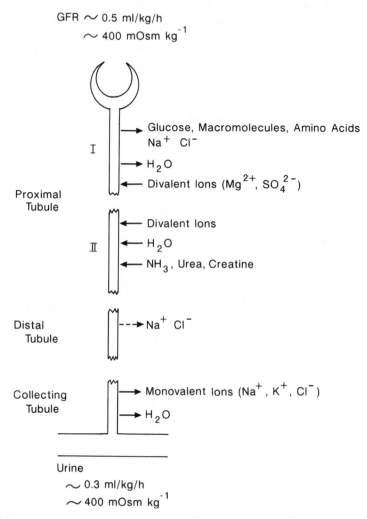

Fig. 7.7 Absorption and secretion during the course of urine formation in a teleost in sea water. Urine formation is shown for a teleost possessing a glomerular kidney. The distal segment is absent from the nephron of the truly marine teleosts, but is present in those species which migrate between the freshwater and marine environments. Thus, in the marine teleosts the more distal parts of the kidney tubules consist of the collecting tubule. GFR, glomerular filtration rate.

urine in the bladder is several hours. In the plaice, *Pleuronectes platessa*, for example, the time between urinations may be 2–3 days or more.

The overall rate of urine production in marine species is very low (approximately 0.3 ml kg^{-1} h^{-1}) amounting to 5–10% of the drinking rate. The osmotic strength of the urine will be similar to that of the plasma and glomerular filtrate (Fig. 7.7). Whilst the overall osmotic strengths of the urine and glomerular filtrate may be similar, the compositions are very different. Compared with the glomerular filtrate the concentrations of organic compounds and monovalent ions in the urine will be low, but the concentrations of the divalent ions will be markedly elevated.

In some species of marine teleosts the urine is produced solely via secretion and the nephrons are aglomerular. Fish having kidneys of this type include the syngnathids, i.e. seahorse and pipefish, and the anglerfish, *Lophius piscatorius*. The kidneys of these fish are also peculiar in that they do not receive a supply of blood via the dorsal aortal route, but the only supply comes from the caudal vein via the renal portal veins.

7.5   ELASMOBRANCHS IN SEA WATER

The plasma of the marine elasmobranchs is almost isosmotic with the surrounding sea water, so unlike the marine teleosts the elasmobranchs are in no danger of losing water to the environment. The body fluids of elasmobranchs tend to be slightly hyperosmotic to sea water, so there is a slight influx of water from the environment and into the fish. Whilst the osmotic strength of the body fluids of the elasmobranchs is very similar to that of sea water, the types of osmolytes present in the plasma, and other body fluids, differ markedly from those found in sea water. The concentrations of inorganic ions in the plasma of elasmobranchs are similar to those seen in marine teleosts, and are, therefore, lower than those found in sea water. Thus, the plasma of elasmobranchs is hypo-ionic to sea water. The osmolarity of the elasmobranch plasma is increased due to the presence of organic osmolytes, notably urea and trimethylamine oxide (TMAO). Thus, the organic osmolytes solve the problem of water balance, but not that of ionoregulation (Fig. 7.8).

In the majority of aquatic organisms, including fish, waste nitrogen derived from protein metabolism is excreted in the form of ammonia. In the elasmobranchs ammonia is sequestered from the blood and converted to urea, $CO(NH_2)_2$, by a series of enzymatic transformations (ornithine–urea cycle) occurring within the tissues of the liver. The urea enters the plasma, and other body fluids, where it acts as an osmolyte. Whilst urea is less toxic than ammonia, plasma concentrations much lower than those found in elasmobranchs would prove fatal to most other animals.

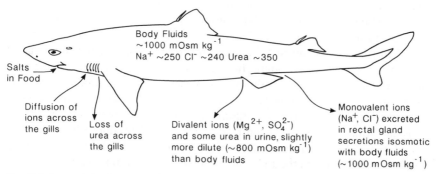

External Environment ~1000 mOsm kg^{-1}

Body Fluids
~1000 mOsm kg^{-1}
Na$^+$ ~250 Cl$^-$ ~240 Urea ~350

Salts
in Food

Diffusion of
ions across
the gills

Loss of
urea across
the gills

Divalent ions (Mg^{2+}, SO$_4^{2-}$)
and some urea in urine, slightly
more dilute (~800 mOsm kg^{-1})
than body fluids

Monovalent ions
(Na$^+$, Cl$^-$) excreted
in rectal gland
secretions isosmotic
with body fluids
(~1000 mOsm kg^{-1})

Fig. 7.8 Summary of ionic exchange and osmoregulation in an elasmobranch in sea water.

There seem to be two major factors contributing to the tolerance of high urea concentrations by elasmobranchs. Firstly, the primary role of the TMAO retained in the body fluids appears to be to counteract the negative toxic effects of urea. A TMAO:urea ratio of about 1:2 seems to give the most effective protection against urea toxicity. Secondly, the enzymes and tissue proteins of elasmobranchs appear to be less sensitive to disruption by urea than those of other species. Thus, the elasmobranchs seem to have developed a tolerance to urea that is greater than that seen in the majority of other animals.

Urea is highly soluble in water, and there will be a tendency both for urea to diffuse out of the fish across the gill epithelia and for the organic osmolytes to enter the kidney tubules as part of the glomerular filtrate. Thus, the fish must solve problems associated with urea retention. Resorption of urea in the kidney is very effective, usually amounting to 95% or more, so very little urea is lost in the urine. There may, however, remain some risk of loss of urea via diffusion over the gills. Urea is not particularly lipophilic and does not, therefore, diffuse rapidly over the gill membranes. In addition, the gill surface areas of elasmobranchs are generally lower than those of teleosts, and the gill epithelium is usually thicker. Thus, the risk of the loss of urea seems to be counteracted by the combination of low membrane permeability, a relatively low gill surface area and a large diffusion distance.

Inorganic ions, largely sodium and chloride, will tend to diffuse into the body fluids of the elasmobranchs across the gills. There may be some additional salt uptake associated with the ingestion and absorption of food. The elasmobranchs, unlike the marine teleosts, are not required to drink sea water, so the levels of salt loading are somewhat lower than in the marine teleosts. Nevertheless, a specialized sodium chloride excretory organ – the rectal gland – has evolved in the elasmobranchs.

The tissues of the rectal gland are very rich in Na^+-K^+ ATPase, so sodium pump activity is high. There is Na^+-K^+-$2Cl^-$ co-transport from the blood to the cells of the rectal gland, and the transport of the monovalent ions from the blood, through the cells to the gland lumen is dependent upon the activity of the sodium pump. The secretion produced by the rectal gland contains little of the organic osmolytes, and consists almost exclusively of a solution of sodium chloride that is isosmotic with the plasma. Thus, the ionic concentration of the fluid produced by the rectal gland is much higher than that of the plasma, and this would seem to provide an efficient method for the removal of excess monovalent ions (i.e. Na^+ and Cl^-). The primary role of the rectal gland may be to extrude the excess salts derived from the diet. Nevertheless, surgical removal of the rectal gland does not appear to have any severe detrimental effects upon the ability of the fish to maintain ionic balance. Thus, it is probable that the role performed by the rectal gland can be taken over by the kidney and gills.

Kidney function in elasmobranchs

Water enters the elasmobranch body over the permeable surfaces. Because water enters osmotically it is freely available for the formation of urine. Electrolytes also tend to enter by diffusion because the concentration of inorganic ions in the body fluids of elasmobranchs is lower than in sea water. The diffusion of ions into the body, along with those ingested with the food create an overall salt loading.

The nephrons of the elasmobranchs are very long and have large glomeruli. Typically, the rates of glomerular filtration in the marine elasmobranchs average 4–5 ml kg^{-1} h^{-1}, so the GFR is much higher than in the marine teleosts. The filtrate comprises monovalent and divalent ions, along with a range of organic compounds. The organic osmolytes, urea and TMAO, are the most abundant solutes in the filtrate. The resorption of urea and TMAO is accompanied by extensive water uptake, and usually only about 15–30% of the water remains unabsorbed. This results in the rate of urine production being about 1 ml kg^{-1} h^{-1} (Fig. 7.9).

There is some absorption of monovalent ions from the filtrate as it passes along the proximal and distal segments of the nephron, but sodium and chloride are the dominant solutes in the urine. Phosphate and the divalent ions magnesium and sulphate are actively transported from the blood to the lumen of the tubule. Consequently, the concentrations of these ions are higher in the urine than those found in the glomerular

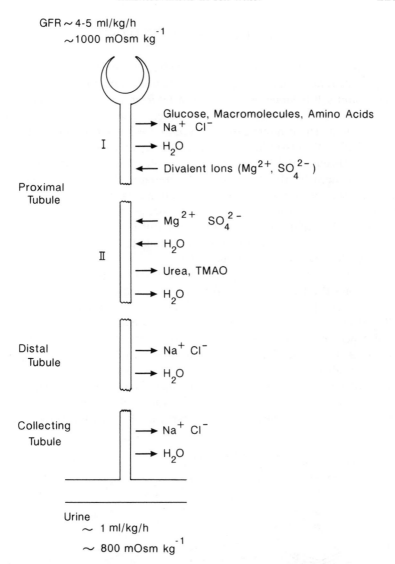

GFR ~ 4-5 ml/kg/h
~1000 mOsm kg^{-1}

Glucose, Macromolecules, Amino Acids
Na$^+$ Cl$^-$
⟶ H$_2$O
⟵ Divalent Ions (Mg^{2+}, SO$_4^{2-}$)

I

Proximal
Tubule

⟵ Mg^{2+} SO$_4^{2-}$
⟵ H$_2$O
⟶ Urea, TMAO
⟶ H$_2$O

II

Distal
Tubule

⟶ Na$^+$ Cl$^-$
⟶ H$_2$O

Collecting
Tubule

⟶ Na$^+$ Cl$^-$
⟶ H$_2$O

Urine
~ 1 ml/kg/h
~ 800 mOsm kg^{-1}

Fig. 7.9 Absorption and secretion in the kidney nephrons during the course of urine formation in an elasmobranch. GFR, glomerular filtration rate; TMAO, trimethylamine oxide.

filtrate. The urine produced by the marine elasmobranchs is hypo-osmotic to the plasma. It contains only very small amounts of the organic osmolytes, and the major solutes are the monovalent and divalent ions which are present at quite high concentrations.

7.6   FRESHWATER TELEOSTS

In contrast to teleosts living in sea water, the freshwater teleosts have body fluids that are more concentrated than the surrounding medium. The water- and salt-balance problems faced by freshwater teleost fish are also, therefore, rather different from those faced by fish living in the marine environment. The osmotic concentration of freshwater fish blood lies in the range 260–330 mOsm kg^{-1} and hence the fish will tend to gain water by diffusion through the body surface. The rate of water influx through the skin is reduced by the scales and the presence of large amounts of connective tissue near the surface. The main site of water influx is the gills, where the great surface area, for gas exchange, allows considerable influx of water. This influx must be counteracted if the fish is to remain in salt and water balance, and the task of pumping out this excess water is accomplished by the kidney (Fig. 7.10).

There is an extensive resorption of ions from the glomerular filtrate as it flows along the kidney tubule. Although the ionic content of the urine is low, the fact that urine is produced in large amounts means that significant amounts of ions are lost. Generally between 2 and 6 ml urine per kg fish per hour may be produced by freshwater teleosts, although values in excess of 12.5 ml kg^{-1} h^{-1} have been reported. Some loss of ions also occurs by diffusion from the body and these losses must be balanced by salt intake in the food and by active uptake of ions across the gills.

The energy required for the active absorption of ions from the gut lumen of freshwater fish species is derived from Na$^+$-K$^+$ ATPase activity located

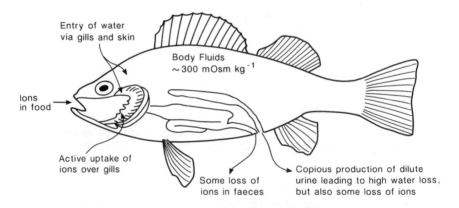

Fig. 7.10 Summary of ionic exchange and osmoregulation in a freshwater teleost.

on the basal membrane of the enterocyte, whilst the entry of monovalent ions over the apical membrane appears to be due to the presence of dual ion-exchangers – Na^+/H^+ and Cl^-/HCO_3^- – in the brush border facing the gut lumen. Thus, the transport mechanisms responsible for the uptake of ions into the enterocyte of freshwater species – double exchange involving Na^+/H^+ and Cl^-/HCO_3^- – seem to differ from that present in marine teleosts, i.e. Na^+-K^+-$2Cl^-$ co-transport. Furthermore, it seems that the ion-absorptive mechanisms present in the intestines of some euryhaline species, such as European flounder, *Platichthys flesus*, and rainbow trout,

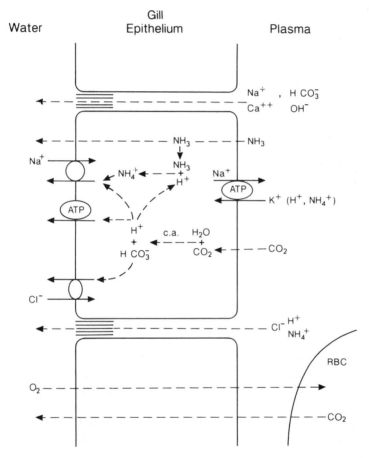

Fig. 7.11 Summary of the possible ionic exchange mechanisms operative in the gills of freshwater fish species. Ion transporting mechanisms are indicated, active transport being indicated by ATP. Carrier-mediated processes are indicated by solid lines and diffusional processes by dashed lines. c.a., carbonic anhydrase; RBC, red blood cell.

Oncorhynchus mykiss, may undergo modification or transformation from the 'freshwater' to the 'marine' form when the fish are exposed to media of increased salinity.

The gills of fish are the site where many of the exchange processes occur between the fish and the environment. Thus, the gills are intimately involved in respiratory gas exchange (i.e. oxygen uptake and carbon dioxide excretion), ion exchange (e.g. active uptake of sodium and chloride ions in freshwater species) and the excretion of waste nitrogenous products (i.e. ammonia and ammonium ions resulting from protein catabolism). These various exchange mechanisms appear to be coupled, and the excretion of carbon dioxide and nitrogenous waste products may result in the uptake of sodium and chloride ions via ion exchange (Fig. 7.11). These ion exchange mechanisms are clearly of benefit to freshwater fish since they serve the dual functions of assisting in the excretion of waste products, whilst, at the same time, aiding in the replacement of ions.

The adenohypophyseal hormone, prolactin, is suspected to play an important role in the regulation of salt and water balance in freshwater teleosts. Some of the effects ascribed to this hormone include inhibition of water resorption from the urinary bladder, stimulation of sodium uptake in the tissues of the gill, bladder and kidney, and stimulation of the co-transport of sodium and chloride in the urinary bladder leading to improved resorption of both ions. Prolactin may also play a role in promoting calcium absorption, and improving calcium retention within the body.

Kidney function in freshwater teleosts

The nephrons of freshwater teleosts are glomerulate, and the glomeruli tend to be concentrated in the posterior lobe of the kidney. The posterior lobe, containing the glomeruli, receives blood from the dorsal aorta, whilst the blood supply to the tubules lying in the more anterior portion of the kidney is derived from the caudal vein.

The primary role of the kidney is the conservation of filtered electrolytes, whilst producing a dilute urine. The urine is often almost devoid of sodium and chloride ions, and it is produced in a quantity that balances that of the water entering the body of the fish from the environment. In order to achieve this, the kidney nephrons must possess effective mechanisms for resorption of monovalent ions accompanied by a low permeability to filtered water.

Glomerular filtration rates in the kidneys of freshwater teleosts are high and resorption of organic solutes (glucose and amino acids) begins as soon as the filtrate enters the proximal tubule (Fig. 7.12). Resorption of the organic solutes from the filtrate occurs, primarily, in the proximal tubule.

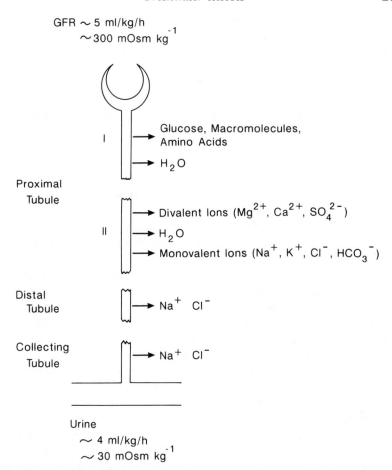

GFR ~ 5 ml/kg/h
~ 300 mOsm kg^{-1}

Proximal
Tubule

I → Glucose, Macromolecules, Amino Acids
  → H_2O

II → Divalent Ions (Mg^{2+}, Ca^{2+}, SO_4^{2-})
  → H_2O
  → Monovalent Ions (Na^+, K^+, Cl^-, HCO_3^-)

Distal
Tubule
→ Na^+ Cl^-

Collecting
Tubule
→ Na^+ Cl^-

Urine
~ 4 ml/kg/h
~ 30 mOsm kg^{-1}

Fig. 7.12 Absorption and secretion in the kidney nephrons during the course of urine formation in a freshwater teleost. GFR, glomerular filtration rate.

Resorption of ions, on the other hand, is largely carried out as the filtrate flows through the distal tubule. The epithelial wall of the distal tubule is relatively impermeable to water so the resorption of ions (particularly the monovalent ions Na^+ and Cl^-) is not accompanied by water uptake.

Although rates of urine production are high, urinary flow is not continuous. The urinary bladder is filled and emptied at regular intervals, i.e. urine is voided in a pulsatile fashion. Urination may occur at 20–30 min intervals. Whilst the residence time of the urine in the urinary bladder reservoir is relatively short, there may be sufficient time for some further resorption of ions to occur. The resultant urine is dilute, from about 15 to 60 mOsm kg^{-1}, and contains low concentrations of inorganic ions.

In freshwater teleosts there is a direct relationship between GFR and rates of urine production, because only a small, and relatively constant, proportion of the water filtered by the glomeruli is resorbed in the tubules. Thus, the absorptive mechanisms in the tubules of the freshwater teleosts differ from those seen in marine species. In other words, the resorptive mechanisms present in the kidney of freshwater teleosts can lead to considerable modification of the composition of the glomerular filtrate, but unlike those of the marine species, do not exert much control over the final urine volume.

7.7   EURYHALINE TELEOSTS

Euryhalinity can be defined as having the ability to tolerate and adapt to conditions of a wide range of salinities. A number of teleosts can be said to be euryhaline species. In considering euryhalinity amongst teleosts it may be pertinent to draw a distinction between those species which experience, tolerate and adapt to relatively large short-term fluctuations in external salinity on a regular basis, and those which migrate between the freshwater and marine environments at different stages of the life cycle.

The former group of species encompass fish living in estuarine, littoral-zone and salt-marsh conditions (e.g. European eel, *Anguilla anguilla*, European flounder, mummichog, long-jaw goby, *Gillichthys mirabilis*, and gilthead sea bream, *Sparus auratus*). These fish may experience changes in salinity ranging from fresh water to full-strength sea water in the space of a single tidal cycle.

The latter group would include a range of salmonid species, many of which undergo a distinct transformation ('parr–smolt transformation') prior to migration from the freshwater to the marine habitat. In other words, these species may change from being relatively 'stenohaline freshwater' fish (the juvenile parr) intolerant of sea water to being 'marine' fish (the smolt, post-smolt and adult) capable of survival and growth under seawater conditions. These represent distinct long-term changes and an abrupt transfer from one habitat to the other at an inappropriate time of the life cycle could compromise survival. For example, juvenile salmonids, prior to the parr–smolt transformation, are relatively intolerant of sea water. The juvenile fish are incapable of maintaining the osmotic strength and ionic composition of their body fluids constant if exposed to full-strength sea water. The fish experience a loss of body water (i.e. they become dehydrated) and there are marked increases in both blood osmolality and the plasma concentrations of ions such as sodium, magnesium and chloride. The osmotic and ionic disturbances are often so severe that mortality results.

Adaptations to fluctuations in salinity

The fish species that are found inhabiting estuaries exhibit varying degrees of euryhalinity, and most are capable of adapting to both fresh water and full-strength sea water. In addition, many of these euryhaline species are able to cope with rapid fluctuations in salinity. This would include abrupt transfer from fresh water to sea water, and vice versa, although they would not normally experience such abrupt changes in nature.

The abrupt transfer of euryhaline teleosts between extreme saline environments leads to physiological changes that can be grouped into two phases, or periods – the *adaptive period* and the *regulatory* period. During the adaptive period there will generally be changes in plasma ionic and osmotic strength but, with time, the plasma values will gradually be restored to those approaching the original. Thus, the direct transfer of the fish from sea water to fresh water will lead to a transitory reduction in plasma osmolality due both to an influx of water and to a loss of inorganic ions. An abrupt transfer from fresh water to sea water, on the other hand, results in a transitory increase in both plasma osmolality and ionic concentrations.

In the *regulatory period*, plasma ionic levels and osmolality are finely regulated, such that there is ionic homeostasis. Thus, during the regulatory period, which follows the adaptive period, the fish can be considered to be well adapted to the new saline environment.

The European eel and European flounder are good examples of species that are capable of tolerating large abrupt changes in salinity. These two fish species have been used in numerous studies of osmoregulation and ionoregulatory mechanisms.

Flounder can be readily adapted to salinities ranging from fresh water to sea water. Acclimation to different salinities leads to some changes in both plasma osmotic concentration and other osmo- and ionoregulatory parameters. The osmotic concentration of the plasma of flounders acclimated to fresh water has been found to be about 310 mOsm kg^{-1}, and plasma osmolality rises almost linearly with increasing salinity to reach a value of 360 mOsm kg^{-1} in fish acclimated to full-strength sea water. Thus, when acclimated to fresh water the euryhaline flounder appears to maintain its plasma osmotic concentration at the top end of the range found in stenohaline freshwater fish species (260–330 mOsm kg^{-1}). However, when flounders are held in sea water the osmotic concentration of the plasma is lower than that normally observed in exclusively marine teleost species (370–480 mOsm kg^{-1}).

As might be expected there is a direct relationship between the salinity of the surrounding medium and the rate of drinking, whereas the relationship between salinity and the rate of urine production is an inverse one.

Some drinking occurs in flounder acclimated to fresh water. Rates of drinking are 3–4 times higher in sea water, where drinking may amount to 8–10 ml kg^{-1} h^{-1}.

In fresh water, the urine production rate amounts to about 5 ml kg^{-1} h^{-1}, but this declines to 1 ml kg^{-1} h^{-1} or less in flounder acclimated to sea water. In flounder held in fresh water the kidney is capable of excreting all of the water that enters the animal osmotically from the environment, and the osmotic concentration of the urine produced in fresh water is much lower than that produced by fish in sea water. The mechanisms for the resorption of monovalent ions from the glomerular filtrate may, however, be less effective in euryhaline species than in strictly stenohaline freshwater fish. This leads to increased urinary loss of these ions in the freshwater-adapted flounder in comparison with stenohaline freshwater fish species.

The poor performance of the kidney in the conservation of monovalent ions appears to be compensated for, and offset, by the changes in the gill epithelium that occur in response to a salinity change. These changes appear to include alterations in both membrane permeability and the mechanisms involved in active uptake of ions. The net results of these changes are a reduction in ion efflux over the gills and a very effective ion uptake in flounder held in fresh water.

When flounder are exposed to fluctuating salinity regimes, simulating those that may occur in an estuary during the course of a tidal cycle, the fish show very rapid changes in both drinking and urine production rates. Both drinking and urine production rates show four-to-fivefold changes during the course of the tidal cycle, with the peak drinking rate occurring at around the time of high tide, when conditions are most saline. The production of urine reaches its nadir at this point, and increases as conditions become less saline. Although there are also changes in the osmotic concentration of the urine produced the changes are transient in nature, since complete adjustments to freshwater and seawater conditions have been reported to require 3–4 days.

The plasma osmotic concentration is relatively little affected by cyclic fluctuations in salinity, having a range of 300–340 mOsm kg^{-1}. Thus, the flounder appears to be able to maintain plasma osmotic concentrations within relatively narrow limits even when experiencing changes in environmental salinity that are so rapid that complete adjustments in the osmo- and ionoregulatory mechanisms are not possible.

The European eel, like the flounder, is capable of surviving abrupt transfer from fresh water to sea water and vice versa. Longer-term adaptation to different saline environments is a biphasic process, involving changes in the osmo- and ionoregulatory mechanisms present in the gills, gut and kidney.

If an eel is transferred from sea water to fresh water there is an almost instantaneous drop in the rate of ion efflux. A return to sea water within the space of a few hours will lead to an almost immediate return of ion efflux rates to the original seawater level. Following a long period in fresh water a different pattern emerges, with a transfer to sea water resulting in a more gradual increase in rates of ion efflux. This appears to be related to a gradual rise in the numbers of differentiated chloride cells, and an increase in the sodium pump (Na^+-K^+ ATPase) activity, that occurs in the course of the first few hours and days after transfer. Thus, long-term adaptation of the fish to freshwater conditions appears to lead to a decline in branchial sodium pump activity and a degeneration of chloride cells. On re-transfer to more saline media a certain amount of time is required before activity can be restored to the levels required for effective ionoregulation in sea water.

Endocrine influences

The changes required for ionoregulation in sea water seem to be, at least in part, under the influence of the corticosteroid hormone cortisol. Growth hormone (somatotropin), secreted by the adenohypophysis, also seems to be implicated in the changes that occur during the period of re-adaptation to seawater conditions. The growth hormone may contribute to seawater adaptation via a number of different actions. Firstly, there appears to be some antagonism between the adenohypophyseal hormones prolactin (the putative 'freshwater-adapting hormone') and growth hormone, so increased secretion of growth hormone may lead to some inhibition of prolactin action. Growth hormone may also prime the tissues so that the effects of cortisol are enhanced. Secretion of insulin-like growth factors (IGFs) is stimulated by growth hormone, and in some studies IGFs have been found to stimulate Na^+-K^+ ATPase production in gill tissue. Thus, growth hormone, cortisol and IGFs may act synergistically, with the net result being an increase in chloride cell numbers and an enhancement of sodium pump activity.

Hormonal influences on the gut are also important during the course of adaptation to media of different salinities. In seawater-adapted eels the oesophagus is highly permeable to monovalent ions, whereas in eels that have been adapted to fresh water the oesophagus is almost impermeable to sodium and chloride. Cortisol administration seems to lead to increased permeability, whereas decreased permeability of the oesophagus to monovalent ions appears to be associated with increased levels of prolactin. In addition to the effects on the oesophagus, cortisol also stimulates sodium pump activity and promotes increased rates of fluid uptake in the intestine.

Glomerular filtration rates

Following abrupt transfer of eels from fresh water to sea water there is a sharp drop in both GFR and urine production (to about 25–30% of the freshwater rate) within the space of a few hours. Later, as the eel becomes adapted to sea water, the GFR tends to show some increase. At the same time there is a change in the permeability of the kidney tubule to water, such that there is an increased resorption of water from the glomerular filtrate. Consequently, the rate of urine production continues to fall as the fish become fully adapted to seawater conditions because the increased resorption of water from the glomerular filtrate more than offsets the increase in GFR.

GFR refers to the total rate of production of glomerular filtrate, and therefore represents the sum of the filtration rates of all the nephrons present. In other words GFR = ΣSNGFR, where SNGFR represents the single-nephron glomerular filtration rate. From this it follows that changes in GFR can result either from changes in the rates at which individual nephrons produce filtrate, from changes in the numbers of functional nephrons (i.e. the numbers of glomeruli that are being perfused with blood and that are actively filtering), or via combinations of both these effects. In teleosts, much of the change in GFR appears to be the result of changes in the number of functional nephrons, a process known either as 'glomerular recruitment' or 'glomerular intermittency'. Single-nephron filtration rate is, however, also known to change when fish are transferred from fresh water to more saline media. The change in SNGFR may not, however, always be in the expected direction, because increases in SNGFR have been recorded in some species on transfer from fresh water to sea water.

When rainbow trout are transferred from fresh water to sea water the rate of urine production decreases to less than 10% of the freshwater rate. This is the result of a decrease in GFR and an increase in the efficiency with which the glomerular filtrate is resorbed in the kidney tubule. GFR declines to about 15% of the freshwater level and resorption efficiency almost doubles (increases from about 45% in fresh water to over 70% in more saline media). Measurements of SNGFR have revealed, however, that SNGFR is almost three times greater in the fish in sea water than in those held in fresh water. Thus, the reduction in GFR must involve a large change in the numbers of filtering nephrons, and it has been suggested that only about 10% of the nephrons present in the kidney of seawater-adapted rainbow trout actively participate in filtration.

There is some evidence that the neurohypophyseal hormone arginine vasotocin (AVT) is involved in the reduction in urine production in seawater-adapted fish. AVT is known to influence the vessels supplying blood to the glomeruli. Hormones of the renin–angiotensin system have also been implicated in the regulation of GFR in euryhaline fish. Angiotensin II is a potent vasoconstrictor and applications of exogenous hormone have been shown to induce a reduced GFR (i.e. 'seawater-adapted' condition) in certain euryhaline species. Other hormones (e.g. urotensins, atrial natriuretic peptide (ANP), cortisol) are almost certainly involved in the changes that occur in the kidney on transferring fish from fresh water to sea water, but the possible interactions between the different endocrine factors have not been elucidated with any degree of certainty.

7.8 PARR–SMOLT TRANSFORMATION IN SALMONIDS

Many salmonid fish, within the genera *Salvelinus*, *Salmo* and *Oncorhynchus*, are anadromous and undergo a distinct transformation prior to seawater migration. Prior to the parr–smolt transformation, juvenile salmonids are relatively intolerant of sea water, and are incapable of maintaining the osmotic strength, and ionic composition, of their body fluids constant if exposed to full-strength sea water. The osmotic and ionic disturbances are often so severe that mortality results. For fish that have successfully completed the parr–smolt transformation, there may also be changes in the osmotic strength and ionic composition of the plasma, but these are of much lesser proportions than those seen in the juvenile fish. In addition any fluctuations tend to be of a transient nature, and the osmotic and ionic disturbances that result from an abrupt transfer to sea water are corrected within the space of a few hours or days.

These differences in the abilities of the juvenile parr and smolt to regulate the osmotic and ionic strengths of the body fluids have been exploited in a simple test (the 'seawater challenge test') that can be used to determine whether or not the fish may be safely transferred from freshwater hatcheries to marine release and ongrowing sites. Because transfer of the fish at an inappropriate time could result in considerable mortality it is important for commercial producers to have information of this type in order to keep economic losses to a minimum.

Seawater challenge test

The 'seawater challenge test' involves the abrupt transfer of a small sample of fish from the hatchery tanks, where they are exposed to fresh water, to tanks containing sea water. Any mortality amongst the fish is registered, and after a given interval of time blood samples are taken for examination of plasma osmolality and the determination of ionic concentrations. If there are no, or very few, mortalities amongst the test fish, and they are capable of regulating the osmotic and ionic strengths of their body fluids within relatively strictly defined limits, then it would be concluded that the fish could be safely transferred to sea water.

The 'seawater challenge test' can take a variety of forms, with both the strength of the sea water used and the duration of the exposure (24, 48 or 96 h) often being decided by the individual investigator based upon previous experience. A commonly used procedure would, however, involve exposure of the fish to full-strength (30–35‰) sea water for a period of 24 h, followed by blood sampling for analysis of osmotic and ionic strength. If measurements show that plasma osmotic strength is about 320–340 mOsm kg^{-1}, and that the fish has been able to maintain sodium and chloride ion concentrations at around 160 and 150 mM, respectively, then it can be safely assumed that the fish may be transferred to sea water. An alternative procedure would involve exposing groups of fish to 40‰ sea water for 96 h, and periodically examining for mortality. Survival of this treatment would generally be taken as indicative that the fish could be transferred to sea water, but blood samples would usually be taken for analysis to provide an additional check of the iono- and osmoregulatory status of the fish.

Smolt characteristics

In the typical parr–smolt transformation there are a series of changes in morphology, physiology and behaviour (Table 7.1). These changes develop over the course of a few weeks, usually during spring, and culminate with a downstream migration followed by a period of residence in the marine environment. Whilst these changes are obviously related they should not be thought of as being tightly linked to each other. Thus, the parr–smolt transformation should not be considered as a single process, in which all the observed changes are strictly controlled by the same limited set of effectors. In other words, different characteristic features of the parr–smolt transformation develop at different rates and can become uncoupled from each other. This has been demonstrated by conducting experiments involving the study of the effects of either environmental manipulations, e.g. photoperiod and temperature, or the application of exogenous hormones.

Morphological and tissue changes

During the course of the parr–smolt transformation the cryptically coloured, stream-dwelling juvenile (parr) typically changes to a silvery, streamlined, pelagic fish (smolt) that is adapted for a life in the marine environment. The change in body colour results from the deposition of purines (guanine and hypoxanthine) in the skin. The purines become deposited in two distinct layers of the skin, one directly beneath the scales,

Table 7.1 Summary of the changes that characterize the parr–smolt transformation in salmonid fish

---

Morphological and physiological changes
- 'Silvering' due to increased deposition of guanine and hypoxanthine in skin and scales
- More streamlined body form and reduced condition factor due to rapid growth of caudal peduncle region
- Development of gill chloride cells, with increased Na^+-K^+ ATPase and succinic dehydrogenase enzyme activity
- Increased salinity tolerance and improved hypo-osmoregulatory ability
- Metabolic changes leading to increased body water content, reduced lipid content and changes in fatty acid composition

Behavioural changes
- Reduced territorial behaviour and increased formation of schools
- Increased activity, negative rheotaxis and downstream migration
- Preference for water of increased salinity
- Decreased ability to hold station against water current

---

and the other deep in the dermis. This leads to a masking of the bars and spots that are characteristic parr markings.

The deposition of purines appears to occur more rapidly in larger fish, and under conditions of increased temperature and long daylength (i.e. increased photoperiod). Application of exogenous thyroid hormones may also lead to increased silvering of the fish. Purine deposition in the skin is indicative of changes in nitrogen and protein metabolism during the parr–smolt transformation, and these metabolic changes may be related to changes in the circulating levels of thyroid hormones. These hormones are known to have effects leading to the general enhancement of metabolic activity, and an increase, or surge, in plasma levels of thyroid hormones has often been reported in fish undergoing the parr–smolt transformation.

In addition to alterations in protein metabolism, salmonids undergoing the parr–smolt transformation also show several changes in carbohydrate and lipid metabolism. These changes may lead to substantial changes in storage reserves and body composition. There may be reductions in liver and muscle glycogen stores, and there are also changes in blood glucose levels. These changes are often found to be associated with seasonal changes in the activities of glycogenolytic enzymes. Total body, and muscle, lipid decreases markedly during the parr–smolt transformation. The greatest reductions are noted in the levels of the storage triacylglycerols in the liver and muscle, but levels of mesenteric fat appear to be little affected. As is common for studies of body composition in fish, the moisture content is found to vary inversely with lipid.

The overall changes in tissue levels of storage glycogen and lipids appear to be due not only to increased rates of breakdown, mobilization and metabolism, but also to a reduction in rates of synthesis. In addition to reductions in the lipid reserves, there is also a reorganization of tissue fatty acid composition.

During the parr–smolt transformation the tissue lipids come to contain relatively high proportions of long-chain, polyunsaturated fatty acids (PUFAs), so that the fatty acid compositions of the lipids gradually change to resemble those of marine fish. These changes occur whilst the fish is still in fresh water, and probably represent a preadaptation for life in the marine environment. The reorganization of lipid fatty acid composition seems to be most extensive in the membrane phospholipids of the gills and gut, in which there is increased incorporation of (n − 3) PUFAs. The exact adaptive value of these changes is not known, but a change towards increased incorporation of PUFAs into the phospholipids would lead to increased fluidity and permeability of the cell membranes. It is also of interest to note that the most radical changes in phospholipid fatty acid compositions occur in the tissues of the gills and gut, both of which are intimately involved in salt and water balance. If the fish are prevented from migrating downstream and reaching the sea, the fatty acid composition of the phospholipids tends to revert to the 'freshwater' type, i.e. the proportions of (n − 3) PUFAs become reduced and there is an increase in the relative amounts of saturated and monounsaturated fatty acids.

The more streamlined body form of the smolt relative to that of the parr is indicative that the body becomes more slender and that there is a decline in weight per unit length. In other words, condition factor (K) is reduced. Condition factor is defined as $[W/L^3] \times 100$, where W is fish weight and L is body length. A reduction in K could be the result of the loss of weight by a fish of given length, a growth in length without any increase in weight, or the growth of the fish in terms of body length being more rapid than the increase in weight. During the period of the parr–smolt transformation the fish show increases in both body weight and length, so the decline in K indicates that the rate of length increase is more rapid than the rate of weight gain. The rapid spurt in terms of growth in length appears to be most marked in the caudal region, leading to an elongation of the caudal peduncle relative to the rest of the body.

Increased rates of growth and body silvering appear to be promoted by combinations of elevated temperatures and long photoperiods during the winter months and, superficially at least, fish exposed to these conditions will come to resemble smolts. The parr–smolt transformation also involves changes in the iono- and osmoregulatory abilities of the fish, enabling survival in the marine environment. Exposure to constant long photoperiods and elevated temperatures may, however, inhibit, rather than promote, the development of the physiological mechanisms required for effective iono- and osmoregulation in sea water. Thus, a silvery appearance and a slender, streamlined body (the crude characters that are typically used to classify salmonid smolts) provide no guarantee that the fish has completed the parr–smolt transformation and is capable of survival in the marine environment.

Ionic regulation and osmoregulation

If the fish is to survive in the marine environment it must be capable of effective hypo-osmoregulation. The parr–smolt transformation will, therefore, involve changes in the osmoregulatory capabilities of the fish, from hyperosmoregulation in fresh water to hypo-osmoregulation in the sea. This transition from a freshwater to a marine existence requires that there be a reversal from net ion influx to net ion efflux, with the gills, gut, kidney and urinary bladder all having a role to play. In most teleosts the changes that can lead to this reversal in ionoregulation are initiated following exposure to the hyperosmotic environment, but in salmonids undergoing the parr–smolt transformation there are structural and functional changes in the iono- and osmoregulatory machinery whilst the fish are still resident in fresh water. In other words, changes occur prior to, and in anticipation of, exposure to a hyperosmotic environment. This means that the fish is, at least in part, preadapted to life in sea water before the commencement of downstream migration.

In a number of euryhaline teleosts sodium pump (Na^+-K^+ ATPase) activity has been shown to change following the transfer of fish from fresh water to sea water, with the transfer to the more saline medium bringing about a marked increase in activity. Much of the sodium pump activity is localized within the chloride cells of the gill tissue, and transfer of the fish to sea water also tends to lead to a proliferation or differentiation of these cells in the euryhaline species. Two- to fivefold increases in gill Na^+-K^+ ATPase occur in salmonids undergoing the parr–smolt transformation, with most of this activity appearing to be associated with the chloride cells. The chloride cells have been reported to increase in number and size, and undergo changes in morphology during the course of the parr–smolt transformation. Numbers of chloride cells peak just prior to the time at which the fish attain maximum salinity tolerance.

Most of the changes occur whilst the fish is still resident in fresh water, i.e. do not require that the fish be exposed to hyperosmotic conditions. If fully functional in fish inhabiting a hypo-osmotic medium, these mechanisms would tend to be counterproductive and result in the fish experiencing iono- and osmoregulatory difficulties (i.e. increased ionic losses and increased water gain). It is, however, unclear whether the hypo-osmoregulatory mechanisms are fully functional in the fish whilst they remain in fresh water, or whether a rapid final induction occurs once the fish become exposed to sea water. For example, the fact that high levels of Na^+-K^+ ATPase activity can be demonstrated in assays carried out on gill homogenates *in vitro* is no guarantee that the enzyme is fuctioning fully as a sodium pump *in vivo*.

There appears to be some experimental evidence that the physiological iono- and osmoregulatory mechanisms that are in place in the fish in fresh water may not become completely functional before the fish has entered sea water. If fish that have apparently completed the parr–smolt transformation are abruptly transferred from fresh water to sea water they may experience a sharp, but transient, rise in plasma ion concentrations. Plasma ion levels begin to fall within a short space of time before stabilizing at concentrations slightly higher than freshwater levels after a period of a few hours. These changes may be, in part, the result of dehydration, because abrupt transfer of the fish from fresh water to sea water may lead to water loss, accompanied by a drop in body weight amounting to 5–8%. Dehydration leads to a rapid onset of drinking, which eventually leads to a restoration of water balance.

The dehydration experienced by the fish in sea water probably stimulates increased activity within the renin–angiotensin system, with the initiation of drinking being a response to changes in circulating hormone levels. The drinking of sea water does, however, lead to an increased salt loading, with plasma concentrations of both monovalent and divalent ions being increased. Excretion of the excess of ions involves both gill and kidney tissues, with the endocrine factor atrial natriuretic peptide (ANP) possibly playing a major role in controlling excretion of the salt load. Plasma levels of ANP have been found to show an increase following the transfer of Atlantic salmon, *Salmo salar*, from fresh water to sea water, and this peptide has been suggested as having a role in salt excretion in marine teleosts.

Hypo-osmoregulatory capacity

The details of the parr–smolt transformation differ between the various salmonid species, both with respect to the extent to which different types of physiological, morphological and biochemical changes occur, and with respect to the ages at which the fish first undertake the migration to the sea. In the pink salmon, *Oncorhynchus gorbuscha*, and chum salmon, *Oncorhynchus keta*, for example, the young fish are capable of tolerating exposure to sea water shortly after hatching, whereas young fish of other *Oncorhynchus*, *Salmo* and *Salvelinus* species may not develop the ability to tolerate highly saline media until they are several months, or even years, old.

Despite the fact that the parr–smolt transformation presents a number of different variations on a theme, there are a number of general conclusions that can be drawn with respect to the transformation and the development of hypo-osmoregulatory capacity in salmonids.

1. Juvenile salmonids resident in fresh water tolerate some change in ambient salinity, with a gradual transfer to increasingly saline conditions leading to a greater long-term tolerance of hyperosmotic media than abrupt transfer.

2. Parr–smolt transformation is usually critical to successful growth in full-strength sea water, although larger individuals of some species (larger fish have a greater salinity tolerance than small) will adapt to sea water and display reasonable rates of growth provided that they are acclimated gradually to these conditions.

3. Salinity tolerance of freshwater residents changes seasonally, usually being found to be greater during the spring and early summer than during autumn and winter.

4. The smolt stage is relatively brief, and fish that do not enter sea water during this period revert to the parr condition. Fish that are transferred to sea water after they have reverted to the parr condition will usually fail to grow normally, and may either become stunted or die. Fish that have been prevented from migrating to sea water and have reverted to the freshwater, parr condition will undergo the parr–smolt transformation again the following year.

5. The seasonal changes in salinty tolerance and the developmental changes associated with the parr–smolt transformation are influenced by environmental factors, particularly photoperiod and temperature, but there is a strong endogenous component.

If fish are held under constant environmental conditions they may show temporal changes in salinity tolerance, and may even undergo regular circannual cycles that appear to resemble the parr–smolt transformation. In the Atlantic salmon, for example, the cycle appears to be of approximately 10 months' duration. Although salmon exposed to conditions of continuous light appear to show seasonal cycles of salinity tolerance, these fish may not develop the complete array of physiological mechanisms required for effective iono- and osmoregulation in the marine environment.

Temporal changes in salinity tolerance have also been recorded in the Arctic charr, *Salvelinus alpinus*, exposed to conditions of continuous light. In the case of the charr, the fish exposed to continuous light do not appear to develop the physiological mechanisms required for hypo-osmoregulation to the same extent as do fish subjected to natural daylength. Thus, current evidence suggests that Arctic charr exhibit marked seasonal changes in hypo-osmoregulatory ability – an increase during spring and early summer, a decline during late summer and autumn, and a nadir in winter – with these changes being substantially modulated by the changes in natural photoperiod that occur during the course of the year.

Endocrine influences

The temporal changes in salinity tolerance and the ability of the salmonids to iono- and osmoregulate efficiently in hyperosmotic media may be a reflection of similar changes in responsiveness to a variety of endocrine factors. The endocrine factors that have been most often implicated in the parr–smolt transformation are the thyroid hormones, growth hormone (and IGFs) and cortisol. There is ample evidence that plasma levels of these hormones become elevated at some stage during the course of the transformation. It is, therefore, these hormones that have been most studied with respect to their influences on salinity tolerance and the development of hypo-osmoregulatory ability in salmonids.

Few studies that have examined the effects of exogenous thyroid hormones on osmotic and ionic regulation have demonstrated that these endocrine factors give rise to increased salinity tolerance, so whether or not the thyroid plays a central role in this aspect of parr–smolt transformation remains open to question. There are, however, clear indications that the salinity tolerances of both brown trout, *Salmo trutta*, and rainbow trout increase following treatment with thyroid hormones.

Applications of exogenous growth hormone, and cortisol, have been shown to produce both increased salinity tolerance and positive effects on the abilities of salmonids to osmo- and ionoregulate in hyperosmotic media. There may be some interactions between growth and thyroid hormones, because growth hormone application has been shown to increase the transformation of thyroxine to the active hormone T_3 in some salmonids. Consequently, it has been suggested that the improvements in salinity tolerance observed in brown trout and rainbow trout following administration of growth hormone may be mediated via thyroid hormones.

Injections of both growth hormone and cortisol have, in several studies, been reported to lead to increased levels of gill Na^+-K^+ ATPase activity in a range of salmonid species. Applications of these hormones may also lead to morphological changes in, and proliferation of, chloride cells. These changes have been assumed to be correlated with the increased salinity tolerance and improved hypo-osmoregulatory abilities displayed by the fish.

Results of some studies with exogenous cortisol and growth hormone are, however, equivocal. Whether or not there is any improvement in salinity tolerance appears to depend upon the hormonal dose applied, the time of year at which the experiments were conducted, the previous holding conditions and the developmental stage of the fish. For example, injections of cortisol administered to juvenile Atlantic salmon in April, at the time they would be expected to be initiating parr–smolt transformation, did not influence the development of gill Na^+-K^+ ATPase activity in fish that had been held under conditions of continuous light since the previous autumn. Cortisol injections, at the same dosage, were, however, effective in inducing a significant increase in sodium pump activity in fish that had been exposed to a simulated natural daylength cycle.

These observations are of interest because it has been reported that exposure of fish to continuous light may lead to a suppression of, or a delay in, the developmental events associated with the parr–smolt transformation. Taken together, the results of experiments with exogenous cortisol applications indicate that the stimulatory effects on gill sodium pump activity are greatest when the fish are near their normal time for undergoing the parr–smolt transformation. The period of maximal sensitivity of the gill tissue to cortisol has been shown to be seasonal in Atlantic salmon and in coho salmon, *Oncorhynchus kisutch* (and the same is probably true for other salmonids). It is probable that these seasonal changes in tissue sensitivity are mediated via changes in daylength.

Species differences in parr–smolt transformation

Amongst the salmonids there is considerable variation in the time of residence in fresh water, the timing of the parr–smolt transformation and the length of residency in the marine environment. For example, pink and chum salmon normally migrate downstream within 2–3 months after hatching, whereas Atlantic salmon may spend 2–7 years in fresh water before undergoing the parr–smolt transformation. Following migration from fresh water to the sea, the fish may spend from one to several years in the ocean before returning to fresh water to spawn. In the anadromous charrs, *Salvelinus* spp., on the other hand, there is an annual migratory cycle in which the fish remain in the sea for no more than a few weeks during the summer, and return to fresh water during the late summer and early autumn.

Thus, the developmental patterns of the parr–smolt transformation, and migratory cycles, in salmonids are extremely heterogeneous, and it is unclear to what extent fish of different species undergo the same sets of physiological changes. Nevertheless, it seems apparent that the suite of physiological changes that occur whilst the fish are still resident in fresh water preadapt the fish for survival and growth in the marine environment, and that the initiation of these changes may be largely mediated via the seasonal change in natural photoperiod.

7.9 CONCLUDING COMMENTS

A high intracellular concentration of K^+ and low concentrations of Na^+
and Cl^- are characteristic of animal cells, and since the ionic compositions
of the ICF differ from those of the ECF some form of active regulation is
required. The ionic composition of the ICF is maintained within strictly
defined limits and any changes in osmolality of the ICF are brought about
by changes in the concentrations of organic osmolytes. Changes in the
intracellular concentrations of organic osmolytes are of utmost importance
for cell volume regulation in response to a challenge resulting from a
change in the osmolality of the ECF. Free amino acids are the organic
molecules most frequently employed as osmotic effectors, but other organic
compounds, such as urea and TMAO, may also function as organic osmolytes.

The ECF of teleosts and elasmobranchs is of similar ionic composition,
with sodium and chloride being the major ions in the ECF. The iono- and
osmoregulatory problems faced by the freshwater and marine teleosts,
and the elasmobranchs do, however, differ from each other. The body
fluids of freshwater teleosts are both hyperionic and hyperosmotic to the
surrounding medium, whereas those of the marine teleosts are hypoionic
and hypo-osmotic to sea water. Thus, freshwater teleosts will tend to lose
ions and have an osmotic water gain from the environment, whereas the
opposite is true for the marine teleosts. The elasmobranchs retain relatively high concentrations of organic osmolytes within their body fluids
and the ECF is almost isosmotic with sea water. There is, therefore, little
osmotic water flow between the fish and the environment, but elasmobranch fish will tend to gain ions because the ECF is hypoionic to sea
water.

Different physiological mechanisms are required for efficient iono- and
osmoregulation in fresh water and sea water, and there are often pronounced structural differences between the kidney, gill and gut of stenohaline species from freshwater and marine environments. The iono- and
osmoregulatory mechanisms in the euryhaline species are extremely labile,
and structural changes in the osmoregulatory organs may be observed
within the space of a few hours or days of transfer from one medium to
another. The anadromous salmonids appear to differ from many other
euryhaline fish species in that many of the changes that lead to improved
hypo-osmoregulatory capacity take place whilst the fish are still resident in
fresh water. In other words, exposure to sea water is not required in order
for initiation of the changes in the organs involved in iono- and osmoregulation, such that the fish appear to become adapted for life in the
marine environment before entering the sea.

FURTHER READING

Books

Hoar, W.S. and Randall, D.J. (eds)(1969) *Fish Physiology, Vol. I,* Academic Press, London.

Hoar, W.S. and Randall, D.J. (eds)(1984) *Fish Physiology, Vol. XB,* Academic Press, London.

Hochachka, P.W. and Mommsen, T.P. (eds)(1991) *Biochemistry and Molecular Biology of Fishes: Phylogenetic and biochemical perspectives,* Elsevier, Amsterdam.

Laverack, M.S. (ed)(1985) *Physiological Adaptations of Marine Animals,* Society for Experimental Biology, Cambridge.

Rankin, J.C. and Davenport, J. (1981) *Animal Osmoregulation,* Blackie, Glasgow.

Rankin, J.C. and Jensen, F.B. (eds)(1993) *Fish Ecophysiology,* Chapman & Hall, London.

Review articles and papers

Bern, H.A. and Madsen, S.S. (1992) A selective survey of the endocrine system of the rainbow trout (*Oncorhynchus mykiss*) with emphasis on the hormonal regulation of ion balance. *Aquaculture,* **100,** 237–62.

Borgatti, A.R., Pagliarani, A. and Ventrella, V. (1992) Gill ($Na^+ + K^+$)-ATPase involvement and regulation during salmonid adaptation to salt water. *Comp. Biochem. Physiol.,* **102A,** 637–43.

Karnaky, K.J. jun. (1986) Structure and function of the chloride cell of *Fundulus heteroclitus* and other teleosts. *Amer. Zool.,* **26,** 209–24.

King, P.A. and Goldstein, L. (1983) Organic osmolytes and cell volume regulation in fish. *Mol. Physiol.,* **4,** 53–66.

Kirschner, L.B. (1993) The energetics of osmotic regulation in ureotelic and hypoosmotic fishes. *J. exp. Zool.,* **267,** 19–26.

Langdon, J.S. (1985) Smoltification physiology in the culture of salmonids. *Recent Adv. Aquacult.,* **2,** 79–118.

Laurent, P. and Perry, S.F. (1991) Environmental effects on fish gill morphology. *Physiol. Zool.,* **64,** 4–25.

McCormick, S.D. and Saunders, R.L. (1987) Preparatory physiological adaptations for marine life of salmonids: Osmoregulation, growth and metabolism. *Am. Fish. Soc. Symp.,* **1,** 211–29.

Nishimura, H. (1985) Endocrine control of renal handling of solutes and water in vertebrates. *Renal Physiol.,* **8,** 279–300.

Sakamoto, T., McCormick, S.D. and Hirano, T. (1993) Osmoregulatory actions of growth hormone and its mode of action in salmonids: A review. *Fish Physiol. Biochem.,* **11,** 155–64.

Shuttleworth, T.J. (1989) Overview of epithelial ion-transport mechanisms. *Can. J. Zool.,* **67,** 3032–38.

Chapter eight

Body form, swimming and movement through the water

8.1 INTRODUCTION

The restraints placed upon a fish moving through water are markedly different from those placed upon an animal moving on land or through the air. Water is a dense, viscous medium that places a premium on effective propulsion mechanisms. Trunk and caudal fin propulsion is mechanically the most efficient form of swimming system. This propulsion system has been retained in almost all groups of fish, and it is the system that enables the achievement of maximum acceleration rates, maximum swimming flexibility and high sprint speeds. Thus, forward swimming is usually achieved by lateral undulations of the main body trunk (Fig. 8.1), but some species rely on the paired and unpaired fins to generate the power for forward swimming (Fig. 8.2).

Even in those species that employ undulations of the trunk as their major swimming mode, there will often be reliance on the paired fins during slow swimming and manoeuvring in confined spaces. Thus, most species will rarely be reliant upon one single mode of swimming, to the exclusion of all others. Nevertheless, most species or groups of fish can be said to be characterized by a particular swimming mode. This has led to swimming being categorized and named after the types or species of fish that typically perform that swimming mode. Thus, fish can be classified as swimming in modes referred to as being rajiform (ray-like), labriform (wrasse-like), thunniform (tuna-like), or anguilliform (eel-like) and so on. The nomenclature defining the different types, or modes, of swimming will give indications both about which fins (e.g. pectoral, dorsal, caudal) are used in propulsion, and about the extent to which the fish relies on lateral

Caudal and Trunk Propulsion

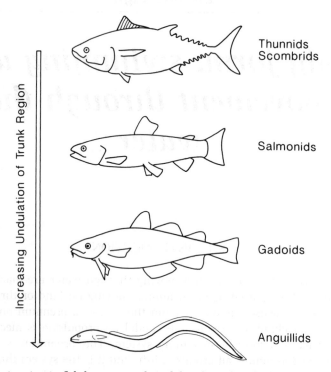

Fig. 8.1 Swimming in fish by means of caudal and trunk propulsion.

undulations of the whole trunk or caudal region during forward swimming (Figs 8.1 and 8.2).

8.2   SWIMMING MODES – FINS VERSUS BODY TRUNK

Swimming in the skates and rays is almost entirely dependent upon the undulatory movements of the greatly enlarged pectoral fins. These fins form a wide lateral extension of the body. In the *rajiform* swimming mode propulsion is achieved by the passage of a series of undulatory waves backwards along the exceedingly flexible fin margins. Under some conditions the holocephalans may use their enlarged pectoral fins in this manner, but usually the pectorals of the holocephalans are used in making rowing movements. Undulatory movements of the pectoral fins are also used by a variety of osteichthyan species, but in the bony fish pectoral fin

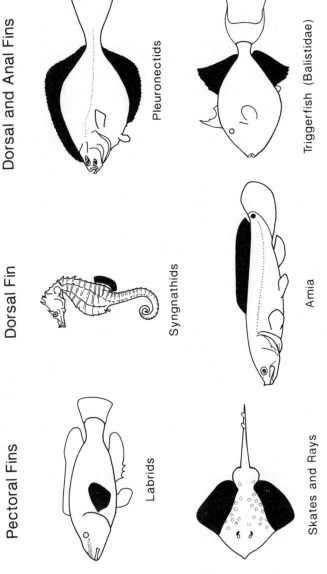

Propulsion using Paired and Unpaired Fins

Pectoral Fins · Dorsal Fin · Dorsal and Anal Fins

Labrids

Syngnathids

Skates and Rays

Amia

Pleuronectids

Triggerfish (Balistidae)

Fig. 8.2 Swimming in fish using various paired and unpaired fins.

undulatory movements are almost invariably combined with some other swimming mode.

When undulations of the fins make a substantial contribution to the generation of the power required for swimming in osteichthyan fish it is usually the long-based median fins (most often the dorsal but, in some cases, the anal fin) that are the locomotor organs. Undulations are passed along the fin due to the contractions of muscles inserted on the fin rays at their bases. Each fin ray can be moved independently of its neighbour on a universal joint, but successive fin rays are joined to each other by a flexible membrane. Contractions of the muscles at the bases of the fin rays move them in the appropriate direction. The extent of the deflection of the fin ray, and the amplitude of the backwardly moving undulatory wave, is, however, limited by the attachment of the fin rays to the body. Thus, the wavelengths of the median fin undulations are usually short, but the frequency of the fin waves can be high (50 Hz or more in some species). Propulsion by means of undulation of the median fins allows the fish to attain only very moderate swimming speeds, but this type of locomotory mode enables precision movements to be made. Median fin undulation usually allows both forward and backward movement, and reversals of direction can be made without turning. In addition, fish employing fin undulation as a means of propulsion usually have the ability to hover in mid-water.

Swimming by means of undulation of the dorsal fin is usually known as the *amiiform* mode (after *Amia calva*, the bowfin). In addition to the bowfin, which typically propels itself slowly by means of dorsal fin undulations, this swimming mode is also displayed by a group of African freshwater electric fish, the mormyrids (Mormyridae), and by the seahorses and pipe-fishes (Syngnathidae). Anal fin undulation is used by a small number of species. This mode of propulsion is used by some South American electric fish, including the banded knifefish, *Gymnotus carapo*, after which this swimming mode – the *gymnotiform* – is named.

The term *balistiform* swimming mode refers to simultaneous undulation of the dorsal and anal fins. This is the swimming mode characteristic of the triggerfish (Balistidae), and is also seen in some cichlids (Cichlidae) and butterflyfish (Chaetodontidae). In many species of flatfish (Pleuronectidae) rapid swimming involves body flexions, but this is often supplemented by rhythmic undulations of the long-based dorsal and anal fins when the fish move slowly over the substrate in search of prey. These fins are also undulated quite rapidly when the fish are lying on the bottom and are making settling, or burrowing, movements.

At the opposite end of the propulsive spectrum from the undulatory movements of the long-based median fins, is the swimming mode based upon the oscillatory movements of the short-based pectoral fins. Fish that use this form of swimming have pectoral fins in which the bases are inserted almost vertically on the body. The fins are also capable of being rotated to some extent on their bases. The pectoral fins are held in such a fashion that they present a broad face to the water when they are moved backward, in a similar manner to oars. The fins are then rotated and present a narrow face when they are moved forward in the horizontal plane. This is the *labriform* swimming mode.

Despite the existence of a range of specialized swimming modes based upon undulation or oscillation of the median or paired fins, trunk and caudal fin propulsion appears to be a common denominator throughout the entire course of fish evolution. Some form of trunk and caudal fin propulsion is found in all fish groups, but the extent to which the body undulates during swimming varies between species.

Thus, there is a continuum of trunk and caudal swimming from the *anguilliform* mode at one extreme to the *thunniform* and *ostraciform* modes at the other. In fish swimming by the anguilliform mode there is considerable undulation of the body, involving almost the entire trunk and caudal region, whereas in those species that swim by the thunniform and ostraciiform modes the only part of the body to show undulation is the caudal region.

8.3 DRAG FORCES AND BODY DESIGN

In order for a fish to swim, its muscles must create sufficient thrust to overcome the forces opposing the motion. In hydrodynamic terms these forces, known as the drag forces, can be expressed as:

$$\text{Drag} = M\,a + 0.5\,d\,S\,(kC)\,U^2 \tag{8.1}$$

where the first term is due to *inertia* and the second due to *friction*. In the expression of inertia, M is the mass resisting acceleration, and a is the rate of acceleration of the body. The friction component can be calculated if information is available about the density of the water (d), the wetted surface area of the body (S) and the swimming speed of the fish (U). Information is also required about the drag coefficient (kC), which describes the drag forces acting on the body in relation to factors such as body shape and the undulatory movements carried out during swimming. On the basis of the above equation it can be seen that the drag forces to be overcome by the fish will differ depending upon whether it is swimming at a constant steady speed or whether it is accelerating rapidly in order to escape from a predator.

Swimming at constant speed

Consider, first, a fish swimming continuously at a steady speed. In this instance the inertia term (M a) drops out of the equation and the friction term is the only one to be considered. The amount of friction to be overcome depends upon the surface area, i.e. the size of the fish, and on the drag coefficient. The drag coefficient is influenced by the overall body shape of the fish and also by the additional friction (drag) introduced by the lateral body movements. This raises the question as to what is the best shape for efficient swimming.

Studies carried out in the early 1900s showed that a cone facing upstream has almost twice the resistance (drag) of a cone facing downstream. The large drag of the upstream-facing cone resulted from the large eddy currents, or turbulence, formed to the rear of the cone. When two cones are placed together, the turbulence is much reduced and this *fusiform* shape gives the least resistance to movement. Other experiments have shown that the shape of the upstream part of the body, channelling the water up to the position of maximum cross section, is less critical than the downstream section.

Active, pelagic species

An examination of the body shapes of the most active fish species shows that there is considerable diversity in the shape of the head and forepart of the body. The hindmost part of the body and tail, however, invariably form a gradually tapering cone. Thus, the resistance to movement in these fish is reduced by them having a streamlined body shape (Fig. 8.3).

The body form giving least drag is one which is 4.5 times as long as its maximum diameter. This relationship between the length of the body and maximum diameter is usually known as the *fineness ratio* (the reciprocal of the fineness ratio – the profile thickness – is used as an expression of body shape by some workers, and drag will be minimized when the profile thickness of the body is 0.2–0.25). In order for drag reduction to be most effective, the maximum diameter should occur at a distance of about one-third the body length from the head. The value of 4.5 for the fineness ratio is not very critical, the total drag only increasing by 10% when the factor is 3 or 7 (Fig. 8.4).

It follows, then, that species that actively hunt for food or spend much of their time swimming pelagically would be expected to have length:maximum diameter ratios of approximately 4.5 and rarely, if ever, outside the range 3–7. This is indeed the case – active pelagic fish, such as the scombrids, salmonids and clupeids have hydrodynamically efficient

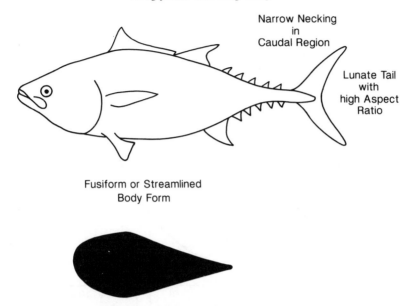

Narrow Necking
in
Caudal Region

Lunate Tail
with
high Aspect
Ratio

Fusiform or Streamlined
Body Form

Fig. 8.3 Diagram of a thunnid fish showing the fusiform body, and other drag-reducing external features. Well-streamlined species, such as thunnids and scombrids, combine a fusiform body shape with lunate tails and narrow necking in the caudal region. These adaptations in the caudal region serve to further reduce turbulence and drag.

body forms, whereas less active fish such as the scorpaenids and cottids show more variation about the theoretical optimum.

The turbulence created by a swimming fish is greater than that produced by a rigid structure of the same shape. This increased turbulence is produced by the lateral body movements which provide the power for swimming. The greater these lateral movements, the greater the turbulence produced. For a given fish the drag increment due to locomotor movements will be largest where the amplitude of the propulsive movements is greatest. Therefore, in order to reduce this turbulence to a minimum a fish should keep the majority of the body rigid and restrict the large-amplitude movements to the tail region. This is the swimming mode adopted by pelagic fish such as tunas and mackerel, but other fish tend to have rather more lateral movements of the body when swimming at steady speed (Fig. 8.1).

Drag can also be minimized by reducing the body and fin area immediately anterior to the trailing edge where the lateral movements are large (i.e. narrow necking in the region immediately anterior to the caudal fin, as seen in the mackerels and tunas). The scooping out of the centre of the

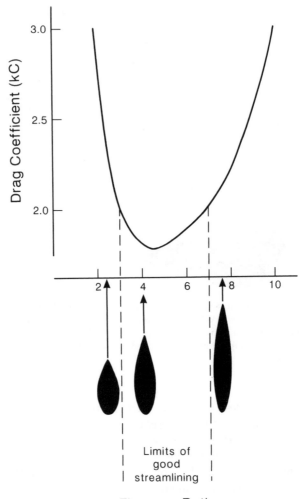

Fig. 8.4 The fineness ratio of the fish body as an indicator of streamlining. The silhouettes indicate body shapes with different fineness ratios.

caudal fin in order to reduce its total area is another means of reducing drag forces (Fig. 8.3). Thus, the continuum from anguilliform to thunniform swimming modes includes trends towards drag reduction via increased rigidly of the anterior part of the body, reductions in the surface area of the flexing part of body and improved streamlining (i.e. more fusiform body shape).

The continuum from anguilliform to thunniform swimming modes also involves a series of other mechanical adaptations. The trends towards narrow necking of the caudal region and scooping out of the tail fin, accompanied by a concentration of the body mass and depth in the foremost part of the body, serve to reduce the magnitude of lateral side forces (i.e. there is reduced risk of the 'tail wagging the head'). Lateral recoil increases the energetic costs of swimming, so any reduction in these forces would be beneficial.

Scombrids and thunnids also typically have caudal fins that are deep in relation to their total area (i.e. the caudal fin has a high aspect ratio [4.5–7.0] – *aspect ratio* is defined as the ratio of the depth of the caudal fin to the mean distance between its anterior and posterior edges)(Fig. 8.3). The large depth of the trailing edge of the laterally moving caudal fin means that a large mass of water is accelerated for each tail beat, something which, in turn, leads to improved swimming efficiency. The muscles and skeletal system in the caudal region exert pressure on the water to generate the propulsive forces required for swimming. In order for swimming to be performed most efficiently the propulsive forces must be transferred effectively with the losses due to elasticity being minimized. The jointing seen in the caudal regions of the scombrids appears to be stiffer than that of other fish species, and this is thought to lead to improved transfer of the muscular forces to the caudal fin without losses due to excessive elasticity.

Costs of swimming

The amount of friction to be overcome by a fish swimming at a steady speed depends upon the speed itself, with friction increasing in proportion to the square of the swimming speed. The power required to be produced by the fish in order to overcome these friction forces is equal to the drag multiplied by the swimming speed, so the power requirements of swimming will increase in proportion to the cube of the swimming speed (U^3). There is, therefore, a curvilinear relationship between the swimming speed of the fish and the energy expended in swimming (Fig. 8.5).

The intercept of the energy expenditure versus swimming speed curve occurs above the origin, since even when not swimming the fish will use a certain amount of energy to maintain other body functions (this is usually called the resting metabolism). The interplay between the resting metabolism and the swimming metabolism has consequences for the amount of energy a fish expends in swimming a given distance. For example, consider a fish swimming a distance of 1 km. If the fish swims slowly, the drag forces the fish must overcome are low and hence the energy costs of swimming are relatively small. However, the slowly swimming fish takes a long time to cover the 1 km and the energy used for other body functions

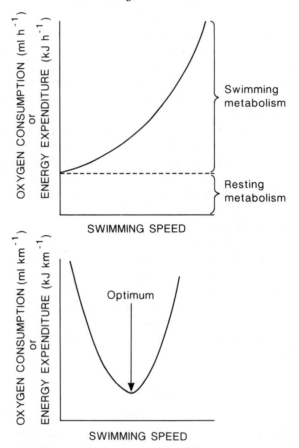

Fig. 8.5 The relationship between oxygen consumption, or energy expenditure, and swimming speed in fish. Note that the cost to swim a given distance is lowest at an intermediate swimming speed.

(resting metabolism) will be considerable. If the fish swims more rapidly, the time taken to swim the distance is reduced and hence the energy required for resting metabolism is less, but the faster swimming speed leads to increased drag forces and hence the swimming metabolism is increased.

When the energy used by the fish to travel a fixed distance at different speeds is calculated and plotted, it can be seen that there is an initial decline in energy use with increasing swimming speed until the *optimum swimming speed* is reached (Fig. 8.5). Optimum swimming speed is that speed at which the energy used to travel a given distance is least. At speeds above the optimum, the drag forces cause increased energy use despite the fact that the time taken to travel the given distance is less than at lower speeds. It is to be expected that a fish swimming a long distance, for example when on migration, would do so by the most economical method, that is, by swimming at the optimum swimming speed. Whilst this is an attractive hypothesis, it is very difficult to get confirmation that fish do swim at optimum speeds during migration.

There are difficulties both in collecting the data required for the calculation of optimum swimming speeds and in recording swimming speeds of fish in the wild. How is information for calculating optimum swimming speeds obtained? Fish are usually made to swim against a water current in a sealed container and, whilst the fish are swimming, the amount of oxygen used by the fish is monitored. The amount of oxygen used is taken as an estimate of the energy expended by the fish. Swimming speed–energy expenditure measurements are made at a number of different swimming speeds, and under different conditions. The data so collected can be used to calculate optimum swimming speeds. It can be seen that the collection of reliable data is extremely time consuming. Studies of this type have shown that both resting metabolism and the energy required for swimming are affected by a number of factors, notably temperature. The energy requirements for swimming at any given speed increase markedly with increasing temperature. In addition, energy use depends upon fish size, resting metabolism for example increasing in proportion to fish weight to the power 0.85 ($W^{0.85}$). Despite the fact that a great deal of information is needed before realistic predictions of optimum swimming speeds of fish can be made, data have been collected for a small number of species under carefully defined conditions. Optimum swimming speeds have usually been found to lie within the range 1–3 body lengths (BL) per second.

There are even greater problems involved in the collection of data on swimming speeds of fish in the wild than are encountered in conducting laboratory studies. Field studies of swimming speeds require extensive long-term tracking and monitoring of the positions of fish using sonar tags. This is both relatively expensive and time consuming so it should come as no surprise that data are scarce. Laboratory and field data are, however, available about the swimming speeds of the plaice, *Pleuronectes platessa*. Optimum swimming speeds calculated from laboratory experiments were close to the maximum speeds that the fish could maintain, and ranged between 1.1 and 1.7 BL s^{-1}, depending upon water temperature. Migrating plaice, observed at sea, were recorded as swimming at an overall mean velocity of 1.2 (range 0.54–2) BL s^{-1} which is very close to the predicted optimum. It seems, therefore, that at least one species of fish adopts the strategy of swimming at optimum speed during migration.

Lateral body movement and undulation give rise to turbulence, and this leads to an increase in drag forces. Thus, the drag forces on a swimming fish are greater than those on a rigid structure of similar shape. These differences in drag can be of major proportions, with the drag forces on an undulating body being 3–6 times those on a rigid structure. This has led to the widespread adoption of the thunniform swimming mode by many actively swimming pelagic fish species. Other species may reduce the negative effects of the drag forces by adopting a *kick-and-glide* swimming mode. In this type of swimming the fish accelerate rapidly during the kick stroke and then glide with the body held rigidly, gradually lose speed and then kick again. This type of swimming is often adopted for prolonged swimming at relatively high speeds (2–2.5 BL s^{-1}). Swimming by means of the kick-and-glide mode can give energetic savings of 40% or so, compared with steady swimming at the same speeds.

Fast-start performance

When a fish makes a sudden dart in order to capture prey or flee from a predator, the acceleration rate is large and speed can increase from zero to 1.0–1.5 m s^{-1} in the space of 50–150 ms. The acceleration is usually brought about by large-amplitude lateral movements of the tail (commonly 0.5 × the fish body length, or more). It is rare that more than a few beats of the tail are involved in producing these rapid accelerations.

Consideration of Equation 8.1, describing the forces to be overcome by a swimming fish, suggests that during fast-start swimming the thrust required to overcome inertia ($M\,a$) is large in comparison with that required to overcome friction. Thus, the drag-reducing mechanisms required for acceleration are those that reduce 'non-essential' body mass. This would involve a reduction in the proportion of non-muscle tissue that has to be accelerated relative to the mass of the 'muscular motor'. In addition, density reductions attributed to buoyancy adaptations would also be expected to contribute to improved acceleration abilities.

In fast-start swimming the amplitudes of the propulsive movements are greatest in the caudal region, and a large caudal fin area would be advantageous. The large-amplitude lateral motions by large fins in the caudal region not only generate thrust forces in the direction of motion, but also give rise to substantial side forces. The potentially negative effects of these side forces can be reduced by the fish having a large body and fin depth in the region of the body close to the centre of mass. Thus, for optimal fast-start performance, the body should be flexible, with a relatively large depth along its whole length, and there should be a large fin area in the caudal region.

Swimming specialists and generalists

On the basis of the above it can be concluded that the fish body design giving optimal fast-start performance is not the same as that for steady swimming, since different forces are operative. For example, good steady-swimming performance is assisted by narrow necking in the caudal region, and by the fish having a large trailing-edge depth to the caudal fin. On the other hand, criteria for fast-start swimming include a large fin area concentrated caudally, where the lateral movements are greatest. Thus, the morphological requirements for good fast-start performance, and those needed for effective swimming at steady speeds, appear to be mutually exclusive. Consequently, the body shapes that lead to maximized acceleration do so at the cost of steady-swimming performance. This does not present large problems for fish that are either highly specialized for continuous steady swimming (e.g. mackerels and tunas), or those that use the body and trunk for 'lunge-swimming' when capturing prey (e.g. pike and other esocids). The latter species are generally relatively sedentary and use alternative swimming modes (e.g. median or paired fin propulsion) when holding station or swimming slowly through the water.

The problem of the dichotomy between the body designs giving good fast-start and steady-swimming performance is greatest for the locomotor generalists, which require to make a balance between both types of swimming activities. Because of the differences in the morphological requirements for the two types of swimming performance, locomotor generalists should, ideally, possess the ability to modify their lateral body profiles. This is possible for the bony fish which can expand and collapse their median fins by means of contractions of the muscles inserted at the bases of the fin rays. Typically such fish expand their median fins before accelerating, and have the fins collapsed against the body when swimming steadily for prolonged periods. Fish, such as the sharks and other elasmobranchs, that are incapable of making significant modifications to their fin areas and lateral body profiles tend to be lacking in locomotor flexibility.

8.4 SWIMMING MUSCLES AND TAIL BEAT FREQUENCIES

In most species of fish the power for swimming is generated by regular contractions of the lateral body musculature and this results in lateral movements of the caudal region. When the tail sweeps from one side towards the centre line of the fish's motion, the tail region is angled obliquely backwards. The thrust exerted by the tail can be divided into lateral and backward components. During a full cycle of tail movement, the lateral components cancel each other out, whereas the backward thrust components serve to drive the fish forwards.

When a fish is swimming at a steady speed the amplitude of the sweep of the tail is usually found to be about one-fifth of the body length, and the fish move forward by about 0.7 BL per tail beat cycle. The step-length of

0.7 BL moved per tail beat cycle refers to fish swimming at steady speeds, but by switching from a steady-swimming to a fast-start mode, it is possible for fish to double the distance moved per tail beat cycle. This increase in step-length obviously contributes to increased speed at any given tail beat frequency.

The swimming speeds capable of being achieved will be dependent upon the contraction properties of the swimming muscles. Data relating to muscle contraction times, assuming a step-length of 0.7 BL, indicate that velocities of 8–15 BL s^{-1} (80–150 cm s^{-1}) could be achieved by 10 cm fish under different temperature conditions. Relative swimming speeds would tend to be lower for larger fish, and a fish of 50 cm in length would be predicted to be capable of speeds of 6–11 BL s^{-1} (300–550 cm s^{-1}) under the same conditions.

Analysis of videotape recordings of fish trained to swim rapidly between two feeding points has suggested that 10 cm long haddock, *Melanogrammus aeglefinus*, and sprats, *Sprattus sprattus*, may reach velocities of over 20 BL s^{-1} (up to 260 cm s^{-1}), whereas Atlantic salmon, *Salmo salar* (25–28 cm in length) can swim at speeds of up to 10 BL s^{-1} (280 cm s^{-1}) under similar experimental conditions. This probably indicates that step-length was greater than 0.7 BL per tail beat cycle for these fish swimming at speeds approaching the maximum.

Further analysis of videotape recordings of swimming fish has revealed that tail beat frequencies vary with fish species and size. Tail beat frequencies generally correlate reasonably well with those predicted from results of physiological experiments carried out in order to examine the contraction properties of isolated muscle preparations. For example, plaice larvae (8 mm body length) may have tail beat frequencies of up to 35 Hz, whereas the tail beat frequency of 25 cm Atlantic salmon may be only 16–17 Hz. A muscle contraction cycle time of approximately 30–50 ms seems to be about average for most species and this implies that fish myotomal muscles are adapted for very rapid cycles of contraction.

8.5 MUSCLE STRUCTURE AND FIBRE TYPES

The source of the driving power for swimming is the lateral body musculature lying on either side of the fish's backbone. In the goldfish, *Carassius auratus*, these muscles make up about 40% of the body weight, increasing to about 60% in salmonids and 70% in scombrids.

The lateral musculature is divided into serially arranged *myotomes* delimited by collagenous connective tissue *myosepta* into which the myotomal muscle fibres insert. The myosepta form a series of overlapping cones which are stacked along axes parallel to the midline, and this gives

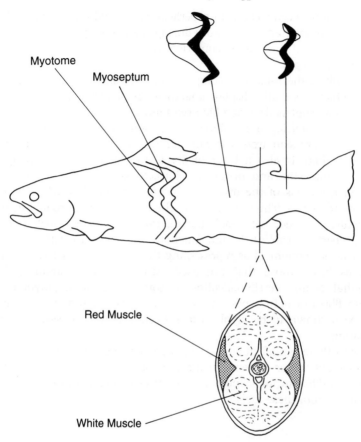

Fig. 8.6 Schematic diagram showing myotomal structure at various distances from the tail (upper and middle figures), and the positions of the white and red myotomal muscles in the cross section of the fish (lower figure).

rise to the characteristic W shape of the myotome when viewed in long-itudinal section (Fig. 8.6).

These lateral muscle myotomes have a variety of roles to play depending upon whether the fish is swimming steadily or is engaged in burst activity. In the former, the muscles are contracting relatively slowly and they must perform over long periods of time. In burst performance, on the other hand, muscle contractions are rapid but the bouts of contraction are of short duration. Several different types of muscle fibres are found in the lateral muscles of fish, and the myotomes can be separated into anatomically distinct regions.

The muscle fibres can be divided into two main types – tonic and twitch.

The tonic fibres occur as a thin superficial layer, and are thought to play a role in the maintenance of posture during the bending of the trunk region. The remainder of the muscle comprises twitch fibres, but the twitch fibres are of several types.

The bulk of the muscle is composed of fast-contracting ('white') twitch fibres, which generally display anaerobic metabolism. A thin superficial band of slow-contracting, aerobic ('red') muscle fibres lies outside the white fibres. The proportion of red fibres in the total muscle cross section increases with progression from the anterior to the more posterior myotomes. Typically, there is also a layer of fibres (the 'pink' fibres), having intermediate contractile and metabolic characteristics. The pink fibres occur between the areas of the myotome occupied by the fast white fibres and the slow red fibres. Thus, the fibres of the myotome can broadly be divided into white and red, with each having rather different distributions (Fig. 8.6), properties (Table 8.1) and functions. As a generalization, the white fibres can be thought of as representing the fast-start muscles, and the red fibres as being those that are used during steady swimming. This is somewhat of an oversimplification as some areas of the myotome may contain fibres of both types and, in some species of fish, the white fibres have been shown to be involved in providing power for slow continuous swimming.

Despite these qualifications, it can be considered that the fish has two very different motor systems in the myotomes. These two motor systems operate on different fuels and use different biochemical routes for liberating the energy from these fuels.

Table 8.1 Comparison of the structure and physiological properties of the red and white fibres found in the myotomal muscles of fish

Red muscle	White muscle
Slow muscle fibres	Fast muscle fibres
'Cruising' or steady-swimming muscle, found in tail region	Burst-performance muscle
Good blood supply; high myoglobin content giving red colour	Poor blood supply
Store and use both lipid and glycogen	Store and use mostly glycogen
Obtain energy by oxidative metabolism: have many and large mitochondria	Respire anaerobically producing lactate
Fibre diameter 100–200 μm	Fibre diameter approx. 300–400 μm

Red muscle

The red muscle has an extensive blood supply, contains relatively large amounts of *myoglobin* (a pigment that is capable of binding oxygen) and the energy from the fuel (respiratory substrate) is released aerobically (using oxygen). The main fuels for aerobic metabolism are the fatty acids derived from the adipose tissues (lipid reserves) of the fish, and glucose from the liver and muscle glycogen stores. These respiratory substrates are usually present in quite large amounts, and have a high energy (ATP) yield per unit substrate when respired aerobically. This makes them ideal substrates for providing the energy required for the performance of continuous work. In general, the rate at which the aerobic pathways can supply energy (ATP) will be limited by the abilities of the respiratory and cardiovascular systems to supply the oxygen and fuels that are transported to the muscles in the bloodstream.

White muscle

The contractions of the white muscle fibres are fuelled anaerobically (oxygen is not used) and the immediate energy supply for contraction comes from the hydrolysis of *phosphocreatine*, present in the muscle. The stores of phosphocreatine are extremely limited and, following a short delay, the energy supply for fibre contraction is supplemented by the activation of *anaerobic glycogenolysis*. Glycogen, which is stored in the muscle, is used as the energy source, but this results in the accumulation of lactic acid (lactate). This method of using fuel is inefficient, since anaerobic metabolism gives just over 5% of the energy yield of the aerobic pathways. Some of the glycogen can be recovered by slow re-oxidation of the lactate, following the completion of a bout of rapid swimming. In addition, the lactate can later be used as a fuel in aerobic metabolism.

The phosphocreatine and glycogen stores are rapidly depleted when the muscles contract anaerobically and, hence, fish have low endurance at high swimming speeds. The relative importance of phosphocreatine and anaerobic glycogenolysis as energy suppliers differs between species, and is related to swimming behaviour. 'Sit-and-wait' predators, such as esocids, that usually perform high-speed swimming over very short distances typically rely on phosphocreatine stores and have limited reserves of muscle glycogen. In contrast, species, such as scombrids and salmonids, with greater endurance at high swimming speeds, tend to have larger muscle glycogen stores, and the activities of the glycolytic enzymes in the muscle tissue are also quite high. In species that rely predominantly upon phosphocreatine hydrolysis the white muscle may become completely exhausted after 1–2 min, or less, of continuous operation. Those fish that have white muscle which is more dependent upon anaerobic glycogenolysis may be able to maintain relatively high swimming speeds for 10–15 min or longer. Under most circumstances, however, a fish would use burst performance for only a few seconds before gliding to a slower swimming speed. If, however, fish are chased to exhaustion it may take 24 h or more before they display complete recovery.

Fibre contraction properties

Burst swimming requires that the myotomal muscles of fish show rapid cycles of contraction and relaxation. However, experiments carried out using isolated muscle fibres have shown that fish muscle fibres contract more slowly than either amphibian or mammalian muscle fibres, at their respective operating temperatures. Since the muscle contraction cycles, at maximum swimming speed, are rapid (cycle time of 30–50 ms, and tail beat frequency of about 25 Hz) this must mean that the individual muscle fibres shorten by very small amounts (2–10% of their resting length) during each contraction cycle. The important conclusion to be drawn from this is that swimming in fish must be, predominantly, an isometric exercise.

Muscle fibre arrangements

If all the muscle fibres were arranged parallel to the long axis of the body they would be required to shorten by a considerable amount in order to produce the large lateral movements of the caudal region seen in swimming fish. It is obvious, therefore, that the fibres must have a special spatial arrangement in order to produce large body flexions for only small amounts of contraction.

Within the myotome, only the most superficial muscle fibres are arranged parallel to the long axis of the fish. The fibres lying deeper in the myotomes are arranged in a kind of helical (spiral) pattern, such that each fibre lies at a different angle. Some of the fibres make angles of up to 40 with the long axis of the body. The advantages of this arrangement appear to be twofold.

Firstly, it probably ensures that, as the fish body flexes, the muscle fibres within any given myotome are all able to contract at the same speed and hence operate most economically (an arrangement in which the muscle fibres were parallel to the long axis of the body would mean that, for any given body flexion, the innermost fibres would have to shorten more, i.e. contract more rapidly, than the fibres lying more superficially in the myotome). This does not, however, imply that all muscle fibres contract at the same rate. Fibres in adjacent myotomes may contract at different rates, with the fibres in the most anterior myotomes seeming to be capable of more rapid contraction than those in the myotomes of the caudal region.

Secondly, the spatial distribution of the fibres within the myotome means that they are only required to shorten by small amounts to produce large flexions of the body. Thus, this spatial arrangement of the fibres enables the swimming movements of fish to be brought about by isometric fibre contractions. When humans engage in exercise it is activities such as weight-lifting and body-building that involve large isometric components and these lead to muscle fibre hypertrophy (increased muscle fibre size) and an increase in muscle fibre strength. On the other hand, training activities such as long-distance running and cycling result in an improved ability of muscles to perform aerobic work leading to increased endurance. When fish are made to swim continuously against a current for periods of several weeks this results in hypertrophy of the muscle fibres but very little change in the capacity to do aerobic work. Thus, although one usually thinks of long-term swimming as being an 'endurance test', the contractions of the fibres within fish myotomes result in changes that are closely allied to those associated with the performance of body-building exercises in humans.

8.6 TERMINOLOGY OF SWIMMING ACTIVITY

In the previous paragraphs a distinction has been made between fast-start, or burst-performance, swimming, and swimming at slower speeds which can be maintained for much longer periods of time. This categorization is useful for drawing a broad dividing line between the extremes of the types of swimming powered by the trunk and caudal region, but the divisions drawn are rather too imprecise to be of value in detailed physiological studies. Consequently, it has become usual for fish physiologists to classify trunk and caudal fin locomotion into a number of categories, each of which is rigidly defined. The definitions of the three main categories – sustained, prolonged and burst swimming – not only indicate the duration over which the fish can perform at a given level, but also give a reflection of the respiratory substrates and metabolic pathways employed in order to provide the energy required for the different types of swimming.

The term *sustained* swimming performance is applied to those swimming speeds that can be maintained for long periods without resulting in muscular fatigue. Sustained swimming can, theoretically, be maintained indefinitely, but for practical purposes sustained swimming speeds are defined as those that can be maintained for a period of 200 min or longer. Metabolism during sustained swimming is aerobic, and there is no accumulation of lactate. The activities that would be encompassed by the term sustained swimming include station holding in mid-water, foraging at low swimming speed and long-distance migration.

Prolonged swimming is of shorter duration (20 s–200 min) than sustained swimming, and results in fatigue. The fish will be said to be fatigued when it collapses and is no longer able to maintain the given swimming speed. The energy supply for prolonged swimming may be provided from both aerobic and anaerobic metabolism. A special category of prolonged swimming is the *critical swimming speed*. The critical speed is a useful term that can be used for comparing the swimming abilities of different sizes or species of fish. Critical swimming speeds are determined under rigidly controlled laboratory conditions, in which the fish are forced to swim against water currents in a flume apparatus. The fish are subjected to stepwise increases in swimming speed until fatigue occurs, and the critical swimming speed is calculated from the maximum speed achieved prior to fatigue. For example, if a fish can maintain a swimming speed of 50 cm s^{-1} for 60 min, but fatigues after 30 min at a speed of 70 cm s^{-1}, the critical swimming speed lies between 50 and 70 cm s^{-1}. In this example, the critical swimming speed would be calculated as $50 + [(70 - 50) \times (30/60)] = 60$ cm s^{-1}, where 50 cm s^{-1} is the swimming speed at the last completed step, $(70 - 50)$ is the incremental increase in speed ('step size'), and 30/60 is the time into the last step at which fatigue occurred (30 min) divided by the prescribed step duration (60 min). Since both the duration and velocity increment of each step will influence the critical swimming speed obtained, it is important that this information be given when presenting the results of studies of this type.

The highest swimming speeds of which fish are capable are termed *burst* swimming speeds. These high speeds can be maintained for only a very short period of time, and fish usually swim at burst speeds for periods of less than 20 s. The energy for burst swimming is provided, predominantly, by anaerobic metabolism. Burst swimming is characterized by an initial acceleration phase followed by a sprint phase of steady swimming. Steady swimming refers to swimming in a given direction at constant speed, whereas swimming that involves either acceleration or direction change is usually termed unsteady swimming.

8.7 RESPIRATORY AND CARDIOVASCULAR ADAPTATIONS DURING SWIMMING

With an increase in swimming activity, the tissue demand for oxygen may increase 5- to 15-fold, with about 90% of this increase being accounted for by elevated muscle metabolism. There are a number of physiological responses invoked in order to meet this demand for increased oxygen supply, with both the respiratory and cardiovascular systems being involved.

Increased swimming results in an increase in the volume of water passed over the gills (i.e. ventilation volume is increased), and this increase is brought about by the combination of a slight increase in breathing rate and quite large increases in ventilatory stroke volume. The increase in ventilation volume, which can be an eightfold increase over that seen in resting fish, results in a substantial decrease in the duration of contact between the water and the respiratory surface. Contact time may decrease from about 300 ms in fish at rest to approximately 30 ms in fish that are swimming rapidly.

Cardiac output may be increased three- to four-fold, largely as a result of increased stroke volume (50–250% increase) of the heart, but there is also some elevation in heart beat frequency (20–40% increase). The increase in cardiac output during exercise results in the blood transit time through the gill secondary lamellae being reduced from about 3 s to 1 s, but there is also an increase in the ventral aorta pulse pressure. Increased blood pressure leads to increased lamellar recruitment (i.e. a greater proportion of the secondary lamellae are perfused with blood), and also results in blood being directed from the basal to the peripheral channels of the lamellae. The net result of this is an increase in the gill area over which gaseous exchange can occur, and there may also be a reduction in the diffusion distance between the water and the blood. Thus, although rates of both blood and water flow over the gills are markedly increased, the efficiency with which the oxygen is extracted from the water is maintained, and oxygen supply is increased to meet increased tissue demand.

Physiological changes during burst swimming

At high swimming speeds there is an increased risk that oxygen supply to the muscle would be unable to meet demand. Under such conditions an increasing proportion of the ATP required for maintaining muscular contraction will be generated by anaerobic glycogenolysis with the production of lactate as an end product. The production of lactate will result in a fall in pH and this will promote an increased offloading of oxygen from the haemoglobin in the muscle tissue. Should, however, lactate be released into the general blood circulation this may be expected to compromise the ability of the haemoglobin to bind with and transport oxygen due to the influence of pH on binding properties (Bohr and Root effects).

The effects of metabolic acidosis on the ability of the blood to transport oxygen are not, however, particularly severe. It appears that the oxygen-carrying properties of the haemoglobin are protected by the release of *catecholamines* during the course of a challenge imposed by a fall in blood pH.

The catecholamines are known to have a number of effects upon the cardiovascular and respiratory systems, and circulating levels of catecholamines are elevated in fish swimming at high speeds. A potent stimulus for the release of catecholamines from the chromaffin tissue of the head kidney appears to be low blood oxygen. The blood reaching the head kidney in the venous return will be en route to the heart from the caudal region, where the bulk of the swimming muscle is concentrated. Thus, when fish increase their swimming speed and there is considerable off-loading of oxygen from the haemoglobin in the swimming muscle, the blood reaching the head kidney will be oxygen-depleted. This will, in turn, trigger the release of catecholamines. Consequently, circulating levels of catecholamines may begin to rise before there has been any large release of lactate into the blood, and prior to the onset of metabolic acidosis.

The release of catecholamines into the blood stimulates an Na^+/H^+ exchange mechanism in the membrane of the red blood cells, leading to the extrusion of protons from the cell (Fig. 8.7). This mechanism can, therefore, serve to maintain intracellular pH in the event of a fall in plasma pH due to metabolic acidosis caused by increasing blood lactate levels. Thus, the maintenance of intracellular pH means that the oxygen-transporting properties of the haemoglobin are not severely compromised under these conditions.

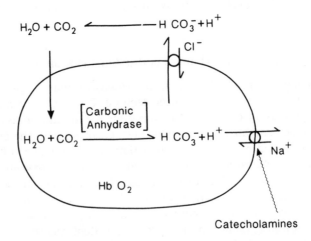

Fig. 8.7 Effects of catecholamines on red blood cell ion exchange mechanisms. The extrusion of protons from the red blood cell serves to maintain intracellular pH in the event of an acidic challenge due to a fall in plasma pH. Hb, haemoglobin.

The catecholamines also exert effects upon the gill vasculature, resulting in both an increase in the recruitment of secondary lamellae and an increase in membrane permeability. Thus, the net result is an improvement in the conditions for gas exchange between water and the blood in the gill tissue.

One further effect of increased levels of circulating catecholamines is the stimulation of contraction of the spleen. Contraction of the spleen results in the release of stored red blood cells into the circulation. Consequently, there is an increase in the ability of the blood to transport oxygen.

Nevertheless, at high swimming speeds oxygen demand will exceed supply and there will inevitably be some ATP production via anaerobic pathways, leading to the accumulation of lactate. Thus, following a bout of strenuous activity, body loadings of lactate may be high, and the fish will require a certain length of time in order to recover.

Oxygen debt

During the recovery period, the rate of oxygen consumption may be quite high, and the fish is considered to be repaying the oxygen debt incurred whilst engaged in high-speed swimming. The oxygen debt can be considered to have two major components – alactacid and lactacid. The repayment of the alactacid oxygen debt requires only a short time, whereas repayment of the lactacid oxygen debt may require several hours. The alactacid oxygen debt is considered to represent the costs associated with the replacement of the tissue reserves of ATP and phosphocreatine, whereas the increased oxygen consumption related to the lactacid oxygen debt is associated with the metabolism of accumulated lactate, regeneration of muscle glycogen and the restoration of acid–base and ionic balance in the body fluids.

8.8 BUOYANCY

If a swimming fish is heavier than water, part of its energy expenditure goes to keep it from sinking and only part is available for forward locomotion. If the body of the fish had the same specific gravity as that of the water, i.e. if the fish were *neutrally buoyant*, more energy would be available for forward motion.

There are a number of ways in which an aquatic animal can reduce its tendency to sink, most notably by reducing the size or weight of the heavy parts of the body, such as the skeleton. Bone is highly calcified, and contains calcium carbonate and phosphate. These calcium salts have specific gravities of approximately 3. The specific gravity of sea water is approximately 1.026 so the replacement of the calcium salts by lighter structural elements could contribute greatly towards the achievement of neutral buoyancy. Amongst fish species, the skeletal elements are usually very much reduced in comparison with those seen in terrestrial animals. In addition, the lipid contents of fish bones are considerably higher than those of mammalian bones. This is most noticeable in the bones of the skull, with some fish skulls containing up to 30% lipid by weight. It is, however, more usual that the skull bones of fish contain approximately 10% lipid, whereas those of mammals rarely contain more than 1% lipid by weight.

Replacement of heavy ions (Mg^{2+}, SO_4^{2-}) by lighter ions such as H^+, Cl^- and NH_4^+ could assist in reducing the overall specific gravity of the animal, but this may not be a very effective buoyancy mechanism. The elasmobranchs employ urea and TMAO as organic osmolytes. These osmolytes appear to reduce the density of the body fluids relative to that of sea water in which the major osmolytes are the inorganic ions Na^+ and Cl^-. TMAO has a lower density than an equimolar solution of urea, and solutions of both TMAO and urea are less dense than equimolar solutions containing inorganic osmolytes. Thus, TMAO and urea accumulation by elasmobranchs may have a role to play in buoyancy regulation. A more effective method for reducing specific gravity than merely changing the composition of body fluids relative to that of the surrounding medium would be the removal or loss of ions from the body fluids without replacement, i.e. the body fluids become more dilute than the surrounding medium.

An increase in the relative quantities of light substances, such as neutral lipids and oils, in the body tissues could make a considerable contribution to specific gravity reduction. A number of the cartilaginous species of fish possess relatively large livers containing high percentages of neutral lipids, oil or *squalene*. Squalene is an extremely light, unsaturated hydrocarbon compound peculiar to elasmobranch fish species. The specific gravities of the lipids and oils are 0.9–0.93, that of squalene is 0.86, whereas the specific gravity of muscle is 1.05–1.1, and that of the skeleton is approximately 2.0. Thus, in the species which have lipid-rich livers the liver functions as an aid to buoyancy.

For example, in bottom-living skates the liver comprises only about 4–6% of the body weight when the fish is weighed in air. The liver tissue contains up to 30% lipid and little or no squalene. The overall specific gravity of the liver is equal to, or greater than, that of sea water. In the skates, therefore, the liver does not function as a buoyancy organ.

In the velvet belly, *Etmopterus spinax*, on the other hand, the liver is about 17% of the body weight in air, and contains 75% oils. Of the oil content about 50% is squalene. Thus, the specific gravity of the liver is less than that of sea water, and the liver of a fish weighing 100 g in air would give an upthrust of about 1.7 g to the fish when it is swimming in sea water.

In pelagic sharks, such as the blue shark, *Prionace glauca*, the liver makes up about 25% of the body weight. Of the liver weight, about 80% is squalene. The liver of a fish of 100 g body weight in air gives an upthrust of approximately 3.5 g to the fish when it is in sea water.

The amounts of upthrust provided by the liver of the pelagic sharks may appear to be small but it must be remembered that the relative weights of the tissues in sea water will be small. For example, in sea water, the relative weight of 1 ml muscle tissue will be $1.1 - 1.026 = 0.074$. Hence, it is easy to see how the upthrust provided by the liver allows pelagic sharks to almost attain neutral buoyancy.

Lipids, being lighter than sea water, can provide some upthrust, contributing to buoyancy. Some osteichthyans, such as scombrids and herring, *Clupea harengus*, deposit large amounts of lipid in the body tissues. These lipid depots, which may be either deposited in the muscle or be accumulated as mesenteric fat, can provide considerable static lift.

In the majority of the bony fish species, however, there may be limited lipid stores, and these will be insufficient to play a major role in buoyancy regulation. Usually the specific gravity of the fish body tissues will be within the range 1.06–1.09, and 1.076 can be taken as representing a reasonable average. Thus, the body lipid stores may contribute relatively little to buoyancy in the majority of bony fish and other mechanisms are invoked.

The osteichthyans are characterized by having a gas bladder. In many of the bony fish species the gas bladder functions as a swim bladder, providing the majority of the lift to give them neutral buoyancy. Whether a particular fish species has a gas bladder that functions as a swim bladder is related more to its lifestyle than its evolutionary position. For example, many bottom-living fish either lack (i.e. have degenerated gas bladders) or have small swim bladders. The ability to achieve neutral buoyancy would seem to be of no particular advantage to a fish living on, or close to, the bottom.

8.9 THE SWIM BLADDER

The gas bladders of fish vary in structure, and they may serve other functions in addition to acting as buoyancy organs. The anatomical origin of the gas bladder is, however, similar in all species. The gas bladder arises as a pouch or outgrowth from the foregut. In many species of fish (e.g. salmonids, clupeids, eels) the connection with the gut is retained as an open tube, the *pneumatic duct*, in the adult. A number of the fish species having *physostomatous* (Greek: physa – bladder; stoma – mouth) gas bladders can fill the swim bladder by gulping air at the surface. This may be the way in which gas is first introduced into the bladder when the fish are in the larval stage of development. The presence of an open connection between the gas bladder and the foregut also enables the release of excess gas via the gut as the fish swim upwards in the water column and the gas expands.

Many teleost fish lose this open connection early in life and, in the adult, the swim bladder is completely closed or *physoclistous* (Greek: kleistos – closed). The rates at which gas can be secreted into or lost from a swim bladder of this type are relatively slow. This is demonstrated by the fact that when fish such as Atlantic cod, *Gadus morhua*, are trawled at depth and brought rapidly to the surface, the decrease in hydrostatic pressure causes the swimbladder gas to expand and push the stomach and gut out through the mouth of the fish. With even more rapid rates of forced ascent the swimbladder gas may expand so quickly that it results in bursting of the swimbladder wall.

Size of the swim bladder and neutral buoyancy

How large must the swim bladder be to give a fish neutral buoyancy? Consider first the size of the swim bladder in a freshwater fish – consider a fish displacing 100 ml fresh water. As the density of fresh water is 1, the upthrust provided by the displaced water is 100 g. This is equivalent to $100/1.076 = 92.9$ ml fish body. Therefore, the gas space in a fish displacing 100 ml fresh water must be $100 - 92.9 = 7.1$ ml.

The size of the swim bladder in a fish in sea water will differ from that of a freshwater fish – consider a fish displacing 100 ml sea water. As the density of sea water is 1.026, the upthrust provided by the displaced sea water is 102.6 g. This is equivalent to $102.6/1.076 = 95.4$ ml fish body. Therefore, the gas space in a fish displacing 100 ml sea water must be $100 - 95.4 = 4.6$ ml.

The above examples show that the size of the swim bladder must be larger in freshwater than in seawater species if neutral buoyancy is to be attained. Thus, freshwater species usually have larger swim bladders than do marine species, but the volume of the swim bladder in bottom-living fish species is, however, often lower than that required for neutral buoyancy (Table 8.2).

A swim bladder can provide a fish with perfect neutral buoyancy but, since the swim bladder is essentially a bubble of gas, it is influenced by hydrostatic pressure. Thus, the volume of the swim bladder will tend to change as the fish swims up and down in the water column. The gas in the swim bladder will obey the gas laws, and Boyle's Law states: pressure

Table 8.2 Swimbladder sizes of a range of fish species occupying different habitats in marine and freshwater environments

Species	Habitat	Swimbladder volume (% total body volume)
Goldfish,	Freshwater,	
Carassius auratus	midwater	7.5–8.0
Dace,	Freshwater	
Leuciscus leuciscus	surface and midwater	8.5–9.0
Brown bullhead,	Freshwater,	
Ameiurus (Ictalurus) nebulosus	bottom-living	5.5–6.0
Wrasse,	Marine,	
Ctenolabrus rupestris	shallow, midwater	4.5–5.0
John Dory,	Marine,	
Zeus faber	midwater	4.0–4.5
Plaice,	Marine,	
Pleuronectes platessa	bottom-living	Absent
Dragonet,	Marine,	
Callionymus lyra	bottom-living	Absent
Atlantic mackerel,	Marine,	
Scomber scombrus	surface, pelagic	Absent

× volume = constant. This means that a swim bladder containing a given amount of gas only gives neutral buoyancy at one particular depth. Thus, if fish are to maintain neutral buoyancy as they swim up and down in the water column they must not only store gas in the swim bladder but they are also required to have the ability to regulate the amount of gas it contains, i.e. fish must be able to secrete and absorb gas.

In aquatic environments the hydrostatic pressure increases by 1 atm for every 10 m increase in depth. If a fish swimming at the surface (with swimbladder gas at atmospheric pressure) swims down to a depth of 10 m, then the volume of the swim bladder will be halved unless the fish can secrete sufficient gas to counteract the pressure changes. If the fish then swims to the surface again the swim bladder will double in volume unless the fish can resorb the gas. Thus, at shallow depths relatively small changes in depth lead to changes in pressure that have drastic effects upon swimbladder volume (Fig. 8.8). The regulatory mechanisms for secretion and absorption of gas may not be able to cope effectively with such large changes in swimbladder volume in cases where the fish move rapidly between different depths close to the surface, and in some surface-dwelling pelagic fish (e.g. many scombrid species) the gas bladder has become degenerated.

At greater depths, changes in hydrostatic pressure have relatively less

Swimming and movement

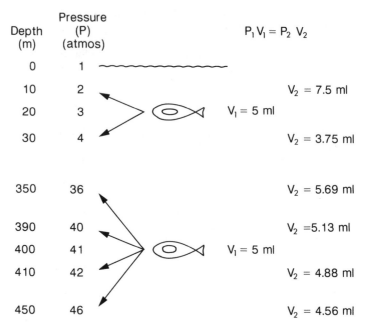

Fig. 8.8 The effects of differences in hydrostatic pressure (in atmospheres) with depth on changes in swim bladder volume experienced by a fish as it moves up and down in the water column. The gas in the bladder is assumed to be subject to changes in accordance with Boyle's law: $P_1V_1 = P_2V_2$, where P is pressure and V is gas volume.

severe effects upon the volume of the gas in the bladder (Fig. 8.8), and a fish changing its depth by 10 m at a depth of 400 m will only alter the volume of its swim bladder by one-fortieth. However, fish living at great depth will face problems with control of swimbladder volume. Consider a fish living at 1000 m depth where the pressure is about 100 atm. To produce and maintain a volume of gas at that pressure entails three major difficulties. Firstly, the gas must be secreted into the swim bladder against a pressure of 100 atm. Secondly, the amount of gas needed to fill the same volume is 100 times greater at 1000 m than at the surface, and finally, once the gas is in the swim bladder it is a problem to prevent it diffusing out again.

8.10 SWIMBLADDER GAS EXCHANGE

The gas contained within the swim bladder is composed mainly of oxygen, although other gases, such as nitrogen and carbon dioxide, are present in lesser proportions. The oxygen is secreted into the swim bladder from the blood supply reaching the bladder via the circulatory system, but there are problems to be overcome in filling, and retaining the gases in, the swim bladder. The gas must be secreted from the blood to the lumen of the swim bladder against a gas pressure gradient, and once secreted into the swim bladder the gas must be prevented from diffusing back into the blood. A further problem arises for species with physoclistous bladders in that they must be able to resorb gas into the blood in a controlled fashion when they ascend in the water column and hydrostatic pressure is reduced.

Gas retention

The explanation of how the gases are retained in the swim bladder once they have been secreted is rather simpler than the description of the method by which gas is secreted and, since both require the structure of the swim bladder to be described, it is easier to begin with the problem of gas retention.

The *gas gland* is found in the wall of the swim bladder, where it can easily be seen because of its bright red colour (Fig. 8.9). The gas gland is supplied with arterial blood transported from the gills to the gas gland via the circulatory system. This arterial blood contains gas at tensions in equilibrium with those in the surrounding water, so one would expect the blood to dissolve the swimbladder gases and carry them away. This does not happen, so the question arises as to how this is prevented.

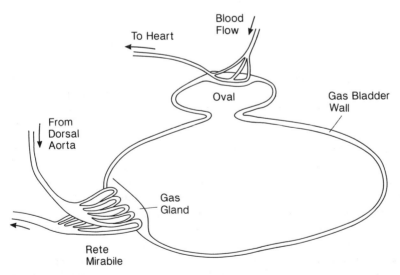

Fig. 8.9 Schematic diagram showing the general structure of the swim bladder of a physoclistous fish.

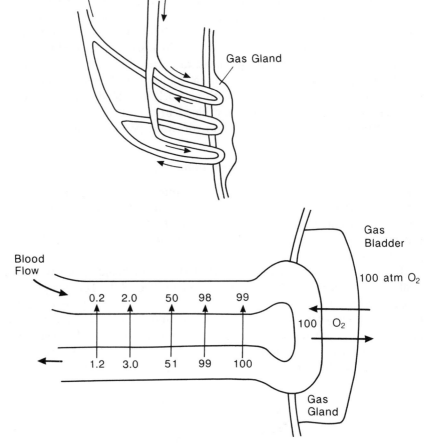

Fig. 8.10 Schematic diagrams showing the general structure of the rete mirabile of the swim bladder. The rete mirabile acts as a countercurrent multiplier system, in that gases and solutes can diffuse from the efferent to the afferent blood vessels.

The gas gland is supplied with blood through a peculiar structure known as a *rete mirabile*, which is essentially a network of blood vessels running parallel to each other and functioning as a countercurrent exchanger (Figs 8.9 and 8.10). It is countercurrent exchange in the rete mirabile that acts to retain the swim bladder gases. For example, assume that the swim bladder contains gas under a pressure of 100 atm. As the blood leaves the swim bladder it will contain dissolved gases at equilibrium with this high pressure. However, the outgoing vessels run parallel to the incoming vessels which carry blood containing gases in equilibrium with lower pressures. Consequently, gases diffuse from the outgoing to the incoming blood. As the blood runs along the venous vessels it loses more and more gas by diffusion to the arterial blood (Fig. 8.10).

The venous blood leaving the rete meets blood which comes directly from the gills and is in equilibrium with the gas pressure in the surrounding water. The venous blood loses gas until it is in diffusion equilibrium with the incoming arterial blood. That is, the venous blood that leaves the rete contains no more gas than the incoming arterial blood. In this way the rete acts as a trap that retains the gases in the swim bladder and prevents losses to the circulating blood. In summary, the rete mirabile is a typical countercurrent exchange system which depends on passive diffusion between two liquid streams running in opposite directions. The exchange of gas is aided by the large surface area for diffusion and by a short diffusion distance.

Gas secretion

The gases found in the swim bladder are the same as those in the atmosphere – oxygen, nitrogen and carbon dioxide – but the gases are present in different proportions. The gases in the swim bladder of fish can contain up to 80% oxygen so it is obvious that oxygen is the gas most commonly secreted.

In a fish swimming at a depth of 1000 m the swim bladder will contain oxygen at a pressure of 100 atm but the oxygen tension in the blood arriving at the gas gland will be the same as that in the surrounding water, that is 0.2 atm. Consider now the oxygen content of the blood rather than the oxygen tension. If oxygen is being removed from the blood and secreted into the swim bladder, it follows that the venous blood leaving the rete of the gas gland must contain less oxygen than the arterial blood that enters the rete. The arterial blood may, for example, contain 10 ml oxygen per 100 ml blood, whereas following secretion of gas into the swim bladder the venous blood leaving the rete may contain 9 ml oxygen per 100 ml blood, so that 1 ml oxygen per 100 ml blood is the amount of oxygen secreted into the swim bladder.

The oxygen in the blood is carried either in solution (a relatively small quantity) or bound to the respiratory pigment haemoglobin (the majority). The ease with which the oxygen binds with the pigment is markedly affected by pH, the binding of oxygen to the haemoglobin being reduced as the conditions become more acid, i.e. with lowered pH. In fish, the effects of acidification of the blood on some forms of haemoglobin are so extreme that the pigments do not bind with oxygen even at high oxygen tensions. This means that the blood haemoglobin does not become fully saturated at high oxygen tensions (the Root effect), and this effect is of fundamental importance in the secretion of oxygen into the swim bladder.

Respiration by cells of the gas gland in the wall of the swim bladder results in the production of lactic acid and carbon dioxide. The lactic acid (lactate and protons) enters the blood in the gas gland and causes a lowering of the pH. Lactic acid secretion by the gas gland cells can cause a lowering of blood pH by almost 1 unit. This reduces the ability of the haemoglobin pigment to bind with oxygen so that oxygen is released from the pigment and the oxygen tension in the blood leaving the gas gland is increased (Figs 8.10, 8.11 and 8.12). It follows that the oxygen tension is now higher in the venous blood than in the arterial blood and, therefore, oxygen will diffuse across into the arterial blood. This process will continue as long as lactic acid is added to the venous blood.

In this system it is important to distinguish clearly between *amount* and *tension*. If oxygen has been secreted into the swim bladder during the passage of the blood through the gas gland it is clear that the amount of oxygen in the venous blood which leaves the rete must be less than that in the incoming arterial blood, but for oxygen to diffuse across from the venous to the arterial blood the tension must be higher in the venous blood. The higher tension in the venous blood results from the offloading of oxygen from the haemoglobin as a result of the reduction in pH caused

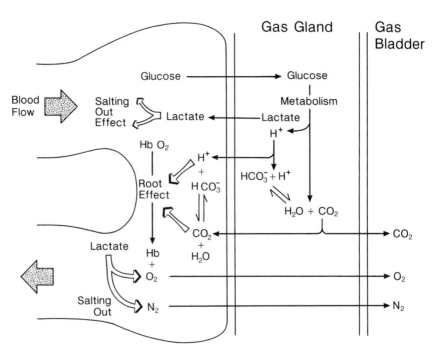

Fig. 8.11 Lactate production and other metabolic activities in the gas gland that contribute to the secretion of gas into the swim bladder. Hb, haemoglobin.

by the acid produced by the gas gland (Fig. 8.11). Since the oxygen tension is higher in the venous than in the arterial blood, oxygen will continuously diffuse from the venous to the arterial blood and gradually accumulate in the rete where it can reach very high tensions (Fig. 8.12).

The lactic acid secreted into the venous blood will also tend to diffuse over into the arterial blood, and this diffusion of lactic acid has two results: (1) the lactic acid is retained within the gas gland and rete, and (2) as the lactic acid diffuses from the venous to arterial blood, the pH of the venous blood will rise, and this results in the oxygen rebinding to the haemoglobin as the blood passes along the rete and re-enters the main circulatory system.

Indications that the gas gland functions in the way described came from the results of experiments carried out during the 1960s. Using micropuncture techniques, blood samples were taken from the arterial and venous vessels of the rete mirabile of the European eel, *Anguilla anguilla*. When these samples were analysed it was found that the venous blood had a lowered oxygen content but an increased tension compared with that of the arterial blood.

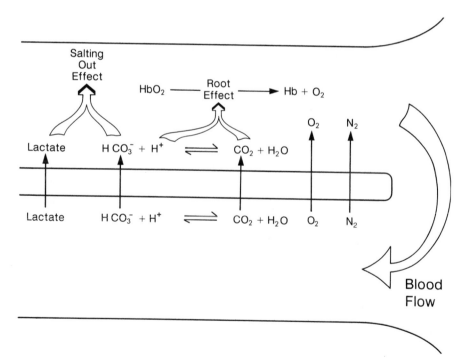

Fig. 8.12 Schematic diagram showing the countercurrent multiplier function of the rete mirabile. Hb, haemoglobin.

The system described above shows how oxygen can be secreted into the swim bladder, but oxygen is not the only gas found in the swim bladder. The secretion of nitrogen and carbon dioxide involves the use of alternative mechanisms. The solubility of gases in a solvent is dependent upon the amount of dissolved solids in solution. Gas solubility declines as the amounts of dissolved solids increase. The secretion of lactic acid by the gas gland not only leads to a decrease in the plasma pH, but also increases the amounts of dissolved solids in the blood. This leads to a reduced gas solubility in the blood, and this colligative effect influences the solubility of all the dissolved gases. This is termed the *salting-out* effect (Figs. 8.11 and 8.12). The solubility of nitrogen will be influenced by the levels of dissolved solutes, and, as is the case for oxygen, the rete system will act as a countercurrent multiplier for nitrogen. The net result is the establishment of gradients that are sufficiently high to promote secretion of nitrogen gas into the swim bladder.

Carbon dioxide produced as a result of respiration by the cells of the gas gland provides additional acid, augmenting the lactic acid. Thus, the carbon dioxide contributes to the induction of the Root effect of the haemoglobin. Carbon dioxide formation also gives rise to gas molecules that can enter the swim bladder because the highest P_{CO_2} occurs in the cells of the gas gland. Thus, carbon dioxide may diffuse both into the swim bladder and into the bloodstream. The carbon dioxide that enters the bloodstream from the gas gland is retained within the countercurrent rete system (Figs 8.11 and 8.12). Thus, the combination of the salting-out effect and the countercurrent multiplier system of the rete can be used to explain how gases such as nitrogen and carbon dioxide can be secreted into the swim bladder.

Gas resorption

When the fish rises in the water column hydrostatic pressure is reduced and swimbladder volume increases. In order to restore volume gas must be removed from the swim bladder. Physostomatous fish can void the excess gas into the gut via the pneumatic duct, which connects the swim bladder to the oesophagus or pharyngeal region. For physoclistous fish, with closed swim bladders, removal of gas is not so simple because the rete mirabile of the gas gland is specifically designed for gas retention. Gas removal occurs, not in the gas gland, but in a different area of the swim bladder which receives a blood supply completely separated from that of the gas gland (Fig. 8.9).

Certain areas of the swim bladder surface are covered with a meshwork of thin blood vessels, and it is here that gas is resorbed from the swim bladder in the physoclistous species. Gas resorption occurs only when the fish is rising in the water column and experiences reduced hydrostatic pressure. At other times these blood vessels tend to be closed off by a series of muscular valves and are not supplied with blood. In some species, such as the gadoids and labrids, the structure of the swim bladder is such that the areas used during resorption of gas are more or less isolated from the remainder of the swim bladder for most of the time (Fig. 8.9). This area of the swim bladder which is specialized for gas resorption is known as the *oval*.

Gas retention at great depth

The system of gas secretion functions efficiently at pressures of up to 150–200 atm, that is, at depths of up to approximately 2000 m. With increasing depth other problems, in addition to those associated with gas secretion, may arise because the swim bladder is not a gas-tight bag. In other words, some gas will diffuse out across the walls of the swim bladder, and the rate of diffusion increases with increasing hydrostatic pressure.

The 'gas tightness' of the swim bladder may be improved by the incorporation of a series of overlapping guanine platelets into the swimbladder wall. The quantities of guanine, and other purines, present in the swimbladder wall tend to be greater in deep-water fish species than in those species living in shallower water, although there are exceptions, e.g. herring, to this general rule. Thus, the brown trout, *Salmo trutta*, which usually inhabits shallow water (depth 0–15 m), has about 24 µg purines cm^{-2} in the swimbladder wall, and the saithe, *Pollachius virens*, a gadoid of surface and midwater habitats (depth 0–200 m), has a swim bladder containing about 47 µg purines cm^{-2}. On the other hand, the swimbladder wall of the cusk or torsk, *Brosme brosme*, which occurs at greater depths (400–1000 m), contains about 236 µg purines cm^{-2}, and the swim bladder of the deep-water (500–1700 m) rat-tail or grenadier, *Coryphaenoides rupestris*, has a very high purine content (393 µg purines cm^{-2}).

The tendency for gas to diffuse out of the swim bladder will increase with increasing depth, and from this it follows that at some particular depth, the rate of diffusion of gas out of the swim bladder will equal the rate at which gas can be secreted by the gas gland. The depth at which this balance between secretion and leakage occurs has been called the *critical depth*. The critical depth will differ between species, depending upon the particular properties of the swim bladder. In the saithe, the critical depth has been estimated to be approximately 78 m, but saithe can often be found swimming at depths greater than this. At depths greater than the critical depth the fish cannot maintain neutral buoyancy by use of the swim bladder alone because gas diffuses out of the swim bladder faster than it can be replaced by secretion. Thus, at these depths the fish must also swim actively in order to maintain their position in the water column.

From this it follows that a gas-filled swim bladder is not the ideal solution to the problem of maintenance of neutral buoyancy at great depth. Numerous deep-water fish species have swim bladders, but many of them fill the swim bladder with lipid rather than gas. High hydrostatic pressure will exert little influence upon the lipid present in the swim bladder, and hence a lipid-rich swim bladder can make a considerable contribution to static lift.

8.11 STATIC AND DYNAMIC LIFT

For fish such as elasmobranchs, which have bodies slightly heavier than water, part of the energy used in swimming is used to provide hydrodynamic lift in order to keep the body from sinking. When a shark or dogfish swims, the side-to-side sweep of the tail fin not only helps to drive the fish forward but also tends to raise the tail. This lift force comes from the large lower lobe of the asymmetrical (heterocercal) caudal fin and, were this lift force not counterbalanced, it would cause the fish to 'nose-dive'. The lift force provided by the tail is counterbalanced by the large pectoral fins which function in more or less the same fashion as the wings of an aircraft. More precisely, the pectoral and caudal fin lift forces are counterbalanced about the centre of gravity of the fish (Fig. 8.13). The total effect is the production of a steady lift force acting through this centre or point of balance enabling the fish to maintain itself level in the water.

Several species of pelagic teleost fish, such as Atlantic mackerel, *Scomber scombrus*, either do not possess or have very much reduced swim bladders. These species also rely on dynamic lift in order to maintain their vertical position. The speeds at which scombrids and thunnids (mackerel, tuna and related species) must swim in order to provide the lift required to overcome their body weight depends upon the density of the fish tissues relative to

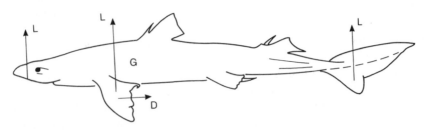

Fig. 8.13 The lift and drag forces operating on an elasmobranch fish during swimming. G, centre of gravity of the fish; D, drag forces; L, lift forces.

sea water, their body shape and the fin area contributing to the dynamic lift. The required speeds can be predicted from hydrodynamic principles and, when compared with the swimming speeds sustained by scombrids over long time periods, there is a close correlation between the two. That is, scombrids and thunnids usually swim no faster than the speed required for them to maintain their position in the water column (Table 8.3).

Other pelagic teleosts may not rely solely upon the swim bladder to remain neutrally buoyant. For example, in the herring, a certain amount of static lift is provided by the relatively high lipid content of the tissues. In this species buoyancy is maintained via a combination of the gas content of the swim bladder, the static lift provided by body lipids and the dynamic lift of swimming.

Many deep-water fish have a degenerated gas bladder and solve the buoyancy problem by a range of alternative methods. The skeletal structures are often considerably reduced, and deep-water fish also often have an increased water content compared with shallow-water species (e.g. shallow-water fish 64–75% water; deep-water fish with swim bladder 70–83% water; deep-water fish lacking swim bladder 88–95% water). In those deep-water fish which possess a swim bladder, it may be filled with lipid. Static lift may be further assisted by the presence of subcutaneous lipid or oil sacs, leading to an overall increase in the lipid content of the body tissues. Finally, deep-water species may display a reduction in the propor-

Table 8.3 The relationships between the density of the body tissues in scombrid and thunnid fish species and their sustained – long-term – swimming speeds

Species	Swim bladder	Body density $(g\ cm^{-3})$	Swimming speed $(BL\ s^{-1})$
Bonito, *Sarda sarda* Skipjack tuna, *Katsuwonus pelamis* Bullet mackerel, *Auxis rochei*	Absent	1.08–1.09	~2.0
Yellowfin tuna, *Thunnus albacares* Big-eye tuna, *Thunnus obesus*	Present	1.05–1.075	~1.3
Atlantic mackerel, *Scomber scombrus*	Absent	1.02–1.06	~1.0
Wahoo, *Acanthocybium solanderi*	Present	1.02–1.03	~0.4

tion of swimming muscle compared with their relatives inhabiting shallower waters.

The way in which these factors can combine to produce a neutral buoyancy in the deep-water species is shown by comparing the body compositions of typical examples of shallow-water and deep-water fish, each of which would weigh 100 g if weighed in air (Fig. 8.14). In

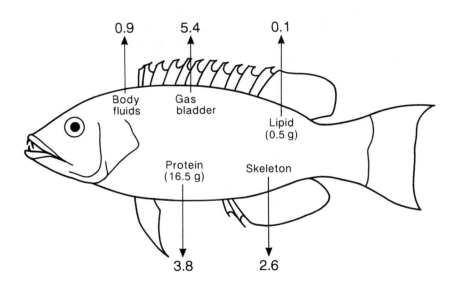

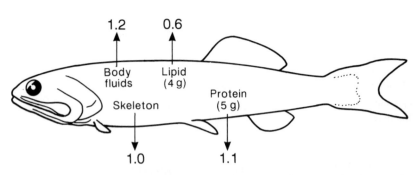

Fig. 8.14 Comparison of the contributions of a range of different mechanisms to the buoyancy of a marine coastal (upper) and deep-water (lower) fish species. Values given in parentheses are the weights of the different chemical components in air. Direction of the arrows indicates how the specific components contribute to the buoyancy balance of the fish in water.

addition to indicating the effects of differences in body composition on the buoyancy of the fish from the different environments, the balance sheets also show that marine teleosts will gain a certain amount of static lift by having their body fluids less concentrated than the surrounding medium, the osmotic concentration of the blood of marine teleosts being 370–480 mOsm kg^{-1}.

8.12   FIN POSITION – STEERING AND BRAKING

The majority of the teleost fish, which have swim bladders, are capable of maintaining neutral buoyancy without having to resort to hydrodynamic lift mechanisms. Consequently, the primary function of the fins is not to produce hydrodynamic lift, and in teleosts the fins fulfil a variety of other roles. The caudal fin, for example, is freed from its role as a producer of lift, and has developed into an effective producer of forward thrust. A symmetrical (homocercal) caudal fin is a common feature of nearly all the more modern bony fish, and the thrust produced by this type of tail is greater than that produced by a heterocercal caudal fin.

In addition to caudal fin symmetry, the acquisition of a swim bladder (and neutral buoyancy) allowed the paired and median fins to develop considerable flexibility. The development of versatile, controllable paired and median fins is just as significant a feature of the evolution of the teleost fish as the change in caudal fin form.

The fins of teleost fish are supported by spines and rays, the fin rays being flexible and capable of being moved by special sets of muscles inserted at their bases. These supportive structures may be moved independently of each other to some extent, allowing for adjustments to be made in fin area, orientation and angling relative to the body surface. The improved flexibility of the fins led to improvements in the capability of the fish to manoeuvre, brake and make fine adjustments in orientation. For example, sharks are capable neither of braking suddenly nor of swimming backwards, but both of these feats are within the repertoire of teleosts.

Three types of body movement can lead to instability. These are *yawing* (movements from side to side about the vertical axis), *rolling* (movements about the horizontal, longitudinal axis) and *pitching* (movements about the horizontal, transverse axis) (Fig. 8.15). Yawing is controlled, largely, by spreading the unpaired, median fins (dorsal and anal) which act as keels. Side-to-side rolling movements are controlled in a similar manner, although spreading of the paired pectoral and pelvic fins may be used to restore equilibrium. Control over pitching movements is more complicated and, in general, the paired fins are most useful.

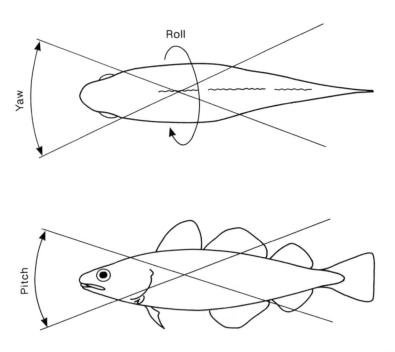

Fig. 8.15 The three main types of unstable body movements in fish are pitching, yawing and rolling. Stability can be restored by use of both the paired and unpaired fins.

Braking

The paired fins are particularly important in the control of pitching during braking. In the teleosts showing the most conservative body characters, the pectoral fins are situated low on the elongated body, below the centre of gravity. This means that when the fish spreads the pectoral fins to brake, it will tend to pitch forward (imagine riding a bicycle downhill and trying to stop using only the brakes on the front wheel – because the brakes are below the centre of gravity of the bicycle and rider, there is a tendency to pitch forward, and if the rider is unfortunate, he will be thrown over the handlebars). In addition, the pectoral fins are set at such an angle to the body that they provide some lift force and hence the fish will tend to rise in the water column. These forces are opposed by those produced by the pelvic fins which are located posteriorly (behind the centre of gravity) and below the centre of gravity (Fig. 8.16). The con-

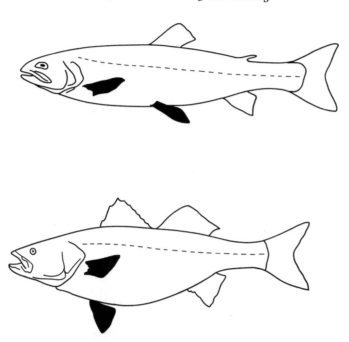

Fig. 8.16 Positions of the paired – pectoral and pelvic – fins used in braking and in making steering movements. Note that the pectoral fins are situated low on, and the pelvics well back on, the body of the salmonid fish (upper). In the percid fish (lower) the pectorals are high on the body close to the centre of gravity of the fish and the pelvic fins are far forward.

tribution of the pelvic fins to braking is relatively slight and their main function is to counteract the pitching movements produced by the pectoral fins.

If the pectoral fins occur higher on the body, as in the acanthopterygians, the forces produced during braking will act directly through the centre of gravity. This improves braking efficiency and also reduces, or eliminates, the tendency to pitch forward. Thus the pelvic fins, released from their role of pitch controllers, can be placed well forward on the body and can begin to function as brakes in combination with the pectoral fins (Fig. 8.16). However, they still function to counteract the lift forces produced by the pectoral fins.

That the pelvic fins have this function was demonstrated in a series of experiments conducted in the USA during the 1930s. It was found that, after the pelvic fins of sunfish, *Lepomis* spp., were removed, the use of the pectoral fins as brakes caused the fish not only to lose speed but also to soar upwards in the water column. There was, however, no pitching movement. The observations showed that the outspread pectoral fins exerted both a drag and a lift, the lift forces being generated due to the fact that the plane of the pectorals is inclined slightly forward from the vertical plane of the fish. The pelvic fins function to counteract this lift, and during braking these fins are held out from the body in such a manner that the trailing edges are held higher than the leading edges. This plane of inclination is, for example, opposite to that of the pectoral fins of a shark, and hence the pelvic fins of these teleosts produce both a drag and a downward force.

The unpaired fins (dorsal, anal and caudal fins) may also be used in braking. For example, during braking in the Atlantic cod, the caudal region is not held straight, but is bent into an S shape. In addition, the unpaired median fins are erected, and this provides increased resistance to forward movement.

Steering

Flexibility of the paired fins, particularly the pectorals, assists not only in improving braking efficiency but also gives the possibility of improved steering and better control over changes of direction. For example, when a shark changes direction little contribution is made by the paired fins. The direction change is achieved by increased contraction of the muscles on one side of the body so that the head is dragged round to face in the new direction. On the other hand, the paired pectoral fins make a considerable contribution to directional changes in the acanthopterygians, such as the percids.

When changing direction, sharks and some bony fish employ a type of controlled yawing. In order for this to be effective it is clearly important that the resistance to turning of the head be less than the resistance to turning imposed by the rest of the body. In the bony fish the median unpaired fins in the tail region can be used to advantage in the control of steering movements because they can be raised to give further resistance to turning of the posterior part of the body. Nevertheless, the distance between the head and the centre of gravity should be relatively short whereas that between the tail and centre of gravity should be greater. This is a general type of body plan shared both by the majority of the sharks and by those bony fish possessing the largest numbers of conservative morphological and anatomical features (Figs 8.13 and 8.16). Changes of direction can be achieved by controlled yawing, but this method has the disadvantage that the fish require quite a large 'turning circle' in order to complete the manoeuvre.

The majority of the acanthopterygians use a more effective method when changing direction. One of the pectoral fins is spread out and functions as a fulcrum around which the body turns. When this method is employed it is clearly an advantage to have the fulcrum (pectoral fin) placed as close to the centre of gravity as possible (Fig. 8.16). Also the resistance to turning of the head and tail should be more or less equal. Thus, the caudal region should be relatively short. From this it follows that the best steering and turning abilities are shown by fish that have relatively short bodies, and that have pectoral fins located high on the body close to the centre of gravity.

In summary, attainment of neutral buoyancy by teleosts has allowed the caudal fin to develop into an organ producing only forward thrust. The pectoral fins are no longer locked into the role of producing lift and have developed in other directions. Pectoral fin development in the teleosts has given rise to improvements in braking (accompanied by a change in the position of the pelvic fins) and steering (along with changes in body form), and has also increased overall manoeuvrability when the pectoral fins are used in slow swimming.

8.13 CONCLUDING COMMENTS

When we think of swimming fish there is a natural tendency to think of darting silvery schools or the impressive performances of salmonids on their migration routes, but this may not be a realistic picture of the role of swimming in the daily life of fish. A number of experiments have been carried out in which various fish species, such as brown trout, pike, *Esox lucius*, and Atlantic cod, have been tagged with sonar tags, re-released and their swimming movements followed. The results from most of these studies suggest that the majority of fish are relatively sedentary and use comparatively little time and energy engaged in active swimming. The majority of the time is spent either resting or in slow swimming, interspersed with short bursts of rapid swimming.

Very slow swimming speeds, and station holding in mid-water, may be achieved by using only the paired – pectoral and pelvic – and unpaired median fins, but higher swimming speeds will require trunk and caudal propulsion. When fish are swimming slowly using caudal propulsion it is only the slow red muscle fibres that are active, but as swimming speed increases the pink, intermediate, and fast white fibres are recruited. In salmonids, for example, the white fibres may be recruited at swimming speeds of $1.5-2$ BL s^{-1}. When swimming at intermediate velocities the fish may adopt one of two strategies. At low speeds most fish species tend to swim steadily at relatively constant speed, but at speeds of $2-2.5$ BL s^{-1} or over they may display swimming of a kick-and-glide (also called burst-

Table 8.4 The proportions of red, slow muscle fibres in cross sections of the caudal regions of various fish species. Note that the muscle of the active, pelagic species contains more red fibres than that of the more sedentary species. Even in the most active species the proportion of red fibres does not, however, exceed 25–30% of the cross-sectional area of the total muscle mass

Species	Habitat	Percentage red fibres in caudal region
Spanish mackerel, *Scomber colias*	Pelagic, continuous swimming	30
Blue shark, *Prionace glauca*	Pelagic, continuous swimming	22
Saithe, *Pollachius virens*	Midwater	11
Pollack, *Pollachius pollachius*	Midwater or bottom-living	3
Rat-tail, *Chimaera monstrosa*	Deep-water	0.6

and-coast) type. Kick-and-glide swimming consists of a short initial acceleration phase followed by a glide phase in which the body is held relatively straight, rigid and motionless. In this type of swimming, energy is used primarily during the acceleration phase. It has been suggested that kick-and-glide swimming can give an energy saving of up to 40–50% when compared with the energy used in covering the same distance, at the same average speed, using continuous steady swimming.

The requirements for optimum performance during steady swimming and burst performance (fast-start swimming) are somewhat different. Power for steady swimming is provided by the red muscle. An examination of the tail regions of various fish species shows that the pelagic species that swim more or less continuously have a higher proportion of red muscle fibres in their caudal region myotomes than do other species (Table 8.4).

Body design criteria also differ, steady swimming being optimized by having the streamlined, or fusiform, body form seen in the pelagic species, whereas fast-start performance is improved by having a deep body profile and large tail-fin area. It is, therefore, easy to see why fish body forms deviate from the optimum predicted for efficient steady swimming. The most generalized fish body form, of a relatively fusiform body and large caudal fin, offers a compromise between reasonable steady swimming ability and good fast-start performance.

FURTHER READING

Books

Blake, R.E. (1983) *Fish Locomotion,* Cambridge Univ. Press, Cambridge.

Day, M.H. (ed)(1981) *Vertebrate Locomotion,* Academic Press, London.

Hoar, W.S. and Randall, D.J. (eds)(1978) *Fish Physiology, Vol. VII,* Academic Press, London.

Hochachka, P.W. and Mommsen, T.P. (eds)(1991) *Biochemistry and Molecular Biology of Fishes: Phylogenetic and biochemical perspectives,* Elsevier, Amsterdam.

Love, R.M. (1988) *The Food Fishes,* Farrand Press, London.

Videler, J.J. (1993) *Fish Swimming,* Chapman & Hall, London.

Review articles and papers

Alexander, R. McN. (1990) Size, speed and buoyancy adaptations in aquatic animals. *Amer. Zool.,* **30,** 189–96.

Fänge, R. (1983) Gas exchange in fish swim bladder. *Rev. Physiol. Biochem. Pharmacol.,* **97,** 111–58.

Johnston, I.A. (1983) On the design of fish myotomal muscles. *Mar. Behav. Physiol.,* **9,** 83–98.

Pelster, B. and Scheid, P. (1992) Countercurrent concentration and gas secretion in the fish swim bladder. *Physiol. Zool.,* **65,** 1–16.

Peters, S.E. (1989) Structure and function of vertebrate skeletal muscle. *Amer. Zool.,* **29,** 221–34.

Sänger, A.M. (1993) Limits to the acclimation of fish muscle. *Rev. Fish Biol. Fisheries,* **3,** 1–15.

Webb, P.W. (1984) Form and function in fish swimming. *Scient. Amer.,* **251,** 58–68.

Chapter nine

Reproduction

9.1 INTRODUCTION

The array of reproductive arrangements found in fish is extraordinarily diverse. Some species are represented only by females, and in other species some fish undergo a sex change during the course of their lives. The fact that sex change occurs naturally in a number of species points to a sexual lability in fish that is not seen in other vertebrate groups.

In most species the sexes are separate and the fish do not normally undergo a sex change, but even here sex may not be irrevocably fixed by the genetic constitution of the individual. In a number of these species it is possible to induce sex change by manipulations of the environment or via exogenous hormonal treatment at certain stages of development.

Reproductive modes differ amongst species, as do the means by which fish find and attract mates, secure breeding sites and care for their eggs and offspring. For example, eggs may be spawned directly into the water column and allowed to drift freely with the currents, they may be deposited on the substrate, or in nests and guarded by the parents, or they may be given added protection by being retained within the body of the female.

Irrespective of the mode adopted, reproductive events will be seen to be cyclic in nature. The cycle will be divisible into distinct phases related to gamete formation, development and maturation, and this culminates in spawning and fertilization of the eggs. Many fish species have distinct breeding seasons, and spawning tends to be confined to a few weeks of the year. The time at which the fish spawn varies little from year to year, which means that the events of the reproductive cycle are locked to, and synchronized by, the environmental changes that occur on a seasonal basis.

9.2 REPRODUCTIVE DIVERSITY

Parthenogenesis, more properly called *gynogenesis*, refers to the reproductive condition when the young develop without there having been fertilization of the egg. Reproduction of this type is rare in fish, but occurs in the live-bearing Amazon molly, *Poecilia formosa*. This fish, of which only female forms are known, is thought to have arisen through hybridization of two closely related *Poecilia* species. The female *P. formosa* are courted by, and mate with, male fish of a related species, it usually being young and in-experienced males that court these females. On completion of the act of mating the sperm serves only to initiate the development of the egg, and gametic fusion does not occur. Therefore, the genetic make-up of the young is wholly dependent upon the genotype of the mother. The offspring are invariably daughters that are genetically identical to their mother.

A number of fish species are *hermaphroditic*, and hermaphroditism is quite common amongst coral-reef fish species and those living in the deep sea. *Synchronous*, or simultaneous, hermaphroditism, refers to the condi-tion in which both parts of the ovo-testis mature at the same time. This gives rise to a fish which is functionally both male and female. *Successive* hermaphroditism, on the other hand, involves the fish changing sex at some stage during its life.

Synchronous hermaphroditism

In species where there is synchronous hermaphroditism the meeting of two individuals could lead to the fertilization of two batches of eggs. This might be of advantage in environments where fish are widely scattered and reproductive contacts are few, or where the capacity of an individual to produce eggs is limited by other factors. Synchronous hermaphroditism also gives the possibility of self-fertilization. This could be advantageous in an environment where contacts between reproductive individuals are few. Whilst this argument may be invoked to explain the development of syn-chronous hermaphroditism in deep-sea environments it does not seem to be a valid explanation for synchronous hermaphroditism in coral-reef fish.

On coral reefs fish may be closely aggregated so there would not appear to be any great difficulty in finding a mate. Thus, reproduction would not be envisaged as being constrained by limited numbers of spawning oppor-tunities. On the contrary, some of the hermaphroditic species of coral reefs seem to form monogamous pairs, thereby ensuring regular access to a mate. This occurs in the hamlet, *Hypoplectrus unicolor*, a synchronous her-maphroditic species found widely distributed on reefs throughout the tropical western Atlantic Ocean. Fish of this species usually form pair bonds with a single partner.

The reproductive behaviour of the hamlet is complex, and a pair may spawn several times during the course of the evening. Courtship and spawning activities usually take place for an hour or so in the period around sunset. Each individual spawns alternatively as male and female.

Shortly prior to spawning the fish ready to assume the male role gives a courtship call, and the fish assuming the female role responds with a different series of sounds. This is accompanied by behavioural displays which eventually result in the fish becoming aligned in the spawning position. This is followed by the release of the eggs, during which the female member of the partnership continues to emit sound. Following fertilization of the eggs the spawning pair break apart and take cover on the reef. In the next spawning event the roles of the two fish are reversed, with the 'original' male acting as the female, and the 'original' female taking over the role of the male.

Successive hermaphroditism

Successive hermaphroditism, in which fish undergo sex reversal, has developed in several species. The form, or direction, of the sex change varies among hermaphroditic species, with both *protandrous* – males change sex to become females – and *protogynous* – females change sex to become males – species being known. Some protogynous species may have a certain proportion of fish born as males, and these males are known as primary males. The males that arise due to a sex change undertaken by a female are termed secondary males. A species that has both primary and secondary males is *diandric*, whereas a species with only secondary males is said to be *monandric*.

Protandry

Since the egg is much larger than the sperm a large body size would be advantageous for the production of eggs. Under reproductive conditions and mating systems where male size is unimportant it might be expected that females would be larger than males. Taking this to an extreme, it could be argued that it would be advantageous for an individual to reproduce as a male when small and then change sex to produce eggs and reproduce as a female as body size increased. Thus, protandry would be favoured.

The deep-sea fish *Gonostoma gracile* is protandric. In this species the males are not only much smaller than the females but they also change sex from male to female after their first breeding season. This sex change from male to female possibly represents an adaptation that can lead to increased fecundity (numbers of eggs produced) in relatively unproductive parts of the ocean.

Protogyny

In some mating systems the males are either territorial, or compete, by some form of physical combat, for the right to court and mate with the females. It is usually the largest males that secure the favours of the females, and mate, so under these circumstances a large male body size is advantageous. Taken to an extreme, the argument would be that such a mating system would favour the development of a pattern of reproduction in which fish first reproduced as females and then changed sex to become males as they increased in size. In other words, protogyny would be favoured.

Many protogynous species have a harem system of mating. This is observed in a number of tropical reef species, such as the pygmy angelfish, *Centropyge potteri*, of the Hawaiian islands. Under the harem system, the sex change of the females may be under some form of social control.

A typical harem will consist of a dominant male and a small number of females with territories or home ranges overlapping with those of the male. If the male dies, or is removed, one of the females, usually the largest, will start to change sex within the space of a few days. On completion of sex-reversal this fish will replace the dominant male.

On some occasions, especially if there are two females of approximately equal size, more than one female will start to change sex. Only one of the females completes sex reversal, sex change in the other being eventually suppressed and regressed. Alternatively, the death or removal of the dominant male from his territory may lead to the arrival of a new male, or to the females being absorbed into the harems of males occupying adjacent territories. In neither of these two cases will any of the females begin to change sex. In other words, the initiation of sex change is suppressed by the presence of the new dominant male fish.

Diandric species

Amongst the parrotfish (Scaridae) and wrasses (Labridae) the reproductive arrangements may be very complex. The population may consist of juveniles (immatures), gonochoristic females (females incapable of changing sex), hermaphroditic females (females that will later change sex), primary males (gonochoristic males incapable of changing sex) and secondary males (derived from sex-reversed females).

In some labrids, the relative proportions of fish within each category appear to be related to the numbers of fish within the breeding group. On small reefs supporting a small breeding population, large secondary males will have a greater mating success with the limited number of females, than will smaller, duller primary males. This arises because the small primary males are incapable of defending a mating territory. On the other hand, on larger reefs supporting large breeding populations, a few large secondary males may be very successful and mate many times a day. The majority of secondary males will not be able to establish a mating territory and will, therefore, probably not be any more successful than the primary males. Both of these types of non-territorial males gather at the fringes of the territories of the successful males and attempt to mate with females attracted to the territory. Thus, secondary males are proportionately commoner and most effective in securing fertilization of eggs in small breeding groups, whereas primary males tend to be more frequent in larger breeding groups.

Gonochoristic species

By far and away the majority of fish species are dioecious and gonochoristic. That is, the eggs and sperm develop in separate female and male sexes, and there is no sex change. Within these bisexual species, however, there are a range of different reproductive patterns, with both group spawning and close pair-bond reproductive behaviours being common. In addition, within a given species, mature individuals may not all display the same types of reproductive behaviours.

Pair mating and sneaking

The butterflyfish (Chaetodontidae) of tropical coral reefs are usually considered to have a reproductive pattern based upon long-term pair bonding. These fish usually mate with a single partner and maintain a monogamous relationship both for reproduction and for the defence of a feeding area. Most species of butterflyfish do not exhibit sexually dimorphic colour patterns, but the abdomen of the female becomes swollen with eggs in the few hours prior to spawning.

In courtship and spawning the female usually swims slightly in front of the male and then adopts a head-down posture immediately prior to the release of the eggs. At this point the male moves alongside the female, and both fish quiver as the eggs and sperm are released simultaneously into the water.

In some cases solitary individuals may follow the courting pair, and these intruders are invariably male fish that attempt to steal fertilizations from the pair-bonded male. Thus, at the culmination of the spawning act the intruder male, or males, will rush in and shed sperm over the eggs released by the female. The number of intruding, or sneaker, solitary males that compete with the pair-bonded males for egg fertilization may range from as few as one to as many as six or seven.

This type of sneaking behaviour is observed amongst males of several fish species. The mature males that perform sneaking often have a different body form, or size, from those males which participate in pair-mating behaviour. Thus, the mature males tend to show size dimorphism, and there may also be differences in colour patterns between the sneakers and the large pair-mating males.

The males that engage in pair mating are usually large individuals. They may either establish distinct breeding territories or actively compete for the right to mate with the females. The sneakers, on the other hand, are usually small, or 'dwarf', males that are incapable of establishing a breeding territory in competition with larger conspecifics. These small, mature males tend, therefore, to congregate around the territories of the larger males.

In salmonids, for example, large males develop distinct morphological characters and breeding colours at the start of the reproductive season. These males compete actively with each other for the favours of the females. Competition often results in physical combat between males. Other males mature at a relatively small body size, and they do not develop the same characters as the large males. The small mature males tend either to retain juvenile coloration, or to develop a darkening of the skin. The small males remain on the periphery during the initial stages of courtship, but at the culmination of the spawning act they dash in and attempt to interpose themselves between the large male and the female. By engaging in this sneaking behaviour they attempt to ensure that their sperm will fertilize at least a portion of the eggs released by the female.

Pair mating, sneaking and cuckoldry

Several alternatives to pair-mating behaviour are seen in some centrarchid species. In the bluegill sunfish, *Lepomis macrochirus*, the males may display one of three distinct mating behaviours – parental, sneaker and satellite (Fig. 9.1). Parental males are larger than the males displaying the other types of mating behaviour. Within the bluegill system the activities of the small mature males not only involve the 'stealing' of fertilizations from the large male, but also result in 'cuckoldry' because the large males build nests, guard the eggs and engage in subsequent parental care of the young.

The large parental males build nests in the littoral zone of ponds or lakes, and the females are attracted to these sites for spawning. After spawning, the males remain at the nest site and provide parental care. Sneaker males are small, young males that congregate around the fringes of the nesting area of the parental male, usually remaining concealed amongst underwater vegetation. The sneakers fertilize eggs by darting into the nest and releasing sperm during the culmination of spawning between a parental male and a female.

The satellite males are intermediate in size between the sneakers and the parental males. The satellite males have a body coloration that closely resembles that of the females, i.e. satellite males can be considered as being

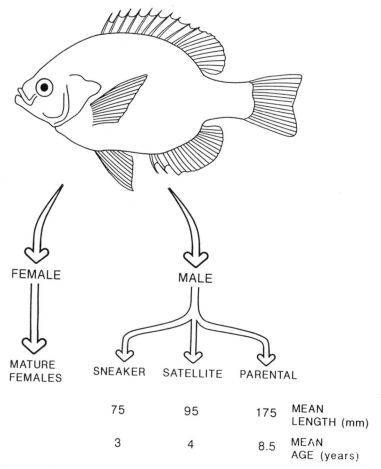

Fig. 9.1 Reproductive development in the bluegill sunfish, *Lepomis macrochirus*, illustrating different developmental patterns in the male.

female mimics. Satellite males do not remain concealed, but actively engage in the courtship and spawning ritual, releasing sperm to fertilize the eggs whilst they 'pair' simultaneously with the female and the parental male.

The large parental males attempt to defend their paternity by chasing any small males detected close to the nesting site. Nevertheless, the specialized reproductive behaviours displayed by the sneakers and satellites ensures that they are successful in fertilizing a certain proportion of the total numbers of eggs produced.

Thus, in many fish species there may be alternative male reproductive behaviours. Males adopting these alternative behaviours both exist in the same environments as, and compete with, males that show the normal type of reproductive behaviour. Whilst some of the alternative behaviours are facultative, i.e. can be changed according to circumstances, others may be associated with distinct and discrete life histories such as early, or precocial, maturation and the development of a sneaker type of reproductive behaviour.

9.3   REPRODUCTIVE SYSTEM – GENERAL ANATOMY

Fish, like all other vertebrates, reproduce sexually, and in the majority of species, but not all, the eggs and sperm are formed in separate individuals. Whilst simultaneous hermaphroditism and sex change do occur in an array of fish species, the discussion of the reproductive system given here will be confined to a general description of the anatomy and developmental events occurring in dioecious, gonochoristic species.

The *gonads* of all vertebrates arise as paired structures in the dorsal lining of the body cavity, and the development of the gonads is intimately associated with that of the kidneys and their ducts. Consequently, it is not unusual to refer to the series of developmental events as leading to the differentiation of the urogenital system. Whilst the gonads originate as a pair, the adults of many fish species possess only a single gonad. This may arise either via a fusion during the course of development, or because one of the gonads fails to develop. In other species two gonads may be present, but one is small, rudimentary and non-functional. There also exist several variations in which partial fusion of the gonads occurs. This may involve a fusion of the posterior region of the gonads, or may be restricted to the gonadal ducts.

The *gonoducts* develop along with those of the nephric system, and the gonoducts are probably derived from a series of embryonic renal tubules in the mesonephros. Evidence for this origin of the gonoducts is seen in the males of the chondrosteans (sturgeons and paddlefish) in which some of the renal tubules appear to form vasa efferentia leading from the testis to the mesonephric duct. In the chondrichthyans the gonoducts draining the testis are thought to have been derived from a group of mesonephric tubules, but, unlike the situation in the sturgeons, the gonoducts cease to have any connection with the ducts draining from the excretory system. In the teleosts there is no connection between the mesonephros and the testis in mature males, and the vas deferens draining the testis is completely separated from the mesonephric duct draining the kidney (Fig. 9.2). In female chondrosteans and chondrichthyans the ova are released from the ovary into the body (peritoneal) cavity and pass to the outside through *oviducts*, or Müllerian ducts. The oviducts run from an anterior collecting funnel to the cloaca (Fig. 9.3). As in the male the gonoducts are thought to be derived from mesonephric tubules and ducts. In the teleosts, on the other hand, the oviducts are formed from posterior outgrowths of the tissues that encircle the ovary, rather than from the Müllerian ducts. Thus, in many teleosts there is no release of ova into the body cavity before they enter the oviduct since the oviducts are continuous with the ovaries. In some teleosts, however, there is either partial or total degeneration of the oviducts so that the ova are released from the ovary into the body cavity before being passed to the exterior. Degeneration of the gonoducts is seen, for example, in the salmonids, anguillids and some cyprinids.

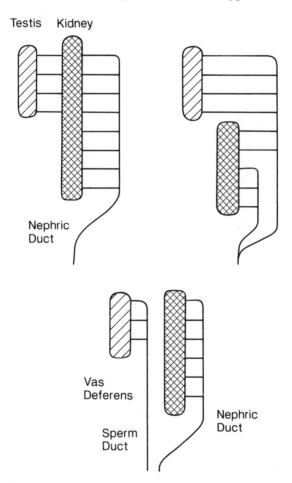

Fig. 9.2 Schematic diagrams showing the structural organization of the male reproductive system in a sturgeon (Chondrostei)(upper left), an elasmobranch (upper right) and a teleost (lower).

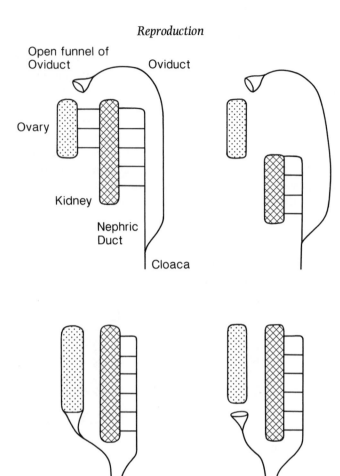

Fig. 9.3 Schematic diagrams showing the structural organization of the female reproductive system in a sturgeon (Chondrostei)(upper left), an elasmobranch (upper
right) and in teleost fish (lower left – no discharge of ova into the peritoneal cavity;
lower right – ova released into the peritoneal cavity before passing to the external
environment).

9.4   MALE REPRODUCTIVE SYSTEM

The *testes* originate as paired structures within the general body cavity. In
the cartilaginous fish they are usually elongated ovoid structures located
anteriorly in the body cavity where they are suspended dorsally by means
of a mesorchium. In most of the bony fish the testes are whitish or cream-
coloured, elongated, lobulate organs attached to the dorsal body wall.
A *vas deferens*, or sperm duct, arises from the posterior region of each

elongated testis and leads to the urinary papilla which is located between the rectum and the urinary ducts.

Testis and spermatogenesis

Spermatozoa, or sperm, are formed from the germ cells via a series of developmental changes known collectively as spermatogenesis (Fig. 9.4). Spermatogenesis involves an initial proliferation of spermatogonia through repeated mitotic divisions, and growth to form primary spermatocytes. Primary spermatocytes then undergo reduction division (meiosis) to form secondary spermatocytes. Division of the secondary spermatocytes results in the production of spermatids, which then undergo metamorphosis to the motile spermatozoa.

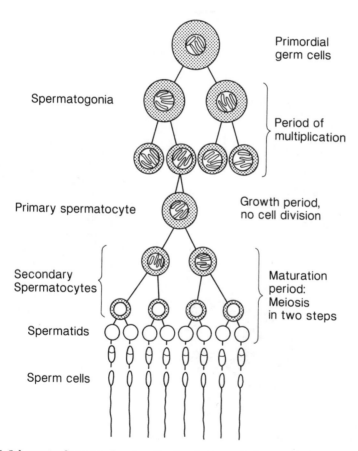

Fig. 9.4 Schematic diagram showing the developmental changes that occur during spermatogenesis.

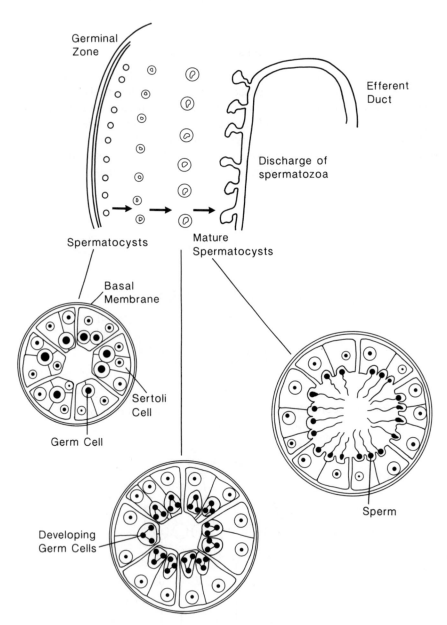

Fig. 9.5 Spermatocysts and the developmental changes that occur in the elasmobranch testis during spermatogenesis.

Elasmobranch testis

The basic structural and functional unit in the testis of the elasmobranchs is the spermatocyst. Spermatocysts consist of clones of germ cells, associated with Sertoli cells. It is the germ cells that ultimately develop into spermatozoa. The spermatocyst forms a closed spherical unit bounded by a basement membrane, and all the germ cells within a given spermatocyst develop synchronously i.e. spermatocysts contain germ cells in the same stage of differentiation.

Spermatocysts arise in the germinal zone of the testis. Groups, or bands, of spermatocysts at the same stage of development extend towards the region where the mature spermatocysts discharge spermatozoa into a system of sperm ducts (Fig. 9.5).

Each spermatocyst may contain several hundred Sertoli cells, and each Sertoli cell may have several germ cell clones associated with it. Thus, each mature spermatocyst may produce and discharge 20–30 000 spermatozoa. Once the spermatocysts have discharged the mature spermatozoa into the sperm ducts they collapse and degenerate.

Sperm discharge in elasmobranchs

The sperm is discharged into a network of ducts that transverses the mesorchium. These ducts coalesce to form the vas deferens. The ducts may be highly coiled so that the sperm traverses a tortuous path from the testis to the point at which the vas deferens emerges into the seminal vesicle. The seminal vesicles, and their associated sperm sacs, open into the urinogenital sinus, which, in turn, discharges into the cloaca (Fig. 9.6).

From the cloaca the sperm passes to the grooves of the claspers. The claspers are extensions of the pelvis fins, and function as intromittent organs via which sperm is transferred to the female. Sperm transfer is assisted by the secretion of fluid from the sperm, or siphon, sacs. The sperm sacs are at a location in close association with the claspers. In many species the fluid from these sacs acts to flush the sperm along, and from, the clasper grooves, thereby ensuring effective sperm transfer from the male to the female.

Teleost testis

The testicular structure of teleosts is somewhat variable, but two basic types can be distinguished – the lobular and the tubular. In the testis of the lobular type, which is seen in most teleosts, there are numerous lobules separated from each other by thin layers of connective tissue (Fig. 9.7). The connective tissue, which extends from the testicular capsule,

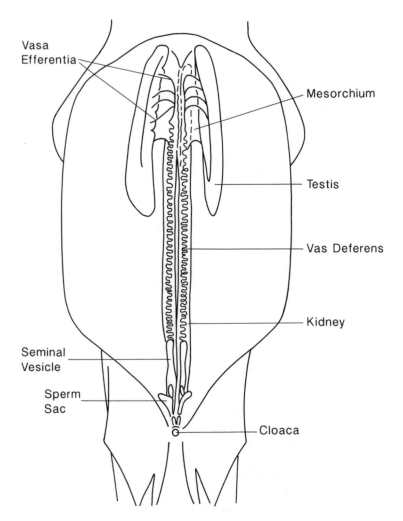

Fig. 9.6 Generalized structure of the major organs of the reproductive system in a male elasmobranch.

effectively divides the testis into lobules, and it is around the perimeter of the lobules that the germinal epithelium develops. This epithelium consists of a layer of Sertoli cells with germ cells embedded within them.

During the course of spermatogenesis the primary spermatogonia undergo numerous mitotic divisions, resulting in the production of cysts surrounded by Sertoli cells. Each lobule of the testis contains numerous cysts. Within each cyst the groups of germ cells divide in synchrony, so all

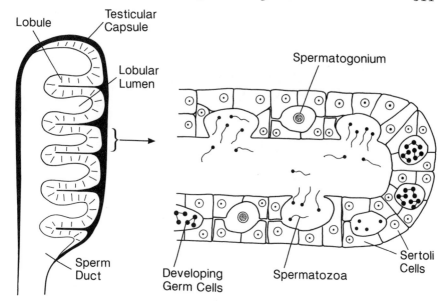

Fig. 9.7 Generalized structure of the lobular testis seen in the majority of teleost species.

of these cells are at the same stage of development at any given time. However, within the lobules the spermatogenetic process need not be synchronous, and germ cells at different stages of development and differentiation may be found within a single lobule (Fig. 9.7).

As spermatogenesis proceeds, the cysts increase in size and they eventually rupture to release the spermatozoa into the lobular lumen. As progressively more cysts reach maturity and rupture, the lumen becomes filled with spermatozoa. Whilst it is usual that it is spermatozoa that are released from the cysts this is not universal.

In some species, rupture of the cysts may occur at an earlier developmental stage. Thus, spermatids, rather than spermatozoa, may be released from the cysts. Differentiation and maturation of the spermatids to spermatozoa then takes place within the lumen of the testicular lobule.

The lobular lumen is continuous with the sperm duct. Following a variable period within the lumen, the sperm are released into the duct. This release of sperm from the lobule into the sperm duct is known as spermiation.

The second type of testicular structure found in teleosts is restricted to relatively few species, such as the guppy, *Poecilia reticulata*. In the tubular testis the primary spermatogonia are only found in cysts located at the blind end of the tubules. As spermatogenesis proceeds the cysts containing

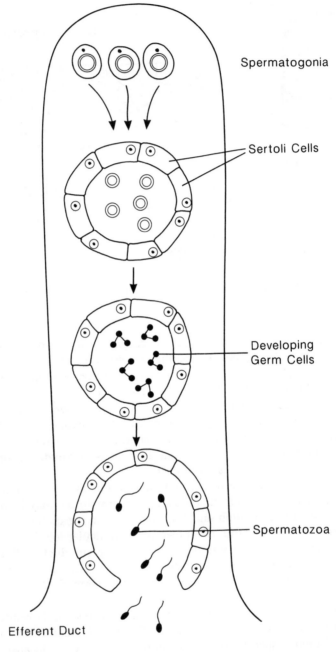

Fig. 9.8 Generalized structure of the teleostean tubular testis.

the germ cells gradually migrate towards the sperm duct. In this type of testis there is no structure corresponding to the lobular lumen (Fig. 9.8).

9.5   FEMALE REPRODUCTIVE SYSTEM

The female *ovaries* are paired, but in some species one of the ovaries may be greatly reduced in size and become non-functional. In the elasmobranchs the ovaries are located anteriorly in the body cavity, each ovary being suspended from the dorsal wall of the body cavity by mesenteric tissue known as the mesovarium. Ova are released into the body cavity from the ovary and then enter the oviduct via a funnel-like structure (Fig. 9.3). The urogenital system of the female holocephalans differs from that of the other cartilaginous fish in that the two oviducts do not fuse to form a common opening to the exterior, but open separately without having any connection with each other or the urinary system.   The reproductive system of the female teleosts is highly variable, reflecting the wide range of reproductive patterns seen in this group of fish. In many teleosts the ovaries are sac-like, hollow paired structures that are continuous with the oviduct, forming a closed system (Fig. 9.3). In some species the paired ovaries become fused early in development to form a single solid structure, and in others there is no direct connection between the ovary and oviduct.

Ovarian follicle

The ovary consists of germ cells, oogonia and oocytes, and their surrounding follicle cells and supportive tissues. The ovarian follicles arise from the germinal epithelium as each germ cell becomes surrounded by a layer of follicle cells early in development.

With growth the follicle cells multiply and form a continuous layer known as the *granulosa* cell layer. At the same time the outer layers of connective tissues become organized to form a follicular envelope, or *thecal* cell layer. Thus, the developing oocyte becomes surrounded by two major cell layers, the inner granulosa layer and the outer thecal layer. The granulosa and thecal cell layers are separated from each other by a distinct basement membrane (Fig. 9.9).

Oogenesis

The earliest stages of the development of the female germ cells are similar to those seen in the initial stages of spermatogenesis (Fig. 9.10). The germ cells first undergo proliferation via a series of mitotic divisions and then they develop to become primary oocytes. There is a period of growth, and a meiotic, reduction division. The final reduction division, with expulsion

Ovarian Follicle

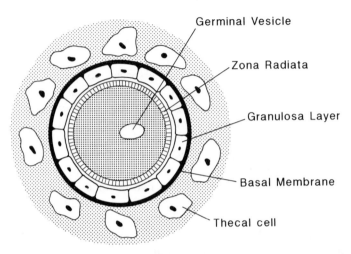

Germinal Vesicle

Zona Radiata

Granulosa Layer

Basal Membrane

Thecal cell

Fig. 9.9 Schematic diagram of the ovarian follicle showing the division into the thecal and granulosa layers.

of the second polar body, does not occur until after fertilization, when the egg has been activated by the sperm.

The growth of the oocytes varies from species to species, with oocyte enlargement being caused largely by the accumulation of yolk. Yolk accumulation occurs during the developmental period known as vitellogenesis. The oocyte enlargement may be considerable, with, for example, the oocytes of salmonids increasing in diameter from 20–50 μm to 4–5 mm during the course of development.

Oocyte development in teleosts

Most teleosts have distinct breeding cycles, with spawning often being limited to particular times of the year. Consequently, the ovaries may vary greatly in appearance depending upon the stage of the breeding cycle. For example, in immature or post-spawned females the ovaries and oviducts may resemble thin strands, having a thread- or string-like appearance. On the other hand, in the period immediately prior to spawning the ovaries may be so large and expanded with oocytes that they almost completely fill the body cavity of the female.

The fact that the teleosts display a range of breeding cycles leads to there being distinct patterns of oocyte development, and three broad categories have been described. For example, all oocytes present within the

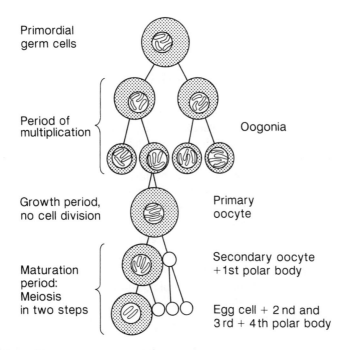

Primordial
germ cells

Period of
multiplication

Oogonia

Growth period,
no cell division

Primary
oocyte

Maturation
period:
Meiosis
in two steps

Secondary oocyte
+1st polar body

Egg cell + 2 nd and
3 rd + 4 th polar body

Fig. 9.10 Schematic diagram showing the developmental changes that occur during oogenesis. In eggs the second meiotic division is postponed until after fertilization. Thus, each mature oocyte (egg or ovum) contains two maternal chromosome sets at the time of spawning.

ovary may be at the same stage of development. This form of synchronous oocyte development is seen in teleosts which spawn only once and then die, i.e. *semelparous* species such as some salmonids and anguillids.

The majority of teleosts spawn more than once during the course of their lives, i.e. are *iteroparous*. Amongst the iteroparous species two general patterns of oocyte development have been identified.

The ovary may contain at least two distinct 'populations' of oocytes at different developmental stages. Oocytes of a given population develop synchronously. This type of developmental pattern is seen in the ovaries of teleosts that spawn once per year within the space of a relatively short breeding season. Thus, the ovary will always contain a reservoir of germ cells at an early stage of development. The oocytes for successive annual spawnings are recruited from this reservoir. The population or populations of oocytes recruited for spawning during the course of a given breeding season will display synchronous, or almost synchronous, development.

In some teleosts the ovaries may contain oocytes at all stages of development, i.e. there is asynchronous oocyte development. This pattern is usually seen in species which spawn many times during the course of a prolonged breeding season.

9.6 SEX DETERMINATION

The reproductive systems and patterns of sexual behaviour displayed by fish are extremely diverse. There are also a number of different genetic systems, and sex determination mechanisms.

As with other animals the genetic material of the fish is located within the cell nucleus in the form of chromosomes. The number of chromosomes varies from species to species, but the chromosome number almost invariably remains constant within a species. In the vast majority of fish the chromosomes occur in pairs – the *diploid* (2N) condition. One chromosome of each pair derives from the male parent and the other from the female. There are some exceptions to the rule that fish are diploid. Some fish, e.g. salmonids and catostomids, may have evolved via tetraploidy, which involved a doubling of the original diploid number (2N) of chromosomes to the *tetraploid* (4N) condition. Some fish have become *triploids* (3N), but the occurrence of natural populations of triploid fish appears to be extremely rare.

There are two basic types of chromosomes, these being named *autosomes* and *sex chromosomes*. Sex chromosomes are those that determine the sex of an individual, and the sex chromosomes are often morphologically different in the two sexes. Autosomes are the other pairs of chromosomes, and these are morphologically the same in males and females.

Sex determination systems have been studied in relatively few species of fish, but nevertheless there are known to be at least nine different systems. Sex is controlled by sex chromosomes in eight of the systems. The sex chromosomes of many species can be identified on the basis of distinct morphology, but this is not always the case.

Sex chromosomes

The most common system of sex determination in fish, as in other vertebrates, is known as the XY system. This is, for example, the way in which sex is controlled in man. In this system individuals that are homogametic (XX) are female, whilst heterogametic (XY) individuals are males.

In a second system, the male is homogametic (ZZ) and the female heterogametic (WZ). This system is known as the WZ system. The designations XY and WZ systems are used in order to prevent the possibility of confusion between the two different systems in which one sex is homogametic and the other heterogametic.

Multiple sex chromosomes

There are a number of sex determination systems in which there are multiple sex chromosomes. One of these systems has multiple X chromosomes, and the female fish are $X_1X_1X_2X_2$, with the males being X_1X_2Y. In another system there are multiple W chromosomes, and here the males are ZZ, whilst the females are ZW_1W_2. In a third system involving multiple sex chromosomes there are multiple Y chromosomes. Males of species showing this type of sex determination are XY_1Y_2 and females are XX.

The number of chromosomes is not constant within species having sex determination systems of any of these types. In the first two cases the female fish have one extra chromosome, and in the third of these systems the males have one more chromosome than the females.

Single sex chromosome

There are two systems of sex determination in which there is only one sex chromosome: the XO and ZO systems. The symbol O is used to denote the absence of a second sex chromosome. In the XO system the female fish are homogametic XX, whilst the males are XO. In the ZO system the males are ZZ and the females ZO.

As with the sex determination systems involving multiple sex chromosomes, the number of chromosomes in species with XO and ZO sex determination systems is not constant. In species with the XO system the females have one more chromosome than the males, whereas in those species with the ZO system it is the males that have the extra chromosome.

WXY system

The WXY system of sex determination is a rather complex one in which the Y chromosome produces males, except when it is paired with a W chromosome. The W chromosome is a modified X chromosome that inhibits, or blocks, the male-determining action of the Y chromosome. Thus, in species with the WXY system of sex determination individuals that are either XY or YY will be males. Fish that are XX, WX or WY will be female.

Autosomal sex determination

Whilst sex determination in fish that have sex chromosomes is usually controlled by these chromosomes, this is not always the case. The sex of individuals of some species can also be influenced by a number of autosomal sex-modifying genes. This further complicates the process of identifying the exact mechanisms by which control over sex determination is exerted.

In some species of fish, sex determination is not controlled by sex chromosomes, but control is autosomal. In these species sex is determined by the number of male or female genes that are located on the autosomes.

Environmental influences on sex determination

Although sex determination can be considered to be largely under genetic control, there are many environmental factors that can play a role in the determination of sex in fish. Factors such as rearing temperature, photoperiod, salinity and social environment have been shown to exert an influence on sex determination and sex ratios in a range of fish species.

The fact that sex determination in fish is rather labile has been confirmed in studies in which sex reversal has been induced by the treatment of larvae or juveniles with sex hormones (either androgens or oestrogens). Gonadal differentiation can be affected by the administration of the steroid sex hormones during certain critical stages of development. Exogenous androgens given to genetic females direct the developing gonad to become a testis, and oestrogens given to the fry either in the food or applied directly via the holding water may cause genetic males to develop ovaries and become functional females as adults. Thus, in some fish species, both the androgenic and oestrogenic steroid hormones are capable of redirecting sexual development in a direction opposite to that of the genetic sex of the individual.

Temperature is also known to influence sexual differentiation in some fish species. For example, in the silverside, *Menidia menidia,* higher proportions of female fish are produced from eggs and larvae reared at temperatures within the range 10–20°C than from those incubated at higher temperature. The increased proportions of females hatching from eggs incubated at low temperature is possibly the result of the inhibition of some form of male-inducing factor at low temperature.

Sexual differentiation

In spite of a plasticity in sex determination, the patterns of sexual differentiation and gonadal development in fish may generally be similar to those described for mammals. In mammals the sex of an individual is determined by the chromosomal constitution established at the time of fertilization – XY individuals become male and XX individuals become female. The male testis-determining gene(s) on the Y chromosome directs the production of certain chemical factors which induce testicular organization of the embryonic gonadal tissue. In the absence of these chemical factors the gonadal tissue develops an ovarian organization. Once the un-differentiated gonad has become a testis or ovary, the subsequent development of the sexual phenotype is dependent upon the presence or absence of gonadal hormones. Female embryos, and castrated individuals of both sexes, develop female urogenital ducts. Thus, development proceeds in a female direction unless testicular secretions provide an overriding influence leading to a redirection towards the development of a male system. Similarly, in the absence of gonadal hormones at around the time of birth, the mammalian pituitary develops the female pattern of gonadotropin release, but if androgens are present during this critical period the individual follows a male developmental pattern.

Thus, in mammals, sexual development seems to be preset to move in the female direction unless it is diverted from doing so by masculinizing influences. There is some evidence to suggest that a similar general developmental scheme exists in a range of fish species, despite the fact that gonadal development is more influenced by exogenous factors than is the differentiation process in mammals. In several gonochoristic species, for example, the gonads appear to pass through a juvenile 'ovarian phase' before differentiating as an ovary or testis. A similar developmental pattern appears to be common amongst the protogynous species. On the other hand, the pattern of gonadal differentiation shown by protandric species would seem to be difficult to place within the simple developmental paradigm of the primacy of female sexual development. There is, however, reason to believe that, even amongst the protandric species the female, or ovarian, developmental scheme is the primary one. Thus, in the protandric species, development of male characters may be inserted into a predominantly female developmental pattern as a 'temporary' phase.

9.7 OFFSPRING NUMBERS AND PARENTAL CARE

Fish may either lay eggs, i.e. be oviparous, or give birth to live young. Those species which give birth to live young can be further divided into the truly viviparous species, in which the developing embryos are supplied with nutrients by the mother, and ovoviviparous species, in which there is egg retention but no additional nutrient supply from the mother.

Within this broad framework of the three basic modes of reproduction – oviparity, ovoviviparity and viviparity – there are large interspecific differences with respect to both the numbers of offspring produced, and the degree of protection and care given to the developing eggs and young by the parents. Within the oviparous species, for example, the variations may range from the production of large numbers of freely floating pelagic eggs at one extreme to nest building and extensive parental care at the other. For example, many marine species produce large numbers of small pelagic eggs, whereas stickleback species (Gasterosteidae) build nests and engage in elaborate courtship and parental behaviours.

By studying the relationships between fecundity (the number of eggs produced per female), egg size and degree of parental care it is possible to make some generalizations about reproductive patterns in fish (Tables 9.1 and 9.2). Thus, it can be said that the following general principles appear to apply.

1. Fecundity tends to be high where eggs and sperm are liberated freely into the water, is less in those species giving some form of protection to the eggs and is lowest in species that show parental care.
2. Fecundity and egg size are inversely related, that is, fish produce many small eggs or a few large ones.
3. The production of pelagic eggs is largely restricted to marine species. Freshwater species usually have non-buoyant, sticky eggs that are deposited on the substrate, or they build some type of nest.
4. Fecundity is higher in oviparous than in either ovoviviparous or viviparous species.

These generalizations refer to interspecific differences, but even within species there can be considerable variations in both fecundity and egg size (Table 9.1). There may, for example, be considerable intraspecific variability in egg numbers, both with respect to differences between females spawning within a given season, and with respect to the fecundity of a given female in consecutive years or breeding seasons. Variations in fecundity of up to 45% have been recorded for plaice, *Pleuronectes platessa*, and variations of about 25%, 34% and 56% have been found for pike, *Esox lucius*, herring, *Clupea harengus*, and haddock, *Melanogrammus aeglefinus*, respectively.

The intraspecific variations in egg size and fecundity can usually be traced either to differences in the sizes, or ages, of the females making up the spawning stock, or to year-to-year fluctuations in the food supply. For example, in many species of fish there is a clear relationship between fecundity and size of the female, larger females tending to produce more eggs than do smaller conspecifics.

Intraspecific variations in fecundity and egg size may also be related to the time of spawning. For example, it is frequently reported that fish spawning late in the season tend to produce smaller eggs than do early spawners.

Table 9.1 Relationships between fecundity, egg and larval size, and reproductive mode for a range of fish species

Species	Reproductive mode		Fecundity	Egg diameter (mm)	Size of larvae (mm)	(% adult length)
Atlantic cod, Gadus morhua	Oviparous	Marine pelagic	$2\text{–}9 \times 10^6$	1.1–1.9	3.5–4.5	0.3
Plaice, Pleuronectes platessa	Oviparous	Marine pelagic	$16\text{–}350 \times 10^3$	1.7–2.2	6–7	0.7
Atlantic mackerel, Scomber scombrus	Oviparous	Marine pelagic	$350\text{–}450 \times 10^3$	1.0–1.4	3–4	0.6
Herring, Clupea harengus	Oviparous	Marine demersal	$5\text{–}200 \times 10^3$	0.9–1.7	5–8	1.6
Carp, Cyprinus carpio	Oviparous	Freshwater demersal	$180\text{–}530 \times 10^3$	0.9–1.6	5–6	0.6
Atlantic salmon, Salmo salar	Oviparous	Freshwater demersal	$1\text{–}10 \times 10^3$	5–6	15–25	1.3
Spurdog, Squalus acanthias	Ovoviviparous		2–16	24–32	250–300	23
Smoothhound, Mustelus mustelus	Viviparous		10–30		~250	20
Guppy, Poecilia reticulata	Viviparous		10–50		6–10	13

Table 9.2 Fecundities of fish species displaying different levels of parental care

Egg type	Degree of parental care	Fecundity	Examples
Pelagic, free-floating	None	$10^4 - 5 \times 10^6$	Atlantic cod, *Gadus morhua* Turbot, *Scophthalmus maximus* Atlantic mackerel, *Scomber scombrus*
Demersal, eggs stick to substrate or vegetation	None	$10^3 - 10^5$	Herring, *Clupea harengus* Carp, *Cyprinus carpio*
Demersal	Eggs buried in substrate, hidden in holes or under rocks	$5 \times 10^3 - 5 \times 10^4$	Capelin, *Mallotus villosus* Salmonids
Demersal	Eggs guarded and cleaned (usually by male); eggs may also be hidden	$10^3 - 5 \times 10^4$	Lumpfish, *Cyclopterus lumpus* Blenny, *Blennius pholis* Painted goby, *Pomatoschistus pictus*
	Nest building; eggs guarded and cleaned (usually by male)	$10^2 - 5 \times 10^3$	Threespine stickleback, *Gasterosteus aculeatus* Bluegill sunfish, *Lepomis macrochirus* Largemouth bass, *Micropterus salmoides*
	Eggs carried by parent either in special brood pouch or in the mouth	50–1000	Sea horse, *Hippocampus ramulosus* Pipefish, *Syngnathus acus* Nile tilapia, *Oreochromis niloticus*

9.8 REPRODUCTION AS A CYCLIC EVENT

In order that survival of young be optimized, the timing of spawning by the mature, adult fish must be closely linked to the cycles of availability of prey consumed by the newly hatched young. The availability of prey of the larvae and juveniles of most temperate and cold-water fish species varies on a seasonal basis. Thus these fish are usually found to have a discrete spawning season timed so that the eggs hatch, and young are ready to consume exogenous food, at a time when prey is abundant.

Thus, in many temperate and cold-water species, spawning is an annual event. Spawning culminates a series of preparatory events during which the gametes develop and the gonads increase in size. In broad terms the annual cycle can be divided into three major periods, or phases:

1. a *postspawning* period when the gonads are small and appear to be in a resting phase;
2. a *prespawning* period in which the gonads begin production of gametes (gametogenesis) and there is production and incorporation of yolk into the oocytes (vitellogenesis); this is accompanied by a gradual increase in gonad size;
3. a *spawning* period, involving final maturation and ripening of the gametes: this phase culminates in the spawning act, with the release of gametes and fertilization of the eggs.

Thus, gonadal development can be considered to consist of a series of interrelated phases, each requiring precise coordination and control if viable gametes are to be produced at the optimum time of the year for the subsequent survival of the young fish. The production of gametes involves a long series of cell divisions (gametogenesis). In the case of the female, gamete development also involves the manufacture and incorporation of yolk into the developing oocytes (vitellogenesis).

During the early part of gonadal development the oogonia become associated with a number of pre-follicular cells, and it is from these that the ovarian follicle is derived. The gametes tend to increase relatively little in size during the earliest developmental phases. In the salmonids, for example, the oocytes increase in diameter from about 50 μm to 500–1000 μm prior to the start of vitellogenesis. The incorporation of yolk that occurs during the vitellogenic phase leads to the diameter of the oocyte increasing to around 5000 μm (5 mm). Thus, there is an enormous increase in oocyte volume during vitellogenesis. The majority of this increase is due to the incorporation of exogenously synthesized yolk into the cytoplasm of the oocyte.

The yolk is derived from vitellogenin, a lipophosphoprotein–calcium complex with a molecular weight of approximately 440 000. Vitellogenin is synthesized in the liver and is then released into the blood system. The vitellogenin is transported to the ovary in the blood from which it is sequestered by the oocyte. Following uptake into the oocyte the vitellogenin is split into two major components – phosvitin (molecular weight 35 000) and lipovitellin (molecular weight 390 000) – which form the yolk stores. The processes of gamete development and gonadal growth usually occur gradually and the prespawning phase of the cycle often commences some considerable time, perhaps several months, before the actual time for spawning. This requires that the fish 'predict' spawning time several months in advance.

Environmental influences on the reproductive cycle

The general cues that trigger gamete growth and development appear to be environmental in nature. These environmental cues are often linked to the annual cycles of daylength and temperature variations. For example, for summer-spawning fish such as carp, *Cyprinus carpio*, and many other cyprinids, gonadal maturation begins in late winter or early spring. The start of the maturation process is thought to be triggered by a combination of increasing daylength and the commencement of the spring rise in water temperature. The spawning season comes to an end during the summer. It is thought that the cessation of spawning activity may, in part, be due to inhibitory effects of high summer water temperatures on gonadal differentiation and maturation. Spawning does not, however, resume as temperature begins to fall with the approach of winter. This indicates that temperature is not the only environmental factor exerting a controlling influence over reproduction. Thus, it is thought that both photoperiod and temperature are major environmental cues responsible for the mediation of the reproductive cycle in several species of fish that inhabit temperate latitudes.

Some species, such as the tobinumeri dragonet, *Repomucenus beniteguri*, do resume spawning activities as water temperature falls during the course of autumn. These fish have two distinct spawning seasons (May–July and September–November) during the course of the year. Spawning commences in spring when the water temperature reaches 18°C, but the spring spawning season comes to an end once temperature exceeds 28°C. Spawning resumes in September as water temperature falls below 27°C, and the autumn spawning season continues as long as temperature stays above 15°C. Thus, gonadal development and maturation appear to be inhibited by temperatures lower than about 15°C and higher than 27–28°C. This suggests that water temperature is probably the major environmental cue for both initiation and termination of the spawning seasons in the to-binumeri dragonet.

In the tropics there may be only very minor seasonal changes in photo-period and water temperature. Consequently, these two environmental factors are unlikely to act as major cues influencing the reproductive cycle, the timing, and the duration of the spawning season. In some tropical species there does not appear to be any distinct spawning season, and some members of the population can be found to be engaged in re-productive activities at almost any time of the year. In other species, spawning activities do seem to vary on a seasonal basis, with factors such as variations in rainfall possibly acting as the environmental cues trigger-ing changes in gonadal development and maturation.

Under natural conditions the seasonally changing pattern of daylength seems to provide the proximate cues that coordinate the timing of gonadal development, oocyte maturation and spawning in many fish species of temperate and high-latitude zones. Temperature is another environmental factor that may have important influences upon the reproductive cycle, but temperature effects seem to be mediated via influences upon physiolo-gical rate processes rather than being used as a proximate cue indicative of the changing seasons.

The overriding importance of the photoperiodic control of the re-productive cycle has been demonstrated in the threespine stickleback, *Gas-terosteus aculeatus*, and some salmonids. The salmonids of the northern temperate zone generally spawn during the autumn and winter months. In salmonid species gametogenesis may commence during the late summer of the year prior to the spawning event. For these species, the environ-mental cue triggering the initiation of gametogenesis, the rate of recruit-ment of previtellogenic oocytes, and the final gamete maturation that occurs shortly prior to spawning, may be decreasing daylength. Vitellogen-esis, on the other hand, is initiated during the spring, a few months prior to spawning, and increasing daylength may be the most important cue triggering the start of yolk production and incorporation into the develop-ing oocytes.

Thus, when rainbow trout, *Oncorhynchus mykiss*, are exposed to condi-tions under which the annual changes in daylength are compressed into periods shorter than 12 months (e.g. 9 or 6 month cycles) the fish can be induced to spawn at intervals of less than 1 year (Fig. 9.11). Similarly, if the seasonal cycle of changing daylength is extended to a period of time longer than 1 year, spawning can be delayed.

Nevertheless, when fish are maintained under experimental conditions of constant photoperiod (e.g. 12L:12D – 12 h light and 12 h dark each day) and temperature the fish may still spawn at approximately yearly inter-vals. This means that there is a strong autonomous component to the

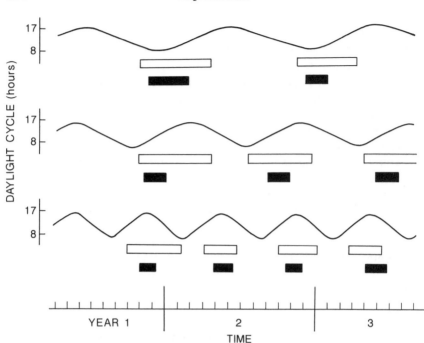

Fig. 9.11 Possible influence of different photoperiodic cycles upon the reproductive development of the rainbow trout, *Oncorhynchus mykiss*. Open bars indicate the times during which the males are in spawning condition, whereas the dark bars indicate the times at which the females can be stripped of ova.

reproductive cycle, and under constant conditions this *endogenous rhythm* will have a periodicity of about 1 year. However, if the fish are held under constant environmental conditions for several cycles it will be found that they spawn at intervals that are approximately, but are significantly different from, 1 year. Thus, the periodicity of the rhythm is only approximately 1 year, i.e. it is circannual.

There is some evidence that there are sexual differences in the periodicity of the endogenous cycle of maturation, with males seeming to have shorter cycles than females. One consequence of this would be that, in the absence of photoperiodic cues, the cycles of the males and females would gradually become desynchronized. The net result would be that the two sexes were sexually active at different times of the year, creating a mismatch between the times at which viable eggs and sperm were produced.

The endogenous rhythm would seem, therefore, to be subject to modification by changes in the ambient photoperiod, leading to corrective delays or advances being made depending upon whether gonadal development is perceived as running ahead of or behind the schedule required for the correct seasonal timing of spawning. Thus, in summary, it can be said that the rate of gamete maturation, and the timing of spawning, in salmonids are dependent upon the entrainment of an endogenous circannual rhythm by seasonal changes in photoperiod.

To what extent temperature can exert a modifying influence on the timing of reproduction has been comparatively little studied. There is, however, a certain amount of evidence indicating that temperature can

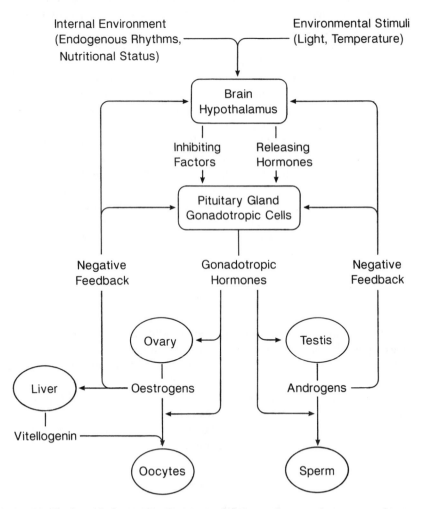

Fig. 9.12 The hypothalamic–pituitary–gonadal axis indicating the major endocrine factors involved in the control of the reproductive cycle.

have a modifying effect upon the photoperiodic control of reproduction in salmonids. For example, many stocks of rainbow trout held on fish farms in constant-temperature water (8–10°C) spawn during the late autumn and early winter (November–January), but if fish are held in river water where temperatures may fall to 1–2°C during autumn and winter the timing of spawning may be delayed until the early spring (March–April). It is suggested that low ambient water temperatures during the course of autumn and winter may reduce the rate of sequestration and incorporation of yolk into the oocytes, leading to an inhibition of oocyte growth and development. Thus, low temperature leads to a slowing of the rate of oocyte development, with the consequence that final oocyte maturation, ovulation and spawning are delayed.

To summarize, many fish species that inhabit temperate regions appear to rely on the seasonally changing cycle of daylength to time their annual cycles of reproduction. These cyclic events will be mediated via a number of endocrine changes. Consequently, it is ultimately the neuroendocrine system which directly controls the different phases of gonadal development and the maturation of the gametes.

A variety of endocrine factors have been implicated in this control, but the hormones most widely studied are those which make up the hypothalamic–pituitary–gonadal axis (Fig. 9.12). Nevertheless, information about the seasonal changes in the endocrine system, and the influences of the various hormones, is confined to relatively few species, notably some salmonids and cyprinids. In addition, much more attention has been directed towards study of the female reproductive cycle than towards study of the male.

9.9   HYPOTHALAMIC–PITUITARY–GONADAL AXIS

Whilst seasonal changes in photoperiod are known to have profound influences upon the endocrine system, and ultimately upon the reproductive cycle, relatively little is known about the endocrine control of the earliest phases of the cycle. Thus, information about the endocrine control of gametogenesis, the initial stages of the ovarian cycle, and the process of transformation of oogonia into oocytes is limited. More is known about the involvement of pituitary and gonadal hormones in the control of vitellogenesis, oocyte maturation and ovulation.

Hypothalamus

Seasonal changes in ambient photoperiod are perceived by the senses, and integration of the sensory inputs at central sites leads to changes in the rates of production and secretion of hormones by the hypothalamus. The hypothalamic hormones include both gonadotropin-releasing (GnRH or GnRF) and gonadotropin-inhibiting (GRiF) factors that have antagonistic actions upon the synthesis and release of *gonadotropins* (GTH) by the gonadotropic cells of the adenohypophysis of the pituitary gland.

The neurosecretory cells of the hypothalamus produce the releasing and release-inhibiting factors which pass down the neuronal connections between the hypothalamus and pituitary in order to exert their controlling effects on the gonadotropic secretory cells. The hypothalamic GnRH appears to be a small peptide hormone consisting of 10 amino acid residues (i.e. it is a decapeptide), whereas *dopamine* is known to exert an inhibitory effect upon GTH secretion by the pituitary. It is, therefore, probable that dopamine, a catecholamine derived from the amino acid tyrosine, is the hypothalamic GRiF.

Pituitary

The next level of control resides in the pituitary gland, and seasonal changes in the morphology and activity of the gonadotropic cells have been reported for a range of fish species.

Stimulation of the gonadotropic cells by GnRH results in the secretion of GTH into the bloodstream, but GnRH is also known to stimulate the increased synthesis and release of growth hormone by the tissues of the pituitary. The GTH and growth hormones secreted into the bloodstream eventually reach the gonadal tissues and bind with specific receptors. Stimulation of the gonadal tissue by these hormones, particularly GTH, leads to increased production and secretion of steroid sex hormones – *oestrogens* and *androgens* (Fig. 9.12).

Hypothalamic–pituitary–gonadal feedback mechanisms

The steroid hormones exert a number of effects, including a feedback control on the production and release of GTH by the pituitary. This feedback control exerted by the steroid hormones is thought to be mediated via an indirect effect upon the antagonistic relationships between GnRH and GRiF (Fig. 9.13). It has been hypothesized that the feedback control of GTH secretion from the pituitary may occur as follows.

Hormones released from the steroidogenic tissues arrive at central sites via the bloodstream, where they are acted upon by a series of enzymes. Transformation of the steroids results in the production of the compound catecholoestrogen. Catecholoestrogen is transformed into 2-methoxyoestrogen by the action of the enzyme catecholoestrogenmethyltransferase (COMT) (Fig. 9.13). COMT can also use dopamine as a substrate, but the

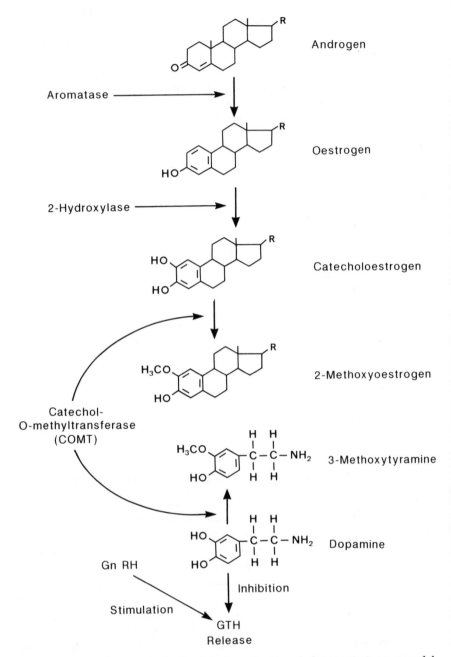

Fig. 9.13 A possible negative feedback loop on the hypothalamic–pituitary–gonadal axis acting to control the rates of production and release of gonadotropins.

affinity of the enzyme is greater for catecholoestrogen than for dopamine. Thus, when catecholoestrogen is present it will be used preferentially as a substrate, and the dopamine will remain undegraded. The dopamine will then be left to exert its inhibitory effect upon the gonadotropic cells of the pituitary.

In the absence of catecholoestrogen the COMT will use dopamine as a substrate and the dopamine will be degraded to inactive 3-methoxytyramine (3-MT) (Fig. 9.13). This, in turn, will lead to the loss of the inhibitory effect of dopamine on GTH release, allowing the stimulatory actions of GnRH to come fully to the fore. Thus, it is envisaged that circulating levels of steroid hormones exert some feedback control on GTH secretion via the enzymatic transformations that occur at central tissue sites. It is thought that the key to the process may reside in substrate switching by the enzyme COMT.

Dopamine may, however, also be metabolized by other pathways and major metabolites may include dihydroxyphenylacetic acid (DOPAC) and homovanillic acid (HVA). There may, therefore, be species differences in the major products of dopamine metabolism. In the brain and pituitary tissues of male goldfish, *Carassius auratus*, DOPAC seems to be a more important metabolite than either 3-MT or HVA. In this species, therefore, it has been proposed that it is the rate of dopamine turnover to DOPAC that exerts a major control over GTH release at the level of the pituitary.

In the goldfish, it has been shown that dopamine may both modulate the release of GnRH centrally and act at the level of the pituitary. Thus, it is not unreasonable to assume that a reduction in dopamine inhibition of the pituitary functions in combination with endogenous GnRH to stimulate the release of GTH from the gonadotrops. Furthermore, there is convincing evidence that both dopamine metabolism and GnRH release can be influenced by sex pheromones. A pheromone-induced reduction in dopamine inhibition, in combination with an increase in GnRH release, may thus be responsible for the surge of GTH observed in the male goldfish following exposure to a ripe female.

Gonadotropins

There are seasonal changes in levels of circulating GTH, with two distinct peaks of serum concentrations of GTH often being observed in female fish. The first measurable increase in serum GTH occurs early in the maturational cycle at around the time that vitellogenesis commences. The inference is that it is the increase in GTH secretion that leads to *oestradiol* production and release by the ovarian tissues. This, in turn, stimulates vitellogenin synthesis and release by the liver (Fig. 9.14).

The second peak of serum levels of GTH is associated with final oocyte

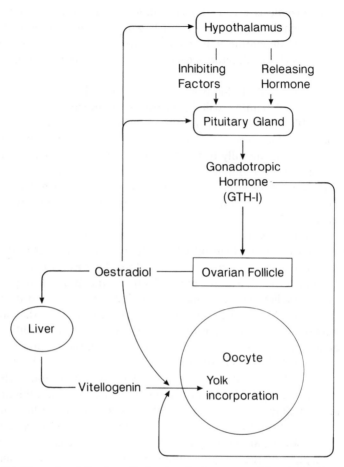

Fig. 9.14 The role of the hypothalamic–pituitary–gonadal axis in the endocrine control of vitellogenesis and incorporation of yolk into developing oocytes.

maturation and ovulation. Once again a cause-and-effect relationship is suspected.

Support for the involvement of the two GTH peaks in the control of vitellogenesis and oocyte maturation comes from experiments in which maturational rates and the timing of spawning have been altered via manipulations of photoperiod. Under these circumstances there are changes in the timing of the GTH peaks which are intimately associated with the shifts in the timing of the other maturational events. Thus, it can be concluded that the changes in circulating levels of GTH are involved in the initiation of vitellogenesis, ovarian growth and in the timing of oocyte maturation and ovulation (Fig. 9.15).

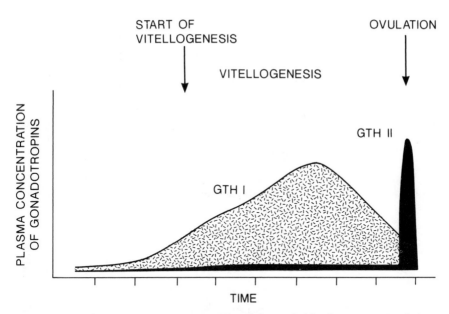

Fig. 9.15 Schematic graphical representation of the probable changes in circulating levels of putative vitellogenic (GTH-I) and maturational (GTH-II) gonadotropins in female fish during the course of the reproductive cycle.

Gonadal steroids

Following closely on the changes in serum GTH levels seen during the early part of the maturational cycle there are increases in circulating levels of the gonadal steroid hormones. In female salmonids, for example, serum levels of oestradiol may increase from a few ng ml^{-1} in immature fish to about 15–20 ng ml^{-1} during the most active period of vitellogenesis. Production of oestradiol occurs in the ovarian follicle, largely under the control of GTH.

Oestradiol synthesis

The ovarian follicle is made up of two distinct cell layers – the *theca* and the *granulosa* – but neither layer is capable of producing oestradiol *de novo*. The cells of the thecal layer can synthesize the androgenic steroid *testosterone* using cholesterol as the precursor. The thecal capacity for testosterone production increases during the vitellogenic phase of the ovarian cycle. This increased production capacity probably results from the stimulatory effects of GTH on the cells of the thecal layer (Fig. 9.16).

The cells of the granulosa layer are incapable of steroidogenesis using

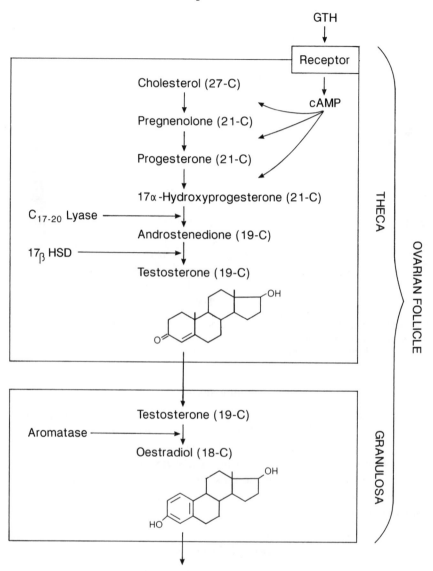

Fig. 9.16 The 'two-layer' model developed to illustrate the involvement of the theca and granulosa of the ovarian follicle in the synthesis of steroids during vitellogenesis.

cholesterol as a precursor. These cells do, however, contain the enzyme aromatase which is essential for the production of oestradiol from testosterone. One of the effects of growth hormone appears to be to induce the gonadal tissue to increase the synthesis of aromatase enzyme. Thus, the effects of growth hormone and GTH secreted by the pituitary will be synergistic.

The synthesis of oestradiol by the granulosa cells is dependent upon a supply of testosterone from the theca, and aromatase is required for the synthesis of oestradiol from testosterone. Thus, both the thecal and granulosa cell layers are required for the production of oestradiol from cholesterol (Fig. 9.16).

Actions of gonadal steroids

The gonadal sex steroids exert control over developmental events in the gonad. In addition, these hormones are implicated in the development of the secondary sexual characters. They may also induce a number of more generalized effects on metabolism and the mobilization of the tissue reserves of nutrients.

In the female, oestradiol secreted by the ovarian follicle induces the tissues of the liver to produce vitellogenin and egg membrane proteins. These materials are then secreted into the blood, and circulating levels of vitellogenin may reach a peak of about 50 mg ml^{-1} serum during the most active phase of yolk incorporation into the developing oocytes. The vitellogenin seems to be taken up by the oocytes via pinocytosis, with uptake appearing to be stimulated by GTH (Fig. 9.14), and possibly also by thyroid hormones and insulin.

Yolk incorporation leads to a displacement of the nucleus, or *germinal vesicle*, towards the periphery of the oocyte. By the end of the vitellogenic stage of development the oocytes will have increased markedly in size, but these post-vitellogenic oocytes are still physiologically immature.

Oocyte maturation

Oocyte maturation requires the rupturing of the nuclear envelopes (*germinal vesicle breakdown*) and the extrusion of the first polar body. It is suggested that the maturational processes are initiated by the second large GTH surge that occurs towards the latter stages of the ovarian cycle (Fig. 9.15).

Post-vitellogenic oocytes do not respond to the presence of GTH by undergoing the final stages of maturation, but they do mature if incubated in a medium that has previously held folliculated oocytes exposed to GTH. In other words, the maturational response of oocytes to GTH requires the production of a secondary effector, or *maturation-inducing hormone*, by the follicular tissue. There are several lines of evidence suggesting that the steroid 17α20β-dihydroxy-4-pregnen-3-one (17α20β-P) is the maturation-inducing hormone in a number of salmonids, but other closely related progestogens are thought to perform this role in other fish species.

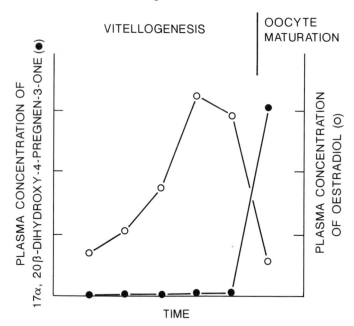

Fig. 9.17 Graphical representation illustrating the changes in circulating levels of oestradiol and maturation-inducing hormone in the female fish during vitellogenesis and the final stages of oocyte maturation.

In the period following the second peak of GTH secretion by the pituitary there is a modification of the steroid biosynthetic pathways of the ovarian tissue towards the production of progestogens. In female salmonids this is accompanied by a steep rise in serum levels of 17α20β-P, with serum concentrations increasing from about 0.5 ng ml^{-1} during vitellogenesis to reach levels of 50–70 ng ml^{-1} at the time of oocyte maturation.

At the same time as progestogen synthesis increases there is a marked decline in the production of oestradiol by the follicular tissues, and circulating levels of oestradiol fall dramatically (Fig. 9.17). Peak levels of the maturation-inducing hormone coincide with the time of germinal vesicle breakdown in the oocyte, which, in salmonids, occurs 2–4 days prior to ovulation.

The initial stages of production of 17α20β-P by the ovarian follicle involve the synthesis of 17α-hydroxyprogesterone (17α OH-P) from cholesterol in the thecal layer. This is followed by the transfer of 17α OH-P to the granulosa, where the 17α20β-P is synthesized in the presence of the enzyme 20β-hydroxysteroiddehydrogenase (20β-HSD)(Fig. 9.18).

Exposure of cells of the granulosa layer to GTH results in an increase in *de novo* synthesis of the enzyme 20β-HSD, thereby increasing the capacity

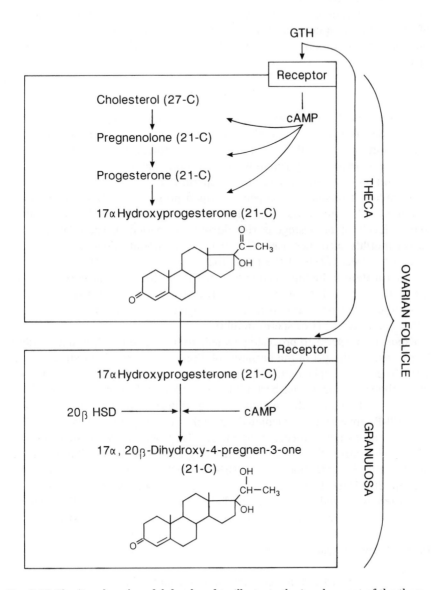

Fig. 9.18 The 'two-layer' model developed to illustrate the involvement of the theca and granulosa of the ovarian follicle in the synthesis of maturation-inducing hormone.

of the granulosa to produce and secrete 17α20β-P. In order for there to be an increased production of the maturation-inducing hormone, the granulosa must, however, also receive increased supplies of the precursor 17α OH-P from the theca. Because 17α OH-P is an intermediate in the synthesis of testosterone from cholesterol in the theca (Fig. 9.16), the production of maturation-inducing hormone involves a switch from testosterone production and secretion, to 17α OH-P secretion by the cells of the thecal layer.

Steroidogenesis in males

The processes of spermatogenesis and spermiogenesis appear to be largely under the control of the androgens *testosterone* and 11-*ketotestosterone*. Production and secretion of androgens is, however, markedly reduced during the period of spermiation (maturation and release of sperm from the germinal cysts and transfer to the sperm duct).

Spermiation seems to be dependent upon progestogens, and in the male salmonids the maturation-inducing hormone seems to be 17α20β-P. Serum levels of progestogens rise during the period of spermiation, and sperm motility also increases. The increase in motility does not appear to be due to direct effects of the progestogens on the sperm. The influence of the maturation-inducing hormone seems to be indirect via effects on the pH of the sperm duct fluids bathing the sperm. Neither testosterone nor 11-ketotestosterone, the two major androgens in teleosts, appear to affect either sperm duct pH or sperm motility.

Exposure of sperm to progestogens fails to influence motility, but motility is increased if the pH of the bathing solution is increased from about 7.4 to 8.0. Exogenous application of progestogens to male fish may lead to the pH of the sperm duct fluid becoming more alkaline (pH increase from 7.4–7.5 to 8.5). Thus, the action of the maturation-inducing hormone, 17α20β-P, appears to be mediated through an increase in sperm duct pH.

The maturation-inducing progestogen is probably synthesized in the sperm cells rather than in the surrounding testicular tissue. The precursor for 17α20β-P production is 17α OH-P, which is acted upon by the the enzyme 20β-HSD. The testicular tissue is able to synthesize androgens, and can, therefore, produce the intermediate 17α OH-P. The 20β-HSD enzyme activity may, however, be limited to the sperm. Thus, the synthesis of 17α20β-P occurs within the sperm cells from 17α OH-P that diffuses from the surrounding testicular tissue.

Feedback mechanisms within the gonadal tissues

When isolated testicular tissue is exposed to GTH the production of androgens is stimulated. The addition of 17α20β-P to the incubation medium leads both to the inhibition of the GTH-stimulated androgen production, and to a marked reduction in the conversion of exogenous 17α OH-P to testosterone and 11-ketotestosterone. Thus, $17\alpha20\beta$-P appears to have an inhibitory effect on the enzymes (e.g C_{17-20}lyase) involved in the biosynthesis of androgens from 17α OH-P.

In the male the switch between androgen and progestogen secretion that precedes, and controls, spermiation may proceed as follows. The GTH surge that occurs prior to spermiation leads to a marked increase in the conversion of cholesterol first to 17α OH-P, and then to androgens, in the testicular tissue. The increased rates of synthesis lead to both 17α OH-P and androgens diffusing out of the testicular tissue. Any 17α OH-P that reaches the sperm cells acts as the substrate for the synthesis of $17\alpha20\beta$-P. The $17\alpha20\beta$-P that is secreted from the sperm cells and enters the testicular tissues will have an inhibitory effect (negative feedback) on the enzymes involved in androgen synthesis. This would, in turn, lead to increased diffusion of 17α OH-P from the testicular tissue, providing greater quantities of substrate for $17\alpha20\beta$-P synthesis. The end result of the operation of this short-loop feedback mechanism would be a rapid rise in progestogen, and a simultaneous drastic decline in androgen synthesis and secretion, in the period immediately following the GTH surge.

If it is assumed that the mechanisms controlling steroid biosynthesis are the same in both the male and the female, then a similar set of arguments can be invoked in order to explain the switch from oestradiol to $17\alpha20\beta$-P secretion by the ovarian tissues prior to the onset of oocyte maturation and ovulation. Thus, it is envisaged that the GTH surge stimulates increased production of 17α OH-P, from cholesterol, in the theca of the ovarian follicle. There is also a GTH-induced increase of 20β-HSD enzyme activity in the granulosa layer.

High rates of production of 17α OH-P in the theca lead to saturation of the enzymes (e.g. C_{17-20}lyase) involved in the biosynthesis of testosterone. Consequently, both testosterone and 17α OH-P begin to diffuse out of the cells of the thecal layer. Testosterone that reaches the cells of the granulosa layer can be used as a substrate for oestradiol production, whereas the 17α OH-P that enters the granulosa layer will result in the synthesis of $17\alpha20\beta$-P. Thus, at this early stage both oestradiol and $17\alpha20\beta$-P will be produced and secreted by the granulosa layer.

Any $17\alpha20\beta$-P that diffuses back to the cells of the theca will inhibit the enzymes (e.g. C_{17-20}lyase) involved in the synthesis of testosterone from 17α OH-P. One result of this enzyme inhibition is an increase in the availability of 17α OH-P for secretion from the theca. The inhibition of testosterone synthesis also leads to a sharp decline in the amounts of testosterone reaching the granulosa cells.

The inhibition of testosterone synthesis, and concomitant decline in diffusion from the theca, results in a marked drop in the production of oestradiol in the granulosa layer. On the other hand, amounts of 17α OH-P reaching the cells of the granulosa layer increase, and rates of synthesis and secretion of the maturation-inducing hormone $17\alpha20\beta$-P increase markedly. The net result is a rapid rise in progestogen production and secretion, accompanied by a sharp decline in the synthesis and secretion of oestradiol by the ovarian follicle (Fig. 9.17).

9.10 REPRODUCTION AND THE MOBILIZATION OF TISSUE RESERVES

Many temperate and cold-water marine species spawn in spring or early summer in order to synchronize the timing of larval hatching with the onset of the spring plankton bloom. One consequence of this is that the majority of the gonadal development, involving vitellogenesis, yolk production and oocyte growth, occurs during the winter months. At this time of the year feeding may be reduced, due both to low water temperature and to limited food availability.

Reduced feeding during winter requires that body reserves be mobilized, both to meet general metabolic requirements of the adult fish and to provide the energy and raw materials for the production of gametes and yolk. Thus, there may be considerable changes in the relative sizes and chemical compositions of the different tissues and body organs during the winter months, as the reserves are mobilized from the storage depots and are either metabolized or are deposited in the gonads.

For example, the Atlantic cod, *Gadus morhua*, in Balsfjord, northern Norway continue to feed in the period October–January, and they obtain sufficient food to enable the continuous build-up of tissue energy reserves. Most of the reserves are deposited in the liver in the form of lipid, there being a steady increase in both the hepatosomatic index (HSI = liver weight / body weight) and the amount of storage lipid during the course of the winter months.

From January until spawning in April these energy reserves are depleted. The reserves accumulated during the autumn and early winter both contribute to gonadal growth and are used as metabolic fuel, but the majority of the liver reserves appear to be used as fuel with there being only a relatively small contribution to the gonad. Gonadal growth appears to continue at a steady rate throughout the period between November and March, with most of the increase being due to the incorporation of yolk proteins into the developing oocytes. The yolk proteins are manufactured in the liver and are then transported to the gonad in the bloodstream before being deposited in the oocyte.

In contrast to the Balsfjord cod, the plaice of the Irish Sea appear to cease feeding in December, and feeding is not resumed until March, after the fish have spawned. Thus, in these plaice, both metabolism and the gonadal development that occurs during the winter are supported by energy reserves accumulated during the previous summer.

The liver of the plaice is relatively small, and most of the reserves are deposited in the muscle as protein and lipid. Chemical analysis of the different organs of plaice collected during the winter months has revealed that about 25% of the muscle protein is mobilized. The majority of the protein is apparently utilized for yolk production and is incorporated into the developing oocytes. Only a small amount of the muscle protein is mobilized for use as metabolic fuel.

About 55% of the muscle lipid is mobilized during the course of the winter and early spring. Very little of this is deposited in the developing gonads, and most is utilized as metabolic fuel.

In most pleuronectids the muscles form the major depots of energy reserves. The mobilization of these reserves during the period of gonad development leads to pronounced changes in tissue composition. There is usually a marked decrease in protein content and an increase in percentage water content of the flesh. In some species the depletion may be so severe that the fish reaches a condition known as 'jellying'. In extreme cases, the muscle of 'jellied' fish may comprise no more than 2.5–3% protein, about 0.05% lipid and over 95% water.

In common with the plaice, and many other temperate fish species, the herring, *Clupea harengus*, tends to cease feeding both during the winter months and during the period of gonadal maturation. The energy stored as body lipids and proteins is mobilized to meet the energy requirements during these non-feeding periods, but the extent to which the different reserves are mobilized depends upon whether or not the fish are undergoing reproductive development. Storage lipids tend to be depleted by all non-feeding fish, but proteins are mobilized only by those fish showing gonadal development. Thus, it seems that the lipids are used as fuel to sustain metabolism whereas the proteins may be mobilized from the soma and utilized to develop the gonadal tissues.

Thus, it is apparent that seasonal cycles of repletion and depletion of tissue storage depots of nutrients are common amongst the species inhabiting temperate and cold-water habitats. The cycles of repletion and depletion may be more pronounced in mature than in immature fish, with the cycles often being observed to be linked with changes in gonadal development. Maturing and immature fish may also rely on the different storage reserves to differing extents. Immature fish tend to mobilize lipids to support metabolism, whereas maturing fish may use lipids as a metabolic fuel and mobilize protein in order to build up the gonads. Thus, the

reproductive growth that occurs during winter and early spring is frequently reliant upon the depletion of the tissue reserves accumulated
during the previous summer when food was abundant.

9.11   THE PRELUDE TO SPAWNING

Even though gametogenesis and vitellogenesis may have been successfully
completed the fish will not yet be in spawning condition. Maturation of the
gametes, ovulation and spermiation are under endocrine control, but
whether or not the reproductive cycle reaches fruition with the performance of the spawning act will be largely determined by external stimuli.

During the early part of the reproductive cycle the rate of development
of the gametes appears to be controlled by an endogenous rhythm which
undergoes modification in response to seasonal environmental cues.
Seasonal cues of this nature can exert a corrective influence on developmental rates in order to bring the fish into reproductive condition at some
specific time of the year. They do not, however, necessarily give sufficient
information to enable the making of precise predictions and fine adjustments to the timing of spawning activities.

If spawning is to be successful, it is obvious that a mating pair must
synchronize gamete production. Since mature gametes may remain viable
for only a limited period of time, the final stages of gamete maturation and
release may be delayed until after pair formation. In other words, the maturation process may be induced by specific aspects of the courtship, or
pair-forming, procedure.

Thus, the external, or environmental, cues that promote the final stages
of gamete maturation may be different from those that exert the greatest
influence at an earlier stage in the reproductive cycle. This may explain,
for example, why many fish species do not successfully complete spawning
in captivity. The captive environment may provide the cues required by
the females to reach the stage of gonadal development where the ovaries
contain postvitellogenic oocytes, but may be deficient in the cues needed
for the induction of the final stages of maturation, ovulation and shedding
of eggs.

The initiation of the final stages of gamete maturation may require the
performance of specific behaviour patterns by potential mates, certain environmental cues or, as is often the case, suitable combinations of several
environmental and behavioural factors. For example, sex hormones in the
ovarian fluids of ripe females may be released into the water and these
hormones can have a pheromonal effect. These sex pheromones may stimulate both spermiation and the performance of courtship behaviour by
the male. The male courtship behaviour may, in turn, act as the trigger

that induces spawning behaviour in the female, culminating in the shedding of gametes and the fertilization of the eggs by sperm.

9.12 SEX PHEROMONES

A number of sex hormones have been shown to function as pheromones. Since different hormones are involved in controlling the various stages of the gonad maturational cycle, these endocrine factors could function as useful chemical signals providing conspecifics with information about the reproductive status of potential mates. Not only may these signals provide information, but the pheromones can also exert physiological effects influencing the final stages of gonadal development in the recipient. Since anosmic fish do not respond to pheromones, the pheromones must be detected by the olfactory sense.

The majority of the compounds that have been shown to have pheromonal activity are sex steroids or their derivatives, and prostaglandins, but the olfactory senses of some species have been shown to be extremely sensitive to releasing hormones (GnRH) and gonadotropins (GTH). It is, however, difficult to envisage how these hypothalamic and pituitary factors would be excreted into the environment and exert pheromonal effects *in situ*. Steroids are relatively insoluble in water, and prior to excretion the steroids are usually converted into a more soluble form, such as glucuronide conjugates or sulphated derivatives. Thus, the compounds that are found to exert pheromonal actions are generally found to be steroid derivatives rather than the sex steroids themselves.

The responses to pheromones can be broadly classified as being either *primer* or *releaser* effects. Primer effects refer to the induction of physiological changes, the results of which may be observed in the longer term. Releaser effects, on the other hand, generally lead to immediately observable changes in behaviour. The types of compounds that produce primer and releaser effects will usually be found to differ. There is also evidence that there are interspecific differences in the nature of the sex pheromones. It should also be remembered that communication between individuals by chemical means can refer to communication either within ($\male$ to $\male$, $\female$ to $\female$) or between ($\male$ to $\female$, $\female$ to $\male$) sexes.

Pheromonal communication within sexes may lead to a synchronization of gamete maturation, or simultaneous ovulation of individuals within the population. For example, exposure to water that has held postovulatory conspecific females induces a GTH surge, and oocyte maturation, in postvitellogenic sharptooth catfish, *Clarias gariepinus*. Thus, it is implied that postovulatory catfish excrete pheromones that are detected by conspecific females.

In the male Pacific herring, *Clupea harengus pallasi,* exposure to fresh milt (sperm and seminal fluid), or testis extracts, can promote spermiation and the release of milt. The maturation-inducing hormone $17\alpha20\beta$-P and some of its conjugates have been found to be present in the milt. It is suspected that these compounds act as sex pheromones, leading to a synchronization of sperm maturation and release in the male herring.

Pheromones released by males may influence females at various stages of the reproductive cycle. For example, in the sharptooth catfish, oocyte recruitment, vitellogenesis and ovarian growth can be promoted by exposure of the females to water that has held sexually mature males. At a later stage in the cycle the pheromones may function as sexual attractants, with postovulatory females being attracted towards a source of water that has held mature males. Neither anosmic nor pre-ovulatory females are attracted by the same water source. This indicates that detection of the male pheromone is via the olfactory sense. The results also demonstrate that the female must be at a very advanced stage in the reproductive cycle before she will respond to the male pheromone by showing pronounced behavioural changes.

In other species, such as the the zebrafish, *Brachydanio rerio,* pheromones released by the male may have stimulatory effects on the final stages of oocyte maturation and ovulation, without acting as attractants. In the zebrafish, the testis is one, if not the only, source of the ovulation-inducing pheromone because exposure of females to testicular homogenates has been shown to evoke a strong ovulation response. In contrast to the male, the female zebrafish does appear to produce pheromones with an attractant effect. Follicular extracts from postovulatory female zebrafish have been shown to be highly attractive to sexually mature males. The exact nature of the sex pheromone is, however, unknown.

The female of the guppy has also been shown to produce a pheromone that is attractive to the males. The male guppy is attracted to a source of oestradiol, either in free or conjugated form. Consequently, it would appear that this oestrogen, rather than a progestogen, exerts pheromonal effects in this species. The guppy is viviparous, and is, therefore, one of the few teleosts in which there is both internal fertilization and sperm storage. In other words, copulation and sperm transfer occur in the absence of ovulated eggs.

Female pheromones may induce pronounced physiological changes in the males of some species. For example, exposure to water that has held postovulatory females leads to a GTH surge in the male sharptooth catfish.

Similarly, a GTH surge, followed by increased levels of circulating testosterone and a rise in milt production, can be induced in male goldfish by exposing them to water that has held postovulatory conspecific females.

The same effects may be induced by exposure of the males to pre-ovulatory females, provided that the females are in the final oocyte maturation stage of the reproductive cycle. Pre-ovulatory females start to release $17\alpha20\beta$-P into the water approximately 6 h prior to ovulation, and it is thought that it is the maturation-inducing hormone which is responsible for inducing the physiological changes seen in the males.

In addition to initiating a sequence of physiological changes, the exposure of male goldfish to water that has held postovulatory females may result in the males displaying immediate behavioural changes. In other words, the males initiate behaviours that resemble the initial phases of the courtship and spawning sequence. It appears that prostaglandins, released from the reproductive tract of the female goldfish in response to the presence of ovulated eggs, act as releaser pheromones eliciting courtship behaviour in the males.

Sex pheromones have been shown to influence courtship and spawning behaviour, and pheromones are also known to be involved in the regulation of the endocrine events related to gonadal development and reproduction in a variety of fish species. There may be distinct primer and releaser pheromones, the former initiating slow developmental processes and the latter evoking rapid behavioural responses.

Primer pheromones can induce a GTH surge in both males and females. In males this results in stimulation of spermiation and increases in milt volume (sperm and seminal fluid), whereas ovulation is stimulated in females.

Releasing pheromones may either act as attractants to the opposite sex, or result in displays of courtship and spawning behaviours. Thus, the priming and releasing actions of the sex pheromones are almost invariably seen during the latter stages of the reproductive cycle. There are, however, some indications that sex pheromones can also influence earlier phases of the gonadal cycle, such as oocyte recruitment and ovarian development in the female.

9.13 ENVIRONMENTAL REQUIREMENTS FOR SPAWNING

Environmental factors may be of considerable importance in governing the reproductive behaviour of fish. In other words, if specific environmental requirements are not met then the fish will not spawn. Thus, if suitable spawning conditions are not found then the eggs may become atresic, degenerate and be resorbed. There may, for example, be specific temperature requirements, requirements for a particular spawning substrate, or photoperiodic influences that result in spawning occurring at a specific times in the light:dark cycle.

Table 9.3 The influence of substrate size on spawning site selection by the Arctic charr, *Salvelinus alpinus*

Substrate type	% Females digging nests
Fine sand	0
Coarse sand	6
Fine gravel	6
Coarse gravel	53
Stones	23
Gravel/sand patches between stones	12

Spawning substrate

Goldfish and related cyprinids may come into spawning condition but will not spawn in the absence of suitable vegetation on which the sticky eggs can be laid. The fish are, however, not totally dependent upon vegetation since, under laboratory conditions, they can be induced to lay eggs on clumps of wool threads or patches of coconut matting. These artificial spawning substrates presumably function as substitute water plants. In the commercial rearing of cyprinids it is not uncommon for artificial substrates, such as coconut matting or plastic turf, to be introduced into broodstock ponds in the period shortly before the fish are expected to come into spawning condition. The eggs are laid onto the artificial substrate, which can then be collected and the eggs transferred to a hatchery for incubation under controlled conditions.

Spawning in some species, such as salmonids, with demersal eggs is dependent upon the fish finding a substrate of suitable particle size, in which to bury their eggs (Table 9.3). Similarly, species such as largemouth bass, *Micropterus salmoides*, herring, and capelin, *Mallotus villosus*, may also delay spawning until a suitable substrate has been found.

Spawning temperature

Plaice, Atlantic mackerel, *Scomber scombrus*, and Atlantic cod are all species which produce free-floating pelagic eggs. In these species spawning

Fig. 9.19 The spawning of the Atlantic cod, *Gadus morhua*, in the Lofoten area of northern Norway. The effects of different climatic conditions upon the depths at which spawning occurs, and the subsequent distributions of the pelagic eggs, are shown. Note differing vertical scales in upper graphs.

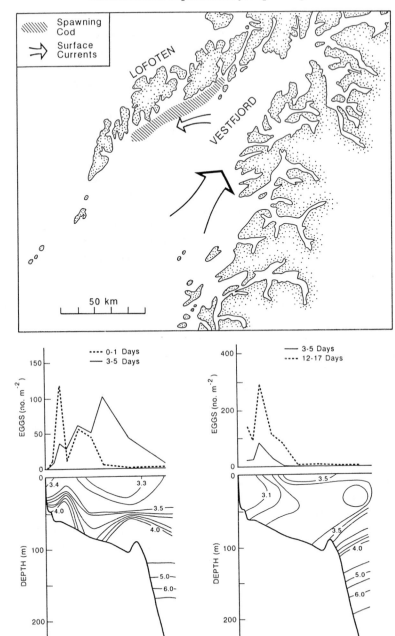

seems to be dependent upon location of water of suitable temperature. For example, the cod of the Barents Sea stock migrate to Vestfjord in the Lofoten Islands, northern Norway, in order to spawn (Fig. 9.19). The cod tend to congregate fairly close to the bottom and spawning occurs in water of temperature 4–5°C. Water of this temperature is found in the transition layer between the waters of the Atlantic and Norwegian Coastal Currents, but the depth at which this transition layer occurs can vary considerably from year to year. For example, in 1961 the transition layer occurred at a depth of 220–240 m but in 1964 it was at a depth of only 30–40 m.

The depth at which the transition layer occurs is largely determined by prevailing wind direction and air temperature conditions during the early part of the spring. South-westerly winds give rise to an instreaming of Atlantic water towards land, and an outstreaming of the overlying Norwegian coastal water. This results in the transition layer being at relatively shallow depth. Under these conditions the cod will spawn close to land and there will be a gradual drift of the eggs away from land in the outstreaming Norwegian coastal water. A north-easterly wind will give the opposite effect with the transition layer being pushed downwards so that the cod spawn at greater depth, and further from land. Under these conditions the eggs rise to the surface offshore and will tend to be carried towards land by the flow of the coastal water.

Light: dark cycle and spawning

Spawning activity in fish species is often observed to be confined to a limited period of day or night. In several cyprinids, the medaka, *Oryzias latipes*, and the tobinumeri dragonet, the timing of spawning is considered to be regulated by the light:dark cycle. In these species the hour of the day at which spawning occurs in captive populations can be influenced by making short-term manipulations to the photoperiod during the course of the breeding season.

That the timing of the light:dark transition can be important in the control of spawning activity has been demonstrated in a series of experiments conducted on the tobinumeri dragonet. When fish were held on a 14L:10D regime with the photophase between 05.00 and 19.00 there was a peak of spawning activity at about 18.00. When the photophase was advanced by 4 h (light period 01.00–15.00) there was a corresponding shift in the timing of the spawning peak, which became advanced by 3–4 h within the space of a few days. A further change in the photophase, leading to the restoration of the 05.00–19.00 light period, resulted in a return of the spawning peak to the period around 18.00.

Additional experiments gave clear indications that it was the timing of 'lights-off' (i.e. the onset of night-time conditions) that played the most important role in the determination of the spawning time in the tobinumeri dragonet. In some other species, however, it may be the timing of 'lights-on' that acts as the major synchronizer of spawning and oviposition.

9.14 COURTSHIP BEHAVIOUR AND SPAWNING ACTIVITY

Many fish species develop special secondary sexual characteristics (Fig. 9.20). These may be permanent dimorphic sexual features, or may be morphological changes that develop in the period prior to, and during, spawning. In many species the development of the secondary sexual characteristics appears to be linked with the endocrine changes that occur in association with reproductive development. For example, the males of some species will develop their spawning coloration, or undergo the morphological changes typical of reproductively active individuals, if treated with exogenous androgens outside of the normal breeding season.

The secondary sexual characteristics of the males may take a variety of forms, dependent upon the type of reproductive activities engaged in and courtship behaviours displayed. Salmonid males, for example, develop a large hooked jaw, or kype, that is used in fighting, whereas the belly of the male threespine stickleback becomes bright red during the spawning season.

Visual signals

The male stickleback shows the bright red belly coloration to advantage during the performance of an elaborate courtship display aimed at attracting ripe females to his nest. The males of several other species, such as the guppy, may develop bright spots or bars on the body and fins, and these are used as visual signals during the performance of courtship behaviours.

There are usually pronounced sexual differences in fish species in which fertilization of the eggs is internal, and the males possess some special type of organ for transferring sperm to the female. For example, the intromittent organ may take the form of the clasper seen in the male elasmobranchs and holocephalans, or the gonopodium present in some of the teleost species with internal fertilization. The intromittent organ may vary in size throughout the course of the year, becoming enlarged during the breeding season, and being partially regressed at other times of the year.

The males of species with internal fertilization generally display some form of courtship behaviour in order to ensure female receptivity. The performance of an elaborate courtship behaviour is, however, not only seen amongst the species which have internal fertilization, but is also the rule

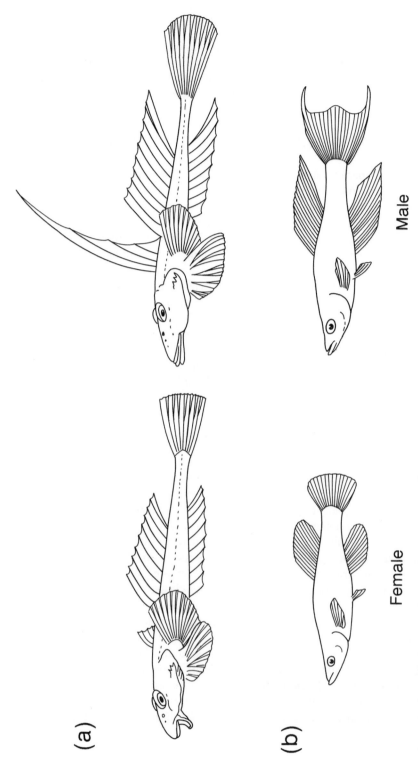

Female

Male

Fig. 9.20 Secondary sexual characteristics in (a) the dragonet, *Callionymus lyra*, and (b) the lyretail, *Aphyosemion arnoldi*.

for fish species that either build nests or defend a spawning territory (e.g. sticklebacks and centrarchids). Visual signals may be used to attract potential mates into the spawning territiory prior to courtship, but chemical or acoustic signals may be more effective over longer ranges, especially in turbid water.

Acoustic signals

Combinations of visual and acoustic signalling are used in the courtship displays of several species of gobies, and males of different species of midshipman also use several types of signals in order to attract a mate. Visual signalling in the midshipman involves the use of photophores, the arrangements of which are species-specific. Acoustic signals also play an important role in mate attraction in the plainfin midshipman, *Porichthys notatus*. The mature male attracts the females to his nesting territory, in the crevices amongst intertidal rocks, by making low-frequency (100 Hz) humming signals. The signals are generated by the contraction of sonic drumming muscles attached to the wall of the gas bladder. The frequency of the acoustic signal is fundamentally equal to the muscle contraction rate.

In the plainfin midshipman there is both intersexual and intrasexual dimorphism in the muscular system that generates the acoustic signals. In large mature males there is considerable hypertrophy of the sonic muscles, with the increase in muscle size occurring near the time of sexual maturity. Only these males produce the humming sounds and use acoustic signals for mate acquisition and territorial defence.

Some small males do become sexually mature, but these males do not establish a nesting territory. The small mature males adopt sneak spawning behaviour in which they either enter the nest of a large male along with the female, or release sperm at the nest entrance and fan it towards the eggs using vibrations of the caudal fin.

Immature fish, small mature males and mature females seldom produce acoustic signals, and their sonic muscles are relatively small. It is thought that the differentiation of the sonic muscles is modulated by the changes in circulating levels of androgens that occur during sexual maturation. The marked elevation of circulating levels of 11-ketotestosterone seen in large males at maturation may be involved in triggering the initiation, and maintenance, of dimorphism of the sonic muscles.

The courtship behaviour of some gadoid species, such as the haddock and the Atlantic cod, includes both a visual display and acoustic signalling by the male. In the haddock, courtship commences when the male approaches a female. The male holds his fins erect, and utters a short series of repeated knocking sounds during the course of his approach. The

sexually active male then swims in a series of tight circles, making exaggerated body movements. During this phase of the courtship display all
the fins are held erect, a heavy pattern of pigmentation appears and the
male issues a long series of calls consisting of rapidly repeated knocks or
grunts.

Towards the latter stages of courtship the male leads the ripe female up
through the water column, his tail moving sinuously from side to side. The
male continues to hold all his fins extended, and he produces a continuous
rasping call.

The male then mounts the female from below, which brings the genital
openings of the two fish into close contact. The continuous call of the male
increases in frequency and reaches a hum.

Sound production ceases as eggs and sperm are released into the water.
Throughout the courtship display the calls of the male consist of series of
repeated pulses, the repetition rate varying with the stage of the behavioural display. In contrast to the male, the female remains silent throughout courtship.

Thus, many fish species may undergo elaborate courtship behaviours
prior to spawning, and several types of signalling may be involved in intersexual communication. Visual, acoustic and chemical signals may be
used in attracting mates, and chemical signals may also be used in order
to convey information about sexual receptivity. Visual and acoustic signals
appear to be those most frequently employed during the final stages of
courtship, with courtship displays culminating in the manoeuvring of the
breeding pair into a position that ensures fertilization of the eggs.

9.15 CONCLUDING COMMENTS

In the majority of temperate-zone fish species spawning is an annual
event, with the fish having a breeding season spanning a period of a few
weeks to a couple of months. Spawning will often be preceded by the fish
making a reproductive migration to specific spawning grounds or sites.
This may involve relatively small-scale movements, such as from a large
water body to feeder streams, or the congregation of the mature fish in the
vegetation-fringed shallows of rivers and lakes. In some marine and diadromous species the distance covered during the course of the spawning
migration may, however, be several hundred kilometres.

The annual cycle of many marine species involves the fish migrating
between geographically discrete feeding and spawning grounds at different
times of the year. This will often be closely linked with distinct cycles of
repletion and depletion of tissue reserves of lipids and proteins. Repletion of
the energy reserves occurs during the summer feeding season, to be

followed by depletion during the course of the winter as the reserves are utilized both as metabolic fuels and for gonad development.

In the female, the gradual accumulation of yolk in the vitellogenic oocytes leads to a pronounced increase in the size of the ovaries. By the completion of vitellogenesis the ovary may account for 10–20% of the body weight of the female fish. The postvitellogenic oocytes must undergo a final maturation prior to ovulation and spawning, and there is often a rapid and marked volume increase in the oocytes at this stage. The maturation-associated volume increases are most pronounced in marine species having buoyant, pelagic eggs, and the primary cause of the increased volume is the osmotic uptake of water.

In the majority of fish species, reproductive development is probably dependent upon the entrainment of an endogenous rhythm by environmental cues. Photoperiodic cues seem to be of greatest importance in the control of rates of gamete development and gonadal growth in fish species of the temperate zone. Thus, the exposure of fish to different photoperiods can be used to influence the reproductive cycle. The use of compressed and extended seasonal light cycles, along with light regimes of different photophase length, provides a method by which the researcher can induce broodstock fish to produce gametes at almost any season of the year. Photoperiodic manipulations of this type currently have some commercial application in the induction of out-of-season production of eggs and fry for the fish-farming industry. At present this form of broodstock management is carried out on very few commercially important species, but these environmental manipulation techniques would seem to offer promise for use with other species of interest to the aquaculture industry.

Whilst photoperiod may be the most important exogenous factor governing rates of reproductive development during the early phases of the cycle, and up until the completion of vitellogenesis, other environmental cues may be required in order for spawning to occur. The processes of gamete maturation, ovulation and spawning may, for example, be delayed under adverse temperature conditions or in the absence of a suitable spawning substrate. In addition, the induction of spermiation and ovulation, and the triggering of spawning behaviour, may require the added stimulus provided by the presence of a potential partner.

The physiological changes that occur during the course of the reproductive cycle are mediated via hormonal factors, and the hypothalamic–pituitary–gonadal axis is the primary route of endocrine control. Interventions have been made at different levels of this endocrine cascade in attempts to manipulate and control various aspects of the reproductive cycle of fish held in captivity. Manipulations, involving the treatment of fish with exogenous hormones, have not been particularly successful in the control of the earliest phases of the cycle, but considerable commercial use is made of hormone treatments for the induction of oocyte maturation and ovulation in broodstock populations of a range of species.

FURTHER READING

Books

Breder, C.M. jun. and Rosen, D.E. (1966) *Modes of Reproduction in Fishes: How Fishes Breed*, T.F.H. Publications, Jersey City.

Hoar, W.S. and Randall, D.J. (eds)(1969) *Fish Physiology, Vol. III*, Academic Press, London.

Hoar, W.S. and Randall, D.J. (eds)(1988) *Fish Physiology, Vol. XIA*, Academic Press, London.

Hoar, W.S., Randall, D.J. and Donaldson, E.M. (eds)(1983) *Fish Physiology, Vols IXA,B*, Academic Press, London.

Keenleyside, M.H.A. (1979) *Diversity and Adaptation in Fish Behaviour*, Springer-Verlag, Berlin.

Potts, G.W. and Wootton, R.J. (eds)(1984) *Fish Reproduction: Strategies and Tactics*, Academic Press, London.

Purdom, C.E. (1993) *Genetics and Fish Breeding*, Chapman & Hall, London.

Scott, A.P., Sumpter, J.P., Kime, D.E. and Rolfe, M.S. (eds)(1991) *Reproductive Physiology of Fish*, FishSymp 91, Sheffield.

Review articles and papers

Balon, E.K. (1975) Reproductive guilds of fishes: a proposal and definitions. *J. Fish. Res. Bd Canada*, **32**, 821–63.

Bass, A.H. (1990) Sounds from the intertidal zone: Vocalizing fish. *BioScience*, **40**, 249–58.

Bromage, N. and Cumaranatunga, P.R.C. (1988) Egg production in the Rainbow trout. *Recent Adv. Aquacult.*, **3**, 63-138.

Bromage, N., Jones, J., Randall, C., Thrush, M., Davies, B., Springate, J., Duston, J. and Barker, G. (1992) Broodstock management, fecundity, egg quality and the timing of egg production in the rainbow trout (*Oncorhynchus mykiss*). *Aquaculture*, **100**, 141–66.

Cochran, R.C. (1992) *In vivo* and *in vitro* evidence for the role of hormones in fish spermatogenesis. *J. exp. Zool.*, **261**, 143–50.

Elgar, M.A. (1990) Evolutionary compromise between a few large and many small eggs: comparative evidence in teleost fish. *Oikos*, **59**, 283–87.

Francis, R.C. (1992) Sexual lability in teleosts: Developmental factors. *Q. Rev. Biol.*, **67**, 1–18.

Grier, J.H. (1981) Cellular organization of the testis and spermatogenesis in fishes. *Amer. Zool.*, **21**, 345–57.

Huntingford, F.A. and Torricelli, P. (1993) Behavioural Ecology of Fishes. *Mar. Behav. Physiol. (Special Issue)* **23**.

van den Hurk, R. and Resink, J.W. (1992) Male reproductive system as sex pheromone producer in teleost fish. *J. exp. Zool.*, **261**, 204–13.

Kime, D.E. (1993) 'Classical' and 'non-classical' reproductive steroids in fish. *Rev. Fish Biol. Fisheries*, **3**, 160–80.

Peter, R.E. (1982) Neuroendocrine control of reproduction in teleosts. *Can. J. Fish. aquat. Sci.,* **39,** 48–55.

Wallace, R.A. and Selman, K. (1981) Cellular and dynamic aspects of oocyte growth in teleosts. *Amer. Zool.,* **21,** 325–43.

van Weerd, J.H. and Richter, C.J.J. (1991) Sex pheromones and ovarian development in teleost fish. *Comp. Biochem. Physiol.,* **100A,** 517–27.

Wourms, J.P. (1977) Reproduction and development in chondrichthyan fishes. *Amer. Zool.,* **17,** 79–110.

Chapter ten

Development of eggs and larvae

10.1 INTRODUCTION

The oocytes pass through a series of developmental stages that ultimately culminate in the production of a mature female gamete. Much of the increase in size of the oocytes is the result of the gradual accumulation of yolk reserves, but the oocytes of many species undergo a rapid and pronounced increase in volume during the final maturation period.

Maturation-associated increases in volume range from slight in most freshwater and euryhaline species, to several-fold in some marine species. Comparisons of the water content of ovaries containing large post-vitellogenic oocytes with the water content of mature eggs have implicated water uptake (hydration) as the primary cause for the volume increases observed during oocyte maturation. Oocyte maturation is usually closely followed by ovulation and spawning.

Whilst external fertilization of eggs is not universal amongst fish species, there are only very few species that have internal fertilization followed by egg retention within the body of the female. The external fertilization of eggs means that all later stages of development occur outside the body of the female.

One consequence of egg laying is that the eggs must contain all the nutrients required by the developing embryo from the time of egg fertilization until the time at which it becomes a young fish and is able to meet its needs by consuming exogenous food. The nutrients utilized from the yolk during development are either incorporated into tissues during embryonic growth, or are used as metabolic fuels.

Young fish may also be dependent upon yolk reserves for some time after hatching, and should the nutrient supply be insufficient, the newly hatched larva risks death from starvation. Consequently, the period of transition from complete reliance on the yolk nutrient reserves to exogenous feeding is deemed to be critical for survival and future growth.

10.2   MODES OF REPRODUCTION AND DEVELOPMENT

The different modes of reproduction can be classified according to the environment in which the embryos develop, and the sources of the nutrients supporting embryonic growth.

Oviparity

Oviparity, or egg laying, refers to the situation where the development of the fertilized egg occurs outside the body of the female. The young hatch when the egg envelope, shell or capsule is broken. Oviparous fish may be further categorized as being either *ovuliparous* or *zygoparous.*

Ovuliparity refers to the release of ova from the reproductive tract of the female followed by fertilization or activation in the external environment. Thus, all organisms that have external fertilization, and this includes most teleosts, are said to be ovuliparous.

Zygoparity refers to the oviparous condition in which the zygotes (i.e. fertilized ova – products of fusion between the eggs and sperm) are retained within the body of the female for a short period of time before being released into the environment. Obviously, zygoparous species display internal fertilization, with there being a transfer of male sperm to the reproductive tract of the female. Zygoparous reproduction characterizes all skates, some sharks and a small number of teleosts.

Irrespective of whether the fertilization of the eggs occurs internally or externally, the nutrients for the developing embryos of oviparous species are provided by the egg yolk.

Internal development

In the majority of fish species with internal fertilization the eggs are retained and undergo development in the maternal reproductive system. Hatching either precedes or coincides with parturition. The result is that the female gives birth to free-living young fish. Retention of the fertilized egg within the body of the female is the dominant mode of reproduction in the cartilaginous fish, being found in over 50% of the extant species. Egg retention is, however, seen in only about 2–3% of teleost species.

The retention of the developing eggs and embryos within the maternal body requires structural modifications to be made to the reproductive tract in order to facilitate this retention. In addition, the mechanisms of endocrinological control over the reproductive cycle undergo some change. In the cartilaginous fish and the coelacanth, *Latimeria chalumnae*, gestation takes place in the oviduct. In the teleosts the developing eggs and embryos are retained within either the ovarian follicle or the ovarian lumen.

Ovoviviparity

In the ovoviviparous species the eggs usually develop within the uterus (modified oviduct) of the female. Fertilization of the eggs occurs internally and the eggs are retained until hatching or beyond. The developing embryos do not receive any supply of foodstuffs from the female but must rely on the yolk of the egg for nutrition. This form of nutrition is known as lecithotrophy. The developing embryos must, however, rely on the female for supplies of oxygen. During pregnancy the wall of the oviduct becomes enlarged and richly supplied with blood vessels.

Viviparity

Viviparous species have eggs which develop either within the uterus (i.e. several species of sharks), or within the ovary (i.e. teleosts). The developing embryos receive some nutrient supply from the female in addition to that provided as yolk in the egg. This form of nutrition is known as matrotrophy. The additional nutrients can be supplied in various forms. In some species the female secretes a nutrient-rich fluid which is taken up by the developing young as a form of 'soup'. In other species some form of 'placenta' may develop, allowing a more direct transfer of nutrients from the blood of the female to the developing embryo.

10.3 REPRODUCTIVE MODES IN ELASMOBRANCHS

Irrespective of the type of reproductive mode shown, the fertilization of the eggs in elasmobranchs occurs internally. This is in distinct contrast to the broadcast spawning employed by the majority of the osteichthyans, in which the eggs and sperm are released freely into the water column. In order to achieve internal fertilization of the eggs the elasmobranchs must engage in courtship and copulatory activities that may be several hours in duration.

The males possess specially modified sexual organs for the transfer of sperm to the female. Parts of the pelvic fins of the male are modified to form elongated structures, known as claspers, and the sperm pass from the male to the female along grooves in the clasper. Associated with each clasper of the male shark is a muscular bladder, or siphon sac, that is filled with water just prior to mating.

During copulation the male flexes one of the claspers at an angle to the main body axis and inserts it into the reproductive tract of the female. Once the clasper has been inserted into the body of the female, contraction of the siphon sac results in the discharge of its contents. The flow of water from the sac carries sperm along a groove in the clasper and into the reproductive tract of the female fish.

The male skates possess a clasper gland, rather than a muscular sac. In skates the flow of sperm along the clasper tends to be slow, since there is no rapid discharge of water from the clasper gland during copulation.

The sperm of the male passes into the reproductive ducts of the female, and begins to ascend. Closely allied to the oviduct is the nidamental, or shell, gland (Fig. 10.1). This gland has several functions, including the

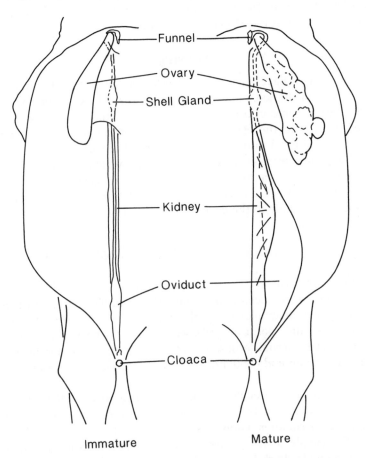

Immature	Mature

Fig. 10.1 Reproductive system of a female elasmobranch, showing both the immature and mature (pregnant) condition.

secretion of both an albumen-like substance and the tough collagenous material that forms the protective coat around the fertilized egg. The nidamental gland also functions in the storage of sperm, which may remain viable for several weeks or months following copulation.

Mature eggs are released from the ovary, or ovaries, and then enter the oviducts. In many species of shark only a single ovary is functional, although most oviparous sharks have both ovaries functional. Whilst fertilization of the eggs may occur in the oviducts in some species, it is more usual that the eggs are fertilized in the nidamental gland. The fact that sperm may be stored for extended periods means that there is no need for copulation to coincide with ovulation. Thus, females can continue to produce fertilized eggs even though they may not have copulated with a male for several days, or even weeks. Following fertilization of the eggs reproductive development may take one of the three main forms – oviparous, ovoviviparous or viviparous.

Oviparous elasmobranchs

All the skates and some shark species are oviparous. The oviparous species tend to be amongst the smallest of the elasmobranchs. The retention of developing embryos within the body of the female would impose strict limits both upon the number and upon the size of the offspring at birth, so oviparity may be an adaptation towards increasing progeny numbers.

The eggs are ovulated and fertilized at regular intervals following copulation. The clearnose skate, *Raja eglanteria*, for example, produces pairs of fertile eggs at intervals of a few days throughout a 6 month egg-laying season. It is usual for the oviparous elasmobranchs to produce ovulated and fertilized eggs in pairs. Following fertilization the eggs are encapsulated in a horny protective egg case.

The egg cases are tough and leathery, and they are often camouflaged by a dark or mottled appearance. Long string-like appendages, or tendrils, on the cases ensure that they become entangled amongst rocks or attached to vegetation. Thus, the egg cases are prevented from being washed away by wave action. The production of a sticky mucous material on the external coating of the egg case may also act as a means of attachment to the substrate.

Despite the fact that the egg case is tough and fibrous it is relatively permeable to ions, urea and other small molecules. Thus, the developing embryo is mechanically protected from the vagaries of the external environment, but soon becomes surrounded with a fluid of similar composition to sea water.

The embryo is, however, capable of maintaining its body concentrations of the major ions below, and urea above, those of the surrounding capsular fluids. Chloride cells appear on the gills at a relatively early developmental stage, and these cells are probably involved in ionoregulation in the embryo. It is unlikely that the rectal gland makes any substantial contribution to ionoregulation at this stage. Although the rectal gland is differentiated, the excretion of ions may be hindered due to rectal closure. Thus, in the oviparous elasmobranchs, it seems most likely that it is the external surface of the embryo, and not the egg capsule wall, that acts as the major barrier to the movements of ions, organic solutes and water.

Ovoviviparous elasmobranchs

Ovoviviparity, involving the retention of the developing embryo within the body of the female, is an effective method of protecting the young from the risks associated with exposure to the external environment. In strictly ovoviviparous species the young develop within the uterus, but without any direct maternal connection or additional nutrient supply over and above that present in the egg. In the ovoviviparous elasmobranchs, the horny egg case is replaced by a thin membraneous sheath or envelope.

Unlike the oviparous species, whose embryos are surrounded by a fluid similar to sea water throughout their development, the eggs and embryos of ovoviviparous species are bathed in fluids that have ionic concentrations intermediate between those of the maternal plasma and sea water. Urea concentrations in the fluids approach those found in the plasma of the female. Later in development the fluids bathing the embryos come to resemble sea water with respect to the concentrations of the major ions. These fluids are, however, more acidic than sea water, having a pH of around 5–6. This ensures that the potentially toxic ammonia, which tends to accumulate with the passage of time, is converted to NH_4^+ and detoxified.

The numbers of eggs produced show considerable intraspecific and interspecific variations. There is, however, usually a close relationship between the size of the female and the number of offspring produced. For example, large female spurdogs, *Squalus acanthias*, generally 'give birth' to more offspring than smaller conspecifics. Mature females may range in length from about 70 cm to 120 cm or longer. The smallest females may sustain a brood of two to three embryos in each of their two enlarged oviducts (or uteri). Large females, on the other hand, may give birth to 12–16 young. In the spurdog gestation lasts about 2 years, and the young are about 20–30 cm in length at birth.

The gestation period of the large (500 cm long) tiger shark, *Galeocerdo cuvieri*, is somewhat shorter than that of the spurdog, being about 1 year in duration. A large female tiger shark may produce 30–55 young, each of which is 70–80 cm in length at birth.

In the ovoviviparous species the developing young are sustained only by the nutrients available from the egg yolk. This is not, however, the case in all of the elasmobranchs.

Viviparous elasmobranchs

The egg yolk may be totally absorbed by the embryo relatively early in development. The young may then emerge from the egg whilst still within the uterus of the female. The female then provides additional nourishment for the developing young. The exact form of nutrient supply varies considerably amongst species.

Aplacental viviparity

Among the lamnoid sharks (e.g. shortfin mako, *Isurus oxyrinchus*, porbeagle, *Lamna nasus*, thresher, *Alopias vulpinus*, and great white shark, *Carcharodon carcharias*) the embryos hatch during the first few months of development. In order to nourish the small numbers of developing young, usually two to eight, the female continues to ovulate throughout much of the gestation period. Thus, the female continues to produce eggs even when pregnant and these arrive in the uterus as suitably sized meals for the developing young. The steady supply of nutritious eggs is consumed by the growing young, a practice known as *oophagy*. The guts of the young fish may become greatly distended due to the presence of these eggs. When ovulation ceases the young are able to continue growth and development by absorbing the yolk reserves accumulated in their guts. In other species, unfertilized eggs ovulated at the same time as those that were fertilized may provide an additional nutrient supply for the young developing in the uterus.

In the sand tiger shark, *Carcharias taurus*, the first embryo to hatch begins to consume the other developing embryos, a form of sibling cannibalism known as *embryophagy*. Once its siblings have all been consumed the remaining embryo in each uterine branch continues to nourish itself via oophagy. Thus, in this species two large offspring are produced per female.

Another form of aplacental viviparity is seen in several species of rays, including the stingrays, *Dasyatis* spp., electric rays, *Torpedo* spp., and devil rays, *Mobula* spp. The egg yolk becomes exhausted relatively early in development and the embryos hatch within the uterus. The inner surface of the uterine wall is lined with a series of string-like projections, or *trophonemata*. These projections secrete a nutritious fluid, which is often termed 'uterine milk'. In none of these species do the embryos become intimately attached to the uterine walls.

Placental viviparity

In two families of sharks (the Carcharhinidae, or requiem sharks, and the Sphyrnidae, or hammerhead sharks), the developing young, after having used up all their yolk supply, become attached to the wall of the uterus by means of a 'yolk sac placenta'. The young receive nutrients and oxygen from the mother via this placental connection much in the same manner as is seen in placental mammals.

Additional means of nutrient uptake are possible in the embryos of certain species of placental viviparous sharks. The placental connection may develop a series of projections, known as *appendiculae*, and these probably aid in the absorption of nutrients from the uterine fluid that surrounds the developing young.

Viviparous sharks usually have a gestation period of about 9–12 months and give birth to 6–14 offspring, although the blue shark, *Prionace glauca*, can produce as many as 40–50 young in a single pregnancy. Having a gestation period of just less than a year, the majority of viviparous sharks tend to produce offspring on an annual basis, but some, such as the sandbar shark, *Carcharhinus plumbeus*, give birth to young only in alternate years.

10.4 OOCYTE MATURATION AND HYDRATION

Hydration is pronounced in the mature oocytes of marine fish that spawn buoyant pelagic eggs into sea water. The water content of such eggs may be 90–92%, whereas the water content of demersal eggs, many of which do not undergo hydration during final maturation, may be only 60–70%.

The physiological basis for the maturation-specific uptake of water by the oocyte remains to be elucidated, but both inorganic and organic osmolytes appear to be involved. There are increases in levels of inorganic ions, such as Na^+ and K^+, during maturation of the oocyte in several teleosts (e.g. medaka, *Oryzias latipes*, mummichog, *Fundulus heteroclitus*, Nile tilapia, *Oreochromis niloticus*, plaice, *Pleuronectes platessa*, and Atlantic cod, *Gadus morhua*), and this may provide at least part of the impetus for the osmotic influx of water.

Organic solutes are also almost certainly involved in the process in some species, because the content of free amino acids is generally found to be higher in the eggs than in the postvitellogenic oocytes of marine species such as the plaice and cod. The source of these free amino acids is probably the yolk protein, which undergoes proteolysis in maturing and hydrating oocytes. Incomplete proteolysis of the yolk proteins would also give rise to a number of small proteins and peptides that may be of osmotic significance in the hydration process.

Thus, there appear to be two mechanisms to explain the hydration of the maturing oocytes. In the first, the influx of inorganic ions during the course of oocyte maturation creates conditions that lead to the osmotic influx of water. This mechanism seems to be present in most species investigated, and probably provides the primary, if not the sole, means of hydration of the eggs of freshwater and euryhaline teleosts. The second mechanism involves the raising of the osmotic pressure of the oocyte during maturation due to the proteolysis of the yolk proteins into osmotically active peptides and free amino acids. This latter mechanism appears to be particularly pronounced in the maturing oocytes of those marine fish species that spawn pelagic eggs.

10.5   CHEMICAL COMPOSITION OF EGGS

Oviparity and external fertilization of eggs is most common amongst fish species. This means that the eggs must contain all the nutrients required by the developing embryo and young fish from the time of fertilization of the egg until the time at which the fish is able to meet its needs via the consumption of exogenous food. The materials required during the developmental period can be broadly divided into those needed directly for the synthesis of the embryonic tissues, and those that are used as metabolic fuels. The total, and relative, amounts of the different nutrients that are needed will obviously vary depending upon such factors as egg incubation time, the size of the young at hatch, and the length of time the young are reliant upon endogenous supplies before they can meet all their needs from other sources.

The egg is bounded by a series of membranes. The outermost membrane, or *chorion*, is relatively tough, and within this is a thinner vitelline membrane that surrounds the yolk (Fig. 10.2). Most of the egg volume is taken up by the yolk mass, eggs with a large amount of yolk being termed *telolecithal*.

At the time of spawning the eggs are composed primarily of water and protein, with a single or series of small oil or lipid droplets (Table 10.1). In addition to the major nutrients – the proteins and lipids – that are used both as building blocks for tissue synthesis and as metabolic fuels, the egg will contain a series of micronutrients – vitamins and minerals – that are required in small doses if the egg is to develop correctly and the young fish grow normally.

It is, for example, known that high concentrations of vitamin C (ascorbic acid) occur in the ovarian tissues of several fish species, including cyprinids, gadoids and salmonids. Whilst the vitamin present in the ovarian tissue seems primarily to be involved in the synthesis and metabolism of

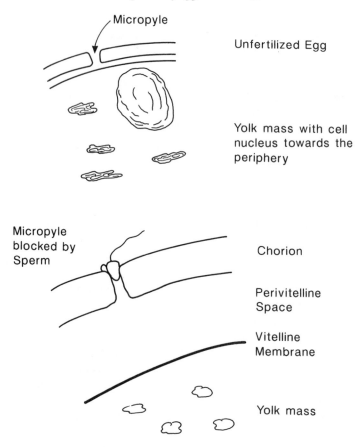

Fig. 10.2 General structure of fish ova prior to and following fertilization, illustrating the changes that occur in the membranes and the development of the perivitelline space. Following fertilization or activation, the egg swells, with the formation of the fluid-filled perivitelline space, and the chorion begins to harden.

the gonadal sex steroids, there is some deposition of ascorbic acid in the eggs. The ascorbic acid deposited in the eggs may play an important role during embryonic development because female broodstock salmonids fed on diets that are deficient in ascorbic acid produce eggs that both contain low levels of the vitamin and have poor hatchability.

Ascorbic acid has a number of biological functions, including acting as a cofactor in the synthesis of collagen, a protein which is abundant in connective tissue and cartilage. Thus, ascorbic acid requirements are likely to be high during embryonic development, which is a period of marked

Table 10.1 Chemical compositions of the eggs of a range of fish species

Species	Approximate percentage composition (% wet weight)				
	Water	Protein	Lipid	Carbohydrate	Ash
Rainbow trout, *Oncorhynchus mykiss*	63.3	28.5	5.5	0.2	1.5
Carp, *Cyprinus carpio*	70.5	20.3	4.7	0.7	1.3
Pike, *Esox lucius*	65.7	28.1	1.7	0.7	1.5
Bream, *Abramis brama*	66.4	23.5	3.2	Trace	1.5
Sockeye salmon, *Oncorhynchus nerka*	59.7	27.4	10.8	Trace	1.5
Sturgeon, *Huso huso*	58.9	25.5	13.6	Trace	1.5

skeletal growth and tissue differentiation. At this stage the only source of the vitamin available to the embryo is the egg yolk.

The proportions of the major nutrients found in fish eggs tend to vary, but some generalizations can be made. For example, lipid content tends to be highest in eggs having long incubation times, and the free amino acid content is usually much higher in the pelagic eggs of marine species than in the eggs produced by other fish species.

Egg lipids

There are certain characteristic patterns of lipid deposition in fish eggs. Depending upon species, total lipid contents of ripe eggs may amount to about 2–10% of the egg weight on a wet weight basis (Table 10.1), and lipid-rich eggs tend to have distinct oil globules that contain high proportions of neutral lipids (triacylglycerols, sterols or wax esters). Thus, higher lipid contents in eggs are usually achieved by increasing the amounts of neutral lipids. As a consequence of this, lipid-rich eggs generally contain lower proportions of phospholipids (less than 50%) than do lipid-poor eggs, in which phospholipids may account for 60–80% of the total lipid content.

As a general rule of thumb it can be said that the neutral lipids will provide metabolic fuel for the embryo during the course of development, whereas the phospholipids will provide the essential fatty acids that are deposited into cell membranes as the tissues become differentiated. Thus, high overall lipid contents, coupled to high levels of neutral lipids, are characteristic of those eggs which have long incubation times. For example, salmonids, the eggs of which may have incubation times of several weeks, tend to produce lipid-rich eggs (Table 10.1), with a high content of triacylglycerols.

Within the phospholipid class, phosphatidylcholines seem to be the major component of the egg polar lipids, followed by phosphatidylethanolamines. The phospholipids almost invariably contain larger proportions of the essential (n − 3) fatty acids than do the triacylglycerols regardless of fish species. Whilst the fatty acid compositions of the egg phospholipids may vary to some extent, the (n − 3) fatty acid 22:6(n − 3) is consistently found to predominate, and 20:5(n − 3) also tends to be present in quite large amounts.

These two fatty acids are required for the normal development of the brain and neural tissues, so any deficiencies in these (n − 3) essential fatty acids might lead to the development and survival of the embryo and offspring being compromised. Such problems are likely to be most severe for those larvae that hatch from eggs with long incubation times, since the egg yolk is their only source of nutrients throughout the protracted developmental period. Deficiencies in the amounts of 22:6(n − 3) and

20:5(n — 3) incorporated into the egg phospholipids are less likely to have severe consequences for the development and survival of larvae which emerge from small eggs with short incubation times. This may be the case because the first-feeding larvae will be able to obtain supplies of the essential fatty acids via the diet only a short time after hatching.

Egg yolk free amino acids

The levels of free amino acids found in the pelagic eggs of a number of marine species are much higher than those recorded in either demersal marine fish eggs or in the eggs of freshwater fish species. For example, the free amino acid pool may account for 20–50% of the total amino acid content (free plus protein-bound amino acids) in pelagic fish eggs, whereas free amino acids rarely exceed 5% of the total in eggs of other types.

The free amino acids, which are derived from the breakdown of yolk proteins, begin to accumulate during final maturation, leading to an osmotic influx of water. Thus, oocyte volume increases markedly around the time of ovulation, and the previously opaque oocyte becomes a transparent and buoyant egg. The osmotic influx of water also seems to be an adaptation that enables the embryo to survive in an hyperosmotic environment, because the embryo appears to be largely dependent upon endogenous water sources during the course of development. Consequently, the free amino acids have been considered to have a primary role as osmolytes in pelagic marine fish eggs. These amino acids may, however, also be a, if not the, major source of metabolic fuel for developing marine fish eggs and larvae.

Amongst the essential amino acids, leucine, lysine, valine and isoleucine tend to dominate in the free pool, whereas alanine and serine are the dominant non-essential acids. During development there is a decrease in the free amino acid pool, and there does not appear to be any preferential utilization of essential amino acids before the non-essential, or vice versa. The free amino acids do not seem to be lost from the egg via diffusion to the surrounding medium, nor do they appear to be used primarily for protein synthesis and incorporation into body tissues. Thus, the amino acids that are lost from the free pool are probably used as metabolic fuel, with perhaps 50%, or more, of the energy requirements of the developing embryo being met by amino acid catabolism.

10.6   FERTILIZATION AND EMBRYONIC DEVELOPMENT

The unfertilized egg is bounded by a series of membranes, the outer of which, the chorion, tends to be relatively thick and tough. The chorion has a small funnel-shaped micropyle through which the sperm enters to fertilize the egg. Fertilization is usually monospermic, the single sperm entering the egg via the micropyle, which is too small to allow the passage of more than one sperm at a time (Fig. 10.2).

Changes in the pH of the fluids of the sperm duct that occur during final maturation may lead to increased motility of the sperm, but in many species the sperm do not become fully active until after they have been shed. The sperm of many marine teleosts, for example, do not begin swimming vigorously until exposed to sea water. The differences in ionic and osmotic strengths between the seminal fluid and sea water seem to be the stimuli required for increased activity.

In some marine species, such as the Pacific herring, *Clupea harengus pallasi*, however, exposure to a hyperosmotic environment is insufficient to induce a marked increase in motility of the sperm. In these species, exposure of the sperm to unfertilized eggs is required in order for the sperm to display maximal swimming activity. Thus, the unfertilized eggs seem to release a substance, probably proteinaceous in nature, that serves to activate the sperm. Contact with the sperm-activating substance leads to an increase in the number of vigorously swimming sperm and lengthens the duration of swimming.

It is also possible that the sperm-activating substance acts as an attractant, guiding the sperm to the micropyle in the outer membrane of the egg. For example, a factor responsible for activation and congregation of sperm has been shown to be localized in the micropylar region of the egg membrane in some cyprinids and salmonids. Thus, there may be a general mechanism by which sperm are fully activated by a sperm-activating substance released from the micropylar region of the unfertilized egg, and the substance may also aid the sperm in locating and penetrating the micropyle to reach the inner membrane of the egg for fertilization.

The fertilization of the egg by the sperm requires the presence of small quantities of divalent ions (Ca^{2+} and Mg^{2+}), and once the egg has been activated by the sperm the micropyle is plugged. Fertilization by the sperm initiates the second meiotic division in the egg, which at spawning contains two maternal chromosome sets. Following the second meiotic division, one of the sets of maternal chromosomes is lost as the polar body is expelled from the fertilized egg. Thus, at this stage the fertilized egg now contains one maternal and one paternal chromosome set.

Egg membranes

Following fertilization the egg absorbs some water and there is separation of the chorion from the vitelline membrane leading to the formation of the perivitelline space. There are also profound changes in the ultrastructure, cytochemistry and histochemistry of the membranes surrounding the egg.

The chorion hardens and, thus, serves to protect the embryo during the earliest stages of development. There are also enzymes present in the egg membranes which may serve to protect the developing egg against bacterial and fungal attack.

The egg membrane enzymes do not appear until after the egg has been activated by fusion with the sperm. The enzyme activity is lost either on death of the developing embryo, or after hatching of the egg.

Bacterial and fungal cell walls consist of various polysaccharides, such as chitin, α- and β-linked glucans and mannans, that are responsible for the strength and integrity of the cell walls. The cell wall polysaccharides often exist in the form of chemical complexes with protein.

The outer membranes of the fertilized eggs of a number of fish species, including some salmonids, are known to possess enzymes that can degrade the cell wall polysaccharides of bacteria and fungi, and this may either cause cell lysis or hinder the growth of micro-organisms that attack the egg surface. Thus, the egg membrane enzymes probably have an important defensive role in protecting the embryo against invaders or pathogens, whereas the thick chorion and enlarged perivitelline space serve to protect the developing embryo against physical damage.

Embryonic development

Following fertilization the egg begins to divide (cleave), but division of the large yolk-containing cells is retarded, resulting in unequal, or meroblastic, cleavage. The result of this is the formation of a layer, the *blastodisc*, on the surface of the yolk at one pole of the egg. It is the blastodisc that undergoes further development to form the embryo (Fig. 10.3).

The periphery of the blastodisc then begins to overgrow the yolk, a process known as *epiboly*, and eventually encloses it as a thin sheet. Thus, the embryo can be envisaged as floating on its yolk, but in practice the embryo lies under the yolk sac because the tissues of the developing embryo are denser than the yolk.

There is a gradual differentiation of the various tissues and with the passage of time the organs become visible. The head region, with pronounced eye cups, soon becomes identifiable. The trunk region elongates and the muscle somites begin to be differentiated. As the trunk lengthens it separates from the yolk sac. The heart becomes visible well before the time of hatching, and there may be the development of a special vitelline circulation on the surface of and within the yolk sac (Fig. 10.3).

The rate at which the embryo develops and uses its yolk supply is very much dependent upon environmental factors, with both temperature and

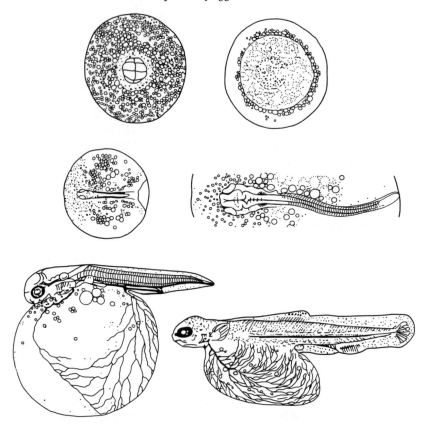

Fig. 10.3 Embryonic development in a salmonid fish, showing an early (8-cell) stage (upper left), epiboly (upper right), an embryo with optic cups (centre), a later stage in which the trunk has separated from the yolk sac (lower left), and a newly hatched fish with yolk sac (lower right).

oxygen concentration being of particular importance. Low oxygen concentrations retard the rate of development and there is usually found to be an increase in the rate of development with an increase in temperature.

10.7   HATCHING

The degree of differentiation of the larva at hatch, and the length of the incubation period depend upon several factors: the fish species, the initial size of the egg and the rearing environment (Table 10.2). For example, in many marine species with pelagic eggs the larvae hatch at very small size. The mouth and jaws may not be fully formed, the eye is unpigmented, the

Table 10.2 The relationship between egg size and the developmental rates of a range of fish species

| Species | Egg diameter (mm) | Hatching length (mm) | Time (days) from fertilization to | | | Temp. range (°C) |
			Hatch	First feed	Meta-morphosis	
Atlantic cod, *Gadus morhua*	1.1–1.9	3.5–4.5	9–25	13–31	30–60	2–12
Plaice, *Pleuronectes platessa*	1.7–2.2	6–7	12–20	18–32	70–100	7–11
Turbot, *Scophthalmus maximus*	0.9–1.2	2–3	5–7	7–9	35–40	13–18
Herring, *Clupea harengus*	0.9–1.7	5–8	7–20	14–32	80–160	6–14
Atlantic salmon, *Salmo salar*	5–6	15–25	50–160	80–300	Not distinct	2–10

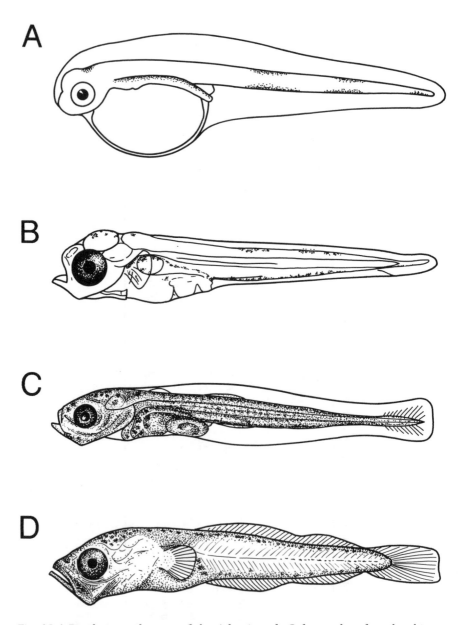

Fig. 10.4 Developmental stages of the Atlantic cod, *Gadus morhua*, from hatching until metamorphosis: (A) newly hatched yolk-sac larva (3.6 mm); (B) exogenous feeding stage (4.5 mm); (C) pre-metamorphosis, with fin fold still present (10 mm); (D) metamorphosis with fin-ray development (14 mm).

larva is transparent with a fin fold running around the trunk and the yolk sac is large (Fig. 10.4). On the other hand, in the salmonids, the newly hatched fish are at a relatively advanced stage of development, and, although the yolk sac is still large, the vascular system and fins are clearly visible (Fig. 10.3).

Hatching stimuli

In the period immediately prior to hatching the embryo becomes active and the body may be flexed rapidly to and fro. The exact stimulus for hatching is not known, but oxygen supply may exert a considerable influence.

Hatching can often be induced if 'full-term' embryos are transferred from oxygen-rich to oxygen-poor water. On the other hand, exposure to hyperoxic conditions may delay hatching in some species. Thus, it seems reasonable to assume that the larvae are able to monitor the environment and time hatching according to external signals.

Oxygen availablity has been shown to influence the time of hatching in species such as the mummichog, medaka, and the two salmonid species rainbow trout, *Oncorhynchus mykiss*, and Atlantic salmon, *Salmo salar*. It has been suggested that hatching may be induced when rates of diffusion of oxygen through the chorion become insufficient to meet the demands of the developing fish. Thus, a critical shortage of oxygen may trigger the hatching response, and this would explain why exposure to hypoxic conditions can lead to accelerated hatching.

Light conditions may also have an important role to play in the induction of hatching in some species, with the eggs of Atlantic salmon, medaka, and zebrafish, *Brachydanio rerio*, tending to hatch during the hours of daylight. In the medaka, exposure to darkness may suppress the hatching of the eggs, but in the Atlantic halibut, *Hippoglossus hippoglossus*, the opposite seems to be the case. When halibut eggs are reared in darkness at 6°C they will tend to hatch at 13–15 days after fertilization, but hatching is delayed or completely inhibited in eggs exposed to light. Hatching can, however, be induced within a short space of time by the transfer of eggs reared in light to conditions of darkness. Exposure to darkness for a period of about 1 h seems to be sufficient to trigger hatching.

Hatching enzymes

Hatching results from the breakdown of the proteinaceous layers of the chorion. Softening of the chorion takes place as a result of the secretion of special hatching enzymes from glands located on the anterior body surface or pharyngeal region of the fish.

The enzymes may not attack the entire surface of the chorion. In the Atlantic halibut egg, for example, there is a specific line of rupture. Thus, in this species the embryo breaks free from the bounding chorion as the 'hatching cap' opens.

In the salmonids, on the other hand, the hatching enzymes appear to exert a more generalized attack on the chorion. This leads to a more widespread weakening of the chorion prior to hatching. Disruption of the weakened chorion is assisted by the rapid flexing movements made by the fish during the process of hatching.

10.8   POSTHATCH DEVELOPMENT

At hatching the larvae of many fish species are more or less transparent, with only scattered melanophores or pigment spots. The notochord and myotomes are usually clearly visible, but there may be little development of cartilage and the skeleton is not ossified. The trunk of the newly hatched larva will usually be encircled by a fin fold, and there may be few signs of fin rays (Fig. 10.4).

The jaws may be seen as a few small skeletal elements, but the mouth and jaw apparatus are not developed at this stage. The gut is generally a straight tube with little or no differentiation into stomach, intestine and associated digestive organs.

The heart is functional in the period prior to hatch, but the remainder of the circulatory system, apart from the specialized vitelline circulation, may be poorly developed. The gills are present as small gill buds and are not functional at this stage, respiratory gas exchange taking place over the general body surface.

Newly hatched Atlantic cod larvae, for example, are approximately 3.5–4.5 mm in length, and there is a large yolk sac. The eyes are unpigmented and the gut appears as a straight tube. After 1–2 days the eyes become pigmented and the alimentary tract begins to become differentiated into distinct regions.

First feeding

As the yolk is resorbed the jaws begin to develop and the mouth opens (Fig. 10.5). The gut and eyes develop further and the larva will start to take exogenous food. In the cod, the larvae may begin by ingesting phytoplankton, and there will be a gradual change from endogenous to exogenous nutrition. At 3–7 days posthatch, depending upon temperature, the cod larvae will begin to consume small zooplanktonic organisms, and during this initial period of exogenous feeding the larvae will be consuming a mixture of phytoplankton and zooplankton.

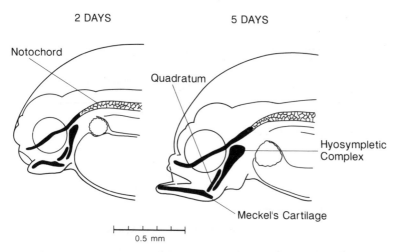

Fig. 10.5 Jaw development in the Atlantic cod, *Gadus morhua*, during the first few days post hatch.

As the phase of first feeding progresses the cod larvae will take copepod nauplii, tintinnids, rotifers and other small organisms, with the size of the prey tending to increase as the larvae grow. The yolk sac will still be visible at this stage, and, in cod, it will not disappear until about 7–10 days post hatch (Fig. 10.4).

This first-feeding stage is critical, and larval mortality is often high. Many larvae either fail to take external food or are unable to capture sufficient prey to ensure continued growth and survival.

The first-feeding larvae of many species locate their prey by sight, but, because they are relatively poor swimmers, the amounts of water the larvae are capable of searching is rather limited. The volume searched by herring, *Clupea harengus*, larvae of length 8–15 mm is, for example, less than 2 l h^{-1}. Consequently, the successful location of areas of high prey density is of vital importance if the fish larvae are to survive the first-feeding phase.

Obviously the length of time a larva can survive without taking external nourishment is dependent upon the amount of yolk remaining in the yolk sac at hatching, and upon the rate at which this yolk is used up following hatch. The interactions between these factors determine the length of time available before the yolk reserves are exhausted, and the larva has become too weak to search for and capture prey (Fig. 10.6). The time at which this occurs has been termed the *point of no return*. Even should the larva encounters prey after this time, it will continue to starve to death because it will be too weak to pursue and capture the prey.

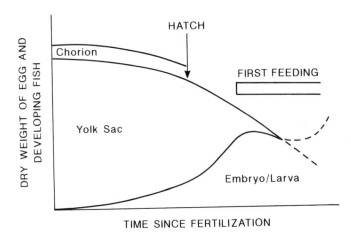

Fig. 10.6 Schematic graphical representation of the development of fish from fertilization until the completion of the transition to exogenous feeding.

Those larvae which manage to capture sufficient prey to survive this critical first-feeding phase rapidly increase the efficiency with which they capture prey. As the larvae grow and increase in experience, the size and type of prey organisms tend to change.

Thus, larvae that survive the first-feeding phase have high rates of food consumption and growth. Atlantic cod larvae, for example, may consume almost their own body weight of food each day, and show rates of weight gain of 10–20% per day. The larval period may last for approximately 35–45 days in cod, during which time they will have increased in length from about 4 mm to 13 mm. Once the cod larvae reach a length of about 13 mm they will begin to undergo the process of metamorphosis (Fig. 10.4).

Metamorphosis

Metamorphosis can be considered to represent the transition from the larval to the adult form, and it involves structural changes in many organ systems (Fig. 10.4). Whilst a clear metamorphosis is seen in many species, the developmental changes may be much more gradual and less pronounced in others. Thus, metamorphosis may take place over a protracted time period in some fish species.

During metamorphosis the median fin fold of the larva degenerates and the unpaired fins become differentiated. The gills develop and increase markedly in size, gradually coming to play the major role in respiratory gas exchange. Metamorphosis is also the time at which the scales develop and the body becomes more deeply pigmented. The lateral line organ may become clearly differentiated at this stage.

In many species, metamorphosis also represents the time at which there is the first appearance of haemoglobin in the blood. There may also be marked changes in the alimentary canal, with increased coiling of the intestine, clearer differentiation of the stomach and pyloric caeca, and development of the gas bladder.

In the flatfish, such as the plaice, dab, *Limanda limanda*, turbot, *Scophthalmus maximus*, and hirame flounder, *Paralichthys olivaceus*, metamorphosis involves pronounced morphological changes. There is both a rotation of the optic region of the skull, and a change in the orientation of the body so that the fish eventually come to lie on one side (Fig. 10.7).

In the pleuronectids, the flatfish family which includes both the plaice and dab, metamorphosis involves the migration of the left eye to the right side of the body, the optic nerves become crossed and the bones of the

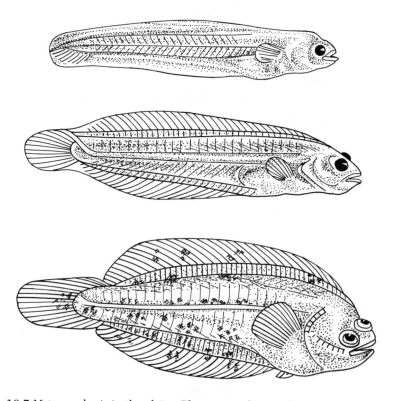

Fig. 10.7 Metamorphosis in the plaice, *Pleuronectes platessa*, showing eye migration. Metamorphosis is completed 60–80 days after hatching when the fish are 12–17 mm in length.

head and cranium become skewed. Thus, on completion of eye migration both eyes are on the right side of the head. Eye migration is usually complete by the time the young fish begin to take up a benthic existence.

The right side of the body becomes darkly pigmented with spots and blotches and the left side of the body is usually white. These dextral flatfish lie on the seabed with the pale left side of the body against the substrate and the darker pigmented, eyed right side uppermost.

In contrast to the pleuronectids, the turbot and hirame flounder are left-sided flatfish. That is, the right eye migrates around to the left side of the body during metamorphosis, and these fish live on the bottom lying on their right side.

In some species of flatfish, 'reversed' specimens occur quite commonly. In addition, abnormal patterns of pigmentation are not infrequently seen, with fish having either dark blotches on the pale side of the body, or white patches on the pigmented surface.

The time between hatching and metamorphosis may range from a few days in some tropical marine species to a few weeks in many species of temperate regions (Table 10.2). In some cases, months, or perhaps even years, may be required for these developmental events to be completed. The great diversity in both the duration of, and the morphological changes that occur during, larval development and metamorphosis make direct comparisons between species extremely difficult. Nevertheless, some attempts have been made to provide a broad classification of the different developmental stages through which a fish passes during the course of the life cycle.

10.9 TERMINOLOGY OF FISH ONTOGENY

There are considerable difficulties involved in devising a standard terminology of developmental ontogeny, and it is open to question whether a single set of terms can realistically encompass all species and patterns of changes. Some workers consider that the entire life of a fish, from the time that development commences with the fertilization of the egg, until death, can be divided into five periods – embryonic, larval, juvenile, adult and senescent. Each period may be divided into phases, as a convenient means of identifying different levels of morphological or physiological development. Early in the life of the fish, developmental events are reflected in rapid morphological and physiological changes, but in each successive period the developmental rate decreases until senescence and death of the fish.

Embryonic period

The embryonic period, which begins with fertilization of the egg, is characterized by endogenous nutrition. This may be either from the yolk or, in the case of the viviparous species, nutrition via special absorptive organs.

The first interval of development within the embryonic period, known as the *cleavage phase*, covers the time between fertilization and the commencement of organogenesis. The *embryonic phase* represents the time of intense organogenesis within the egg membrane. This phase continues until hatching has been completed. On hatching the embryo is considered to have entered the *eleutheroembryonic phase* (named after the Greek – eleutheros meaning free). The phase commences with hatching and continues until most of the yolk has been utilized and the fish begins to take exogenous food.

Larval period

The larval period commences when the transition to exogenous feeding has taken place. This period is usually considered to last until there has been ossification of the axial skeleton.

The larval period is characterized by the persistence of some embryonic organs. There may also be development of special larval organs which are later lost or replaced by other organs performing the same function (e.g. surface blood vessels on fins and filamentous appendages which are used in gas exchange).

The larval period may be divided into two phases – the *protopterygiolarval* and the *pterygiolarval phases*. The protopterygiolarval phase encompasses the interval between the transition to exogenous nutrition and the commencement of the differentiation of the fins. This is followed by the pterygiolarval phase, which is deemed to terminate when the median fin-fold is no longer visible. The larval period may be extremely long in some species, such as the eels, *Anguilla* spp. In other species, such as the salmonids, it is difficult to distinguish a distinct larval period, and this period may be considered to be absent.

Juvenile period

The juvenile period begins once the fins are fully differentiated, and the temporary larval organs have regressed or have been replaced. The transition from larva to young fish sometimes involves extensive changes from an unfishlike appearance to one resembling the adult. In other words, commencement of the juvenile period can be considered to coincide with the completion of metamorphosis.

The juvenile period lasts until the fish mature and the first gametes are produced. This period is usually characterized by rapid growth, and there is often a distinct juvenile body colour or pigmentation pattern.

Adult period

The adult period commences with onset of gonad maturation. This culminates in the production of the first gametes. Thus, the onset of the adult period is almost invariably accompanied by a spawning run or migration, specialized reproductive behaviour, and by changes in external morphology and colour.

Spawning may be repeated for a number of years, or seasons, or be performed only once, as is the case with some eels and salmonids. During the adult period, available resources are usually directed more towards the development of the gonads and reproductive products than towards somatic growth. Consequently, rates of growth are generally substantially lower during this period than during the juvenile period.

Senescent period

The senescent period, or old age, can be considered to cover the time period when the growth of the mature fish has become exceptionally slow, or may even have ceased. Production of gametes is usually also markedly reduced.

The period of senescence can last for several years, during which there is a gradual decline in the numbers of fertile gametes produced (e.g. sturgeon, *Acipenser sturio*). Alternatively, the senescent period may last for no more than a few days or weeks, during which time the body undergoes rapid degenerative changes (e.g. Pacific salmon, *Oncorhynchus* spp.).

Other developmental transitions

Whilst this classification provides a useful indication of the ontogenetic changes that occur during the course of the life cycle of a fish, the terminology has not been universally accepted. The classification scheme may also fail to highlight some special developmental transitions.

For example, according to the classification given above, the larval period is defined as commencing at the time when the fish undergoes the transition from endogenous to exogenous feeding. The hatching of the fish is not deemed to be of sufficient significance to merit being used as a criterion for demarcation of the transition between developmental periods. Many workers, however, consider the larval period to commence at the point at which the fish hatch, and will refer to the newly hatched fish as a yolk-sac larva.

In addition, there can be little doubt that the parr–smolt transformation shown by the salmonids is an important developmental transition in the life cycle of these fish. The parr–smolt transformation involves a range of morphological and physiological changes associated with the migration from fresh water to the marine environment. This transitional event is not, however, encompassed by the terminology given above.

10.10   RATES OF DEVELOPMENT

Whilst rates of development, and the timing of the various transitional events, are, to a certain extent, determined by species-specific character- istics, environmental factors may have a marked modifying influence. Ambient temperature, for example, is known to be one of the most potent factors influencing both rates of development, and the overall survival of the eggs and larvae.

Temperature effects on development

Rates of development are generally found to increase with increasing tem- peratures. Incubation of eggs at high temperature may, however, result in increased incidence of malformations and abnormalities in the embryo, or the death of the eggs. The majority of temperate marine species of fish, for example, have eggs capable of normal development and hatching through- out a temperature range of about 10°C.

At one time it was thought that the product of the incubation tempera- ture (T in °C) and the time (D in days) required to reach any particular stage of development was constant, i.e. $T \times D$ = constant, and this became established as the concept of 'degree-days'. A formula of this type would allow developmental times to be predicted very easily, and the degree-day (°day) concept was rapidly adopted by fish hatchery managers for planning purposes. For example, it has been suggested that Atlantic salmon require approximately 500 °days to complete development from fertilization to hatch. This would mean that the eggs would hatch after 125 days if incubated at 4°C, but hatching would occur after just over 60 days if the incubation temperature were raised to 8°C.

Subsequently, simple tables have been prepared giving information about the number of degree-days required by various salmonid species in order to reach specific developmental stages. Information is usually given about the number of degree-days required from the time of fertilization for the eggs to reach the 'eyed' stage. This is the stage at which the eggs can be safely handled and moved without massive mortalities. The tables also provide information about the number of degree-days required between fertilization and hatch (Table 10.3).

There has been increasing criticism of the degree-day concept because plots of developmental times against temperature yield relationships that

Table 10.3 The 'degree-day' concept and the development rates of different salmonid species

Species	Egg diameter (mm)	Degree-days (°C × days) to	
		'Eyed egg' stage	Hatching
Atlantic salmon, *Salmo salar*	~5.5	230	500
Brown trout, *Salmo trutta*	~5.0	220	490
Rainbow trout, *Oncorhynchus mykiss*	~4.5	175	375
Arctic charr, *Salvelinus alpinus*	~4.2	200	450

are not well described by this concept (Fig. 10.8). The degree-day concept builds upon the principle of thermal summation (i.e. $T \times D$ = constant), in which the plotting of developmental times against temperature would result in a curvilinear relationship in the form of a rectangular hyperbola.

Nevertheless, there may be a close approximation to this form of relationship over the relatively limited range of water temperatures encountered in salmonid hatcheries. Thus, under these conditions, use of the formula is unlikely to give estimates that are seriously in error. Consequently, the degree-day concept may still have application in practical hatchery management to give estimates of the times at which eggs can be safely moved or shipped, or the dates on which eggs might be expected to hatch under different incubation conditions.

The relationship between incubation time and temperature is a curvilinear one, and a number of attempts have been made to describe the relationship in mathematical terms. Power law (i.e. $D = a\ T^{-b}$) and exponential (i.e. $D = a\ \exp^{-bT}$) equations have been quite widely used in order to describe the relationship, and these equations have been found to give a good description of the effects of temperature on the rates of development of several fish species.

When data have been fitted to the power law function, the constant b, which describes the inverse relationship between incubation time and temperature, has usually been found to be within the range 1–1.5 (in the exponential the value of b will usually be 0.1–0.15). Both the power law and exponential functions have the weakness that they do not take into account the fact that there is a maximum temperature above which egg incubation is impossible, and mortality will be almost complete.

For example, maximum incubation temperatures have been reported to be 19 and 23°C for pike, *Esox lucius*, and rainbow trout, respectively, with

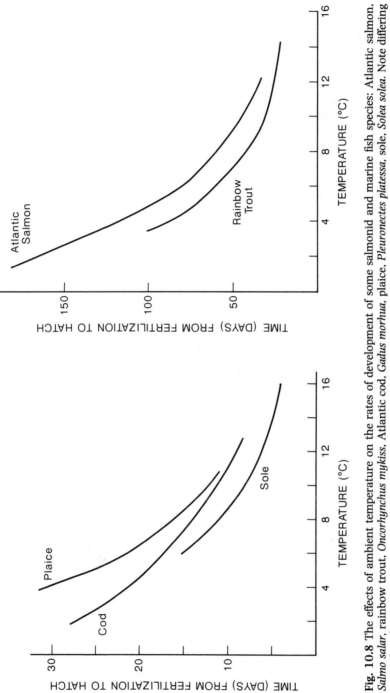

Fig. 10.8 The effects of ambient temperature on the rates of development of some salmonid and marine fish species: Atlantic salmon, *Salmo salar*, rainbow trout, *Oncorhynchus mykiss*, Atlantic cod, *Gadus morhua*, plaice, *Pleuronectes platessa*, sole, *Solea solea*. Note differing vertical scales.

the time to hatch being 4 and 15 days for eggs incubated at their respective maximum temperatures. These maximum incubation temperatures are several degrees lower than the maximum temperatures tolerated by larger fish of the same species. Eggs and larvae are often found to have a more restricted temperature tolerance range than larger individuals. For example, eggs of the Arctic charr, *Salvelinus alpinus*, experience considerable mortality when incubated at temperatures of 12°C or over. On the other hand, juvenile and adult charr may tolerate temperatures of 20°C or over, and the larger fish are capable of displaying relatively high rates of growth at temperatures that would prove fatal to eggs and start-feeding fry. Similarly, mortality may be 80% or more in Atlantic cod eggs incubated at temperatures of 12°C and over, yolk-sac larvae should not be exposed to temperatures above 16–17°C, but juvenile and adult cod seem to be able to tolerate maximum temperatures of up to 23–24°C.

Other environmental factors

Other environmental factors may also influence rates of development. Low levels of dissolved oxygen, for example, tend to have a retarding effect on development, especially at high incubation temperature. The salinity of the incubation medium can also affect both developmental rates and the survival of the eggs and larvae. Changes in the salinity of the incubation medium do not, however, seem to produce consistent changes in the acceleration or retardation of development.

Endocrine factors

Rates of development of the eggs and larvae of several species have been shown to be influenced by the addition of exogenous hormonal factors to the incubation medium. Thyroid hormones, which have been implicated in various aspects of amphibian metamorphosis, are known to be transferred to the developing oocyte by the female fish. Consequently, it is the thyroid hormones which have been most studied with respect to influences on rates of development and growth of fish eggs and larvae. The thyroid hormone content of the eggs can be increased by administering these hormones to the female fish, and this form of treatment has been reported to result in increased larval survival and growth in the striped bass, *Morone saxatilis*. Similar findings have been reported following the injection of female broodstock rabbitfish, *Siganus guttatus*, with various doses of thyroxine. It has been suggested that the thyroid hormones of maternal origin enter the oocytes bound to vitellogenin. Incorporation of thyroid hormones into the oocytes may not, however, be solely limited to the period of most active vitellogenesis, because injection of the female fish in the period shortly prior to ovulation also leads to the production of eggs containing elevated levels of thyroid hormones.

Immersion of eggs and larvae in media containing thyroid hormones has been shown to lead to accelerated growth in several species, including the Mozambique tilapia, *Oreochromis mossambicus*, Nile tilapia, *O. niloticus*, carp, *Cyprinus carpio*, and milkfish, *Chanos chanos*. This form of treatment also led to enhanced survival amongst the tilapia and carp larvae.

There is evidence that thyroid hormones play an important role in the regulation of rates of morphogenesis and metamorphosis in species such as the conger eel, *Conger myriaster*, the black sea bream, *Acanthopagrus schlegeli*, and the goldfish, *Carassius auratus*, but the most convincing demonstration of the role of these hormones in metamorphosis derives from studies carried out on the hirame flounder.

Developmental rates of larval hirame flounder are increased by administration of exogenous thyroxine to the incubation medium. Hormone treatment leads to increased rates of gut differentiation and more rapid development of fully functional gastric glands. Similarly, the treated larvae initiated and completed metamorphosis much more rapidly than did larvae which had not been exposed to an exogenous source of thyroxine.

10.11 SPAWNING SEASONS AND MIGRATION

The majority of fish larvae hatch with variable, but limited, reserves in their yolk sacs and, if they are to survive, they must find suitable food in sufficient quantity before their own reserves are depleted. For the larvae of most marine fish species their first food consists of small zooplanktonic organisms.

In the marine environment of tropical regions plankton production is more or less continuous, but at relatively low levels. Thus, suitable food for fish larvae will be available over a long period. Consequently, spawning may be spread over a period of several months.

By way of contrast, the discontinuous production cycles in temperate and polar latitudes are associated with substantial, but short-lived, standing stocks of plankton. These plankton blooms could sustain large numbers of fish larvae if spawning and hatching were timed to coincide with these periods of abundance. Thus, in the temperate and polar latitudes reproductive success is greatest for those individuals, breeding groups or species whose eggs hatch in the right place at the right time.

The adoption of a migratory lifestyle might allow the exploitation of the resources of different areas on a seasonal basis. Thus, migratory behaviour would appear to be a life history feature directed towards improved reproductive success and increased abundance. The belief that migration is an adaptation leading to increased abundance is supported by an examination of fishery statistics. Of the 25–30 most important commercial fish species almost all are migratory. For example, in northern waters, all of the top commercial species, which include capelin, *Mallotus villosus*, Atlantic cod, plaice and herring, are migratory species. Thus, commercially important species are migratory, and they are commercially important because they are abundant.

Since migratory behaviour is viewed as being an adaptation to exploit seasonal productivity in different areas, the cycles of migration should be *regular* and *predictable*. Thus, the implications are that migrations occur along more or less defined routes, at specific times of the year, and that the migratory species have specific feeding, wintering and spawning areas.

In general terms the migratory movements of marine fish are related to the water currents. In the simplest case a migratory cycle could be thought of as following a specific water circulation system, or gyral. The adult fish would move, or be carried, around the system and would eventually return to the starting point after some length of time. The drift of the eggs and larvae from the spawning ground might occur at a different rate from that of the movements of the adults. These differences in rates of movement would lead to the distributions of the various life history stages becoming discontinuous.

The gyral can be thought of as a closed specific regional water circulation so that fish migrating within this circulation system will become more or less isolated from other fish of the same species. In other words, the fish migrating within a given gyral system can be thought of as being an independent breeding unit, which can be described in specific terms of reproductive characteristics, recruitment, growth and mortality.

In many cases, some form of environmental barrier, such as unsuitable water temperatures, may prevent the fish going around the whole gyral system. Thus, the older, mature fish would be required to make some compensatory movement in the direction opposite to that in which the eggs and larvae are carried by the drift currents. This does not mean, however, that the adult fish must swim directly against the current in order to reach the spawning grounds. The eggs and larvae of marine fish will generally be carried pelagically by the *surface drift* currents, but in deeper waters there is often a *countercurrent stream*. Therefore, the adults could move in the opposite direction to the surface drift merely by migrating at a different depth to their offspring.

Migratory cycles can usually be described by simple triangular patterns in which the movements of the young and adults are linked (Fig. 10.9). More complex cycles are, however, displayed by a number of fish species. Irrespective of the complexity of the migratory pattern, the point to be made is that migratory behaviour appears to be an adaptation allowing

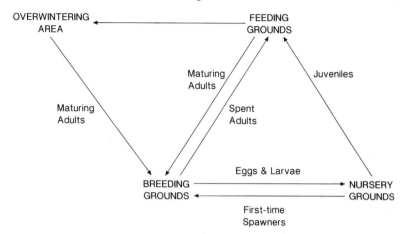

Fig. 10.9 Schematic diagram illustrating possible migratory patterns in fish species.

the exploitation of resources that become available in different geographic locations at different seasons of the year.

10.12 CONCLUDING COMMENTS

There are three basic reproductive modes shown by fish species – oviparity, ovoviviparity and viviparity. Using all three reproductive modes, the elasmobranchs have adopted the strategy of giving birth to relatively few young at any one time, and each of the offspring represents the investment of a great deal of maternal energy. This stands in marked contrast to the situation observed for the vast majority of the teleosts. Most teleosts are oviparous, there is external fertilization of the egg, and large numbers of gametes are usually produced.

Nevertheless, there are some viviparous teleosts – about 500 out of a total of 22 000 species – but viviparity in the teleosts takes a markedly different form from that seen in the elasmobranchs. Since teleosts do not develop 'true' uteri or oviducts, there is intraovarian gestation of the developing eggs and embryos, sometimes within the follicle, but within the ovarian cavity in other species.

Amongst the oviparous teleosts there is great diversity, and it is almost impossible to make simple generalizations about patterns of development. Nevertheless, there are certain discrete transitional events that occur during ontogeny, and the fish may be considered to pass through five life periods in the time between the fertilization of the egg and death – embryonic, larval, juvenile, adult and senescent periods.

FURTHER READING

Books

Breder, C.M. jun. and Rosen, D.E. (1966) *Modes of Reproduction in Fishes: How Fishes Breed*, T.F.H. Publications, Jersey City.

Hoar, W.S. and Randall, D.J. (eds)(1969) *Fish Physiology, Vol. III*, Academic Press, London.

Hoar, W.S. and Randall, D.J. (eds)(1988) *Fish Physiology, Vol. XIA,B*, Academic Press, London.

Kamler, E. (1992) *Early Life History of Fish: An Energetics Approach*, Chapman & Hall, London.

Keenleyside, M.H.A. (1979) *Diversity and Adaptation in Fish Behaviour*, Springer-Verlag, Berlin.

Potts, G.W. and Wootton, R.J. (eds)(1984) *Fish Reproduction: Strategies and Tactics*, Academic Press, London.

Review articles and papers

Balon, E.K. (1975) Reproductive guilds of fishes: a proposal and definitions. *J. Fish. Res. Bd Canada*, **32**, 821–63.

Balon, E.K. (1975) Terminology of intervals in fish development. *J. Fish. Res. Bd Canada*, **32**, 1663–70.

Balon, E.K. (1984) Reflections on some decisive events in the early life of fishes. *Trans. Am. Fish. Soc.*, **113**, 178–85.

Craik, J.C.A. and Harvey, S.M. (1984) Biochemical changes occurring during final maturation of eggs of some marine and freshwater teleosts. *J. Fish Biol.*, **24**, 599–610.

Elgar, M.A. (1990) Evolutionary compromise between a few large and many small eggs: comparative evidence in teleost fish. *Oikos*, **59**, 283–7.

Hamlett, W.C., Eulitt, A.M., Jarrell, R.L. and Kelly, M.A. (1993) Uterogestation and placentation in elasmobranchs. *J. exp. Zool.*, **266**, 347–67.

de Jesus, E.G., Hirano, T. and Inui, Y. (1993) Flounder metamorphosis: its regulation by various hormones. *Fish Physiol. Biochem.*, **11**, 323–8.

Kjørsvik, E., Magnor-Jensen, A. and Holmefjord, I. (1990) Egg quality in fishes. *Adv. mar. Biol.*, **26**, 71-113.

Kormanik, G.A. (1992) Ion and osmoregulation in prenatal elasmobranchs: Evolutionary implications. *Amer. Zool.*, **32**, 294–302.

Rønnestad, I. and Fhyn, H.J. (1993) Metabolic aspects of free amino acids in developing marine fish eggs and larvae. *Rev. Fish. Sci.*, **1**, 239–59.

Schindler, J.F. and Hamlett, W.C. (1993) Maternal–embryonic relations in viviparous teleosts. *J. exp. Zool.*, **266**, 378–93.

Wourms, J.P. (1977) Reproduction and development in chondrichthyan fishes. *Amer. Zool.*, **17**, 79–110.

Wourms, J.P. (1981) Viviparity: The maternal–fetal relationship in fishes. *Amer. Zool.*, **21**, 473–515.

Wourms, J.P. and Lombardi, J. (1992) Reflections on the evolution of piscine viviparity. *Amer. Zool.*, **32**, 276–93.

Chapter eleven

Recruitment and population fluctuations

11.1 INTRODUCTION

The fact that all natural populations fluctuate in size hardly needs stressing. It is, however, often assumed that the size of an unexploited fish population oscillates around some equilibrium level. In effect it is assumed that the environment has a certain *carrying capacity*, i.e. an ideal number of fish that the environment can support, and that this carrying capacity is stable over time. This is obviously an oversimplification because environments do change with the passage of time. Such changes will mean that the carrying capacity of an environment for a given species will tend to be on a sliding scale rather than rigidly fixed. In addition there are often short-term fluctuations in environmental parameters superimposed upon the longer-term trends. Such short-term fluctuations are usually relatively unpredictable, and represent, therefore, stochastic events that influence the resource base of the fish population.

The concept of regulation of population size around some equilibrium level implies that there are control mechanisms in operation. The control mechanisms would act to return population numbers to the equilibrium level should any deviations occur. Thus, should numbers increase to exceed the equilibrium level the regulatory factors will act to reduce numbers. On the other hand, a decline in the numbers of individuals within the population would be expected to introduce a series of changes that would lead to the rapid restoration of the size of the population to the equilibrium level. In other words, it is assumed that the numbers within the population are regulated in some density-dependent way.

11.2   POPULATION SIZE AND DENSITY-DEPENDENT
REGULATION

Density-dependent regulation can be thought of as working via both reproductive and mortality mechanisms in order to return population size to the equilibrium level. There are a number of possible ways in which this could be brought about, and density-dependent mechanisms could act at several different stages of the life cycle.

At high population density there may be intraspecific competition for food, leading to reduced growth rates. This may result in individuals reaching maturity, and entering the spawning stock, at a high age. Slow growth and late recruitment of individuals to the spawning stock would tend to have the effect of reducing the reproductive rate of the population. This would, in turn, lead to a gradual reduction in numbers to the equilibrium level.

On the other hand, a low population density could lead to there being adequate food for all members of the population. Under such circumstances individual growth rates would be high and fish would be expected to mature, and recruit to the spawning stock, at a comparatively young age. Rapid growth and early recruitment of fish to the spawning stock would result in an increase in reproductive rate. This would lead to a rapid restoration of population size to the equilibrium level.

At high population density, competition for food may reduce spawning stock fecundity. Some of the adult population could fail to secure sufficient resources to enable completion of the reproductive cycle, so the relative size of the spawning stock could be reduced, i.e. not all the mature fish produce eggs and spawn each year. In addition, reduced levels of food available to each individual within a dense population could result in reduced individual relative fecundities, i.e. lower egg numbers per unit body weight of spawning fish.

Inadequate food supply resulting from high levels of competition could also influence reproductive success via effects on egg quality. That is, poor egg quality could result from a reduced energetic investment in the yolk deposited in each egg.

All of these factors would act to reduce the reproductive rate of the population to a level below its true potential.

Under conditions of low population density each mature individual would be expected to secure sufficient food supplies to enable it to reproduce. Furthermore, large numbers of good-quality eggs should be produced at each spawning season. Thus, each individual would be able to display its maximal reproductive potential, thereby ensuring a high reproductive rate for the population.

It is far from easy to obtain evidence that density-dependent regulatory factors are operative in fish populations. Nevertheless, field studies have shown that there may be very large fluctuations in reproductive parameters within populations. For example, fecundity has been found to vary markedly between years, variations of up to 45% having been found for plaice, *Pleuronectes platessa*, 25% for pike, *Esox lucius*, 34% for herring, *Clupea harengus*, and 56% for haddock, *Melanogrammus aeglefinus*. Above-average fecundity appears to be associated with low population density and vice versa. It is tempting to suggest that this association could have a basis in a close connection between fecundity and food supply. In other words, less food would be expected to be available for each individual at high population densities, and this would, in turn, lead to reduced egg production per female.

Food supply and breeding success

Laboratory studies carried out on a range of fish species have demonstrated that, in addition to affecting fecundity, food supply may also influence a number of other characteristics that play a role in governing overall reproductive success. For example, in studies with brown trout, *Salmo trutta*, and rainbow trout, *Oncorhynchus mykiss*, it has been shown that a higher percentage of well-fed, than poorly fed, fish come into spawning condition. In addition, the poorly fed fish that spawn may have lower fecundities than well-fed fish.

Similar results have been reported in studies carried out on threespine sticklebacks, *Gasterosteus aculeatus*, where it was also found that the inter-spawning interval was shorter for the well-fed fish. In other words, well-fed fish spawned more eggs and more often than did fish eating less food.

There have been few controlled laboratory experiments carried out to examine the effects of food supply on reproductive success of marine fish, but some information is available for the Atlantic cod, *Gadus morhua*, and haddock. In these two gadoids, food supply may influence the proportion of the population reaching spawning condition, the number of eggs produced per spawning individual, and egg size (which is a measure of egg quality or energy invested per egg).

Availability of breeding sites

When population densities are high, certain individuals within the population may either be prevented from spawning or be forced to spawn under conditions that are far from ideal. For example, when there are a limited number of suitable nesting sites, dominant individuals may maintain spawning territories, whereas other members of the population will fail to secure a nesting site and will not spawn. Even when there is no apparent overt intraspecific competition for spawning territories and nesting sites, density-dependent factors may still play a role in governing reproductive success.

For example, species such as salmonids and the grunion, *Leuresthes tenuis*, bury their eggs but do not guard the nests. Thus, if spawning occurs in waves over an extended time period, fish spawning later in the season would be free to use the same nest sites as those spawning earlier. These late-spawning fish could then destroy the nests and eggs of the earlier spawners. Such nest destruction has been observed in salmonid species, and this would be expected to be more frequent when population densities are high.

The demersal eggs of fish such as herring often form thick layers or clumps on the substrate. Eggs at the centre of such clumps may experience conditions under which oxygen levels become depleted. This may result in the death of some eggs due to suffocation, and the degeneration of these eggs can lead to increased bacterial growth and further egg mortalities. Problems of this type are likely to be more severe when there are large, rather than small, spawning aggregations of fish.

Survival of eggs and larvae

The aforementioned density-dependent factors affect the reproductive performance of the adults in the population, but larval growth and survival is also likely to be density dependent. For example, at high population density under severe competition for a limited food supply, the amount of energy invested in the yolk of each egg may be reduced. An egg containing relatively little yolk would result in a reduced food reserve for the newly hatched larva. The consequences of this might be expected to be severe. Larvae derived from eggs containing little yolk will be totally dependent upon exogenous food supplies at an earlier age, and at a smaller size, than individuals that hatch from larger, more yolk-rich eggs. Thus, a small yolk reserve could lead to an increased risk of death of larvae by starvation before they have learnt to take external food. Even should such larvae learn to take exogenous food, the mortality risk may still be high because their developmental rate is insufficiently rapid to enable them to search sufficiently large volumes of water to capture adequate numbers of prey.

When larval density is high, there may be a rapid depletion of the available food supply so that density-dependent mortality is generated by a lack of suitable prey. Even if the depletion in the food resources is not sufficient to result in starvation of larvae *per se* the reduction in the numbers of suitable prey available may be sufficiently severe to cause reduced growth rates of the larvae. Slower growth will mean that the larval period is extended, with the possibility that a large proportion of the vulnerable larvae will be consumed by predators.

Spawning stock and recruitment

It is the rate at which individuals recruit to the population that determines how rapidly the numbers within a disturbed population will return to the equilibrium level. The absolute numbers of fish recruiting to the population will be influenced by the interplay between the different density-dependent factors. How might such interactions be expected to influence quantitative aspects of the recruitment process?

When there is a high population density and the spawning stock is large, density-dependent factors related to the amount of food available to individuals might be expected to lead to reduced individual fecundity (Table 11.1). Nevertheless, despite reduced individual fecundity, the population fecundity may be higher for a large than a small spawning stock. The population fecundity gives an indication of the potential numbers of recruits immediately after spawning (N_0). The eggs and larvae are vulnerable to predation, and rates of mortality (M) may be relatively high, so that there is a relatively rapid decline in numbers with time. The change in numbers with time (D in days) may be described by $N_D = N_0 e^{-MD}$, where N_D is the number of eggs or larvae remaining after a time D days has elapsed.

When initial numbers (N_0) are large, the newly hatched larvae may deplete their food base and, consequently, individual rates of growth may be low. The consequences of this would be that the slow-growing larvae

Table 11.1 Possible influences of the size of the spawning stock on population fecundity, rates of larval development and recruitment of juvenile fish to the population. See text for further details

	Spawning stock size			
	5	10	20	50
Individual fecundity	10^5	10^5	8×10^4	6×10^4
Population fecundity, N_0	5×10^5	1×10^6	1.6×10^6	3×10^6
Mortality rate, M	0.08	0.08	0.08	0.08
Time to metamorphosis, D	45	50	55	65
Recruitment, N_D	13661	18316	19644	16550

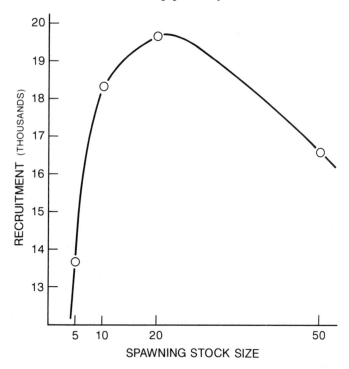

Fig. 11.1 Graphical representation illustrating the possible relationship between the size of the spawning stock and recruitment of fish to the population. See Table 11.1 and main text for further details of assumptions and calculations.

will require a longer time to reach the size of metamorphosis than more rapidly growing individuals. Slow-growing larvae will, therefore, be vulnerable to predation for a longer period of time than rapidly growing individuals. If it is assumed that recruitment takes place at metamorphosis, then the possible influences of these density-dependent factors on recruitment can be examined in a simple simulation exercise (Table 11.1). The results clearly demonstrate that there may be fewer recruits to the population when the spawning stock is large than when it is small, the relationship between the number of recruits and the spawning stock taking the form of a dome-shaped curve (Fig. 11.1).

Density dependence and inherent variability

Density-dependent regulation might be expected to operate upon the population in order to return fish numbers to the equilibrium level, and intuition would suggest that the regulation of numbers by density-dependent factors should lead to stability. Whilst density-dependent regulatory processes generally introduce a trend towards long-term stability, the nature of the regulatory processes themselves may lead to fluctuations in population numbers in the short term. This is particularly evident in populations with potentially high reproductive rates.

Complications arise because populations do not respond immediately to changes in their own density, but there is a time lag before any response occurs. Suppose, for example, that the reproductive rate is determined by the amount of food available to individuals within the population. The total amount of food available at any given time is likely to be determined by the amount of food consumed by the population at some earlier time. This will mean that reproductive rate displayed at one particular time is governed by the density of the population at an earlier date, and there is a time lag in the density-dependent response.

Alternatively, some organisms may respond to density-dependent factors at one point in their life cycle, but only mature and reproduce some time later; the period that elapses between the time at which the density-dependent response occurs and the time of reproduction is also a form of time lag. The point to be made is, that by the introduction of a time lag, the population is responding to a situation that no longer exists. This time lag in response is likely to result in increased fluctuations in numbers about the equilibrium level. In summary, it can be concluded that density-dependent regulation will usually lead to long-term stability in populations, whilst at the same time permitting, or even inducing, short-term fluctuations.

Despite the fact that fluctuations are to be expected in population sizes, one can nevertheless consider multi-age fish stocks as being relatively stable, with overall numbers fluctuating comparatively little. Populations which contain many year classes, and several overlapping generations, tend to show damped oscillations around the equilibrium population size. In such populations, there may be two- to fivefold differences between population maxima and minima. On the other hand, populations of fish species that reproduce rapidly, and in which there are only a single or few generations, may show fluctuations in numbers of almost two orders of magnitude.

11.3 STOCHASTIC EVENTS AND RECRUITMENT VARIABILITY

Within multi-age populations there may be quite large fluctuations in the sizes of the recruiting year classes. In other words, the recruitment to fish stocks varies considerably from year to year. It must be remembered that the fish are not exposed to constant conditions, but are living in a changing environment. The year-to-year variations in recruitment arise due to the responses of the population to the short-term changes in the environment. Thus, fluctuations in recruitment, or year class strength, reflect the short-term success of the population in exploiting the changes in the environment. The numbers within a given year class are largely determined at a relatively early stage so it is egg and larval survival that is particularly important in determining fluctuations in recruitment.

The environmental factors most likely to influence egg and larval survival are those related to feeding conditions, disease outbreak and predation rates. Year-to-year fluctuations in environmental conditions may lead to changes in the food base and thereby influence the time (D in days) required for individual larvae to reach the size at which they can undergo metamorphosis. Disease outbreaks and predation will be the prime determinants of mortality rates (M). Thus, it will be the year-to-year variations that influence the interactions between mortality rates and the time taken to reach metamorphosis that govern the extent of the fluctuations in year class strength.

As has been stated previously, the numbers of recruits (N_D) may be estimated from $N_D = N_0 e^{-MD}$. The performance of a simple simulation exercise (Table 11.2) reveals that relatively small changes in mortality rates (M) and time to metamorphosis (D in days – reflecting changes in the food base) can have important consequences for the numbers of fish recruiting to the population. The results of the simulation exercise suggest that year-to-year variations in year class strengths can easily be an order of magnitude or more (Table 11.2).

In practice, 100- to 200-fold differences in recruitment have been observed between the poorest and strongest year classes in gadoids such as the haddock and bib, *Trisopterus luscus*. Fluctuations in year class strength

Table 11.2 Possible influences of different feeding conditions and rates of predation and mortality on the developmental rates of fish larvae and the recruitment of juveniles to the population. See text for further details

Conditions	Mortality rate, M	Time to metamorphosis, D	Recruitment, N_D
Good	0.08	50	18316
Poor feeding conditions	0.08	65	5517
High mortality rates	0.1	50	6738
Poor feeding conditions, high mortality rates	0.1	65	1504

Population fecundity (N_0) is 1×10^6 in each case.

in other species of gadoids, pleuronectids and clupeids are, however, usually smaller, perhaps amounting to 10- to 30-fold differences between maxima and minima.

The effects of large fluctuations in year class strength on overall population size will, however, be damped in a long-lived stock, with 16–25 or more year classes. Fluctuations in total stock size will be only one-fourth to one-fifth as variable as those of the annual recruitment.

Spawning season and year class strength

Some of the observed variations in year class strength can be attributed directly to density-dependent regulation of the overall population size. Much of the variability is, however, probably more a reflection of the effects of unpredictable, short-term stochastic events.

Many fish species spawn at particular times of the year, and these spawning seasons generally ensure that the hatching of the eggs and growth of the larvae occur at times when larval food is abundant. For example, spawning in many temperate-zone marine fish species occurs during spring, with the entire spawning season lasting for a period of a few weeks to 3 months. The spawning season of the fish will more or less coincide with the spring plankton production cycle. Within the protracted spawning season there occurs a spawning peak, and the timing of this peak varies little from year to year. For those species for which long-term records are available, e.g. some stocks of plaice, herring, and Atlantic cod, the timing of the spawning peak seems to have a standard deviation of about 1 week. This means that the majority of the eggs spawned by a population hatch, and the larvae begin to seek exogenous food, within a very restricted time period.

The time of the onset of the spring phytoplankton bloom, which indicates the start of the annual production cycle, may vary quite considerably from year to year. For example, the interannual variation in the timing of the start of the plankton bloom may be as much as 6 weeks in the coastal waters of northern Europe. This variability is related to the degree of stability in the water column, which influences nutrient supply. Water column stability is, in turn, dependent upon climatic factors. In summary, it can be said that the onset of the spring phytoplankton bloom in temperate waters varies from year to year at a given place, may vary in timing from place to place within any given year and may display a general trend towards being earlier or later over longer intervals of time.

The net result of this is that the time at which the zooplanktonic secondary producers increase in numbers also fluctuates considerably from year to year, and from place to place. Thus, there may be considerable year-to-year variations in the time at which larval food is most abundant.

By contrast, the time at which the majority of fish larvae hatch, and are ready to accept exogenous food, varies little from year to year. Furthermore, the time interval over which the larvae are ready to accept exogenous food is extremely restricted. In other words, the variation in the peak date of spawning of many temperate-zone marine fish populations is low in comparison with the variability in the timing of the spring production cycle. It might be expected, therefore, that larval growth and survival, and subsequent recruitment to the stock, would increase or decrease depending upon whether or not the peak hatching period of the fish larvae coincided with the time at which most larval food was available.

Match–mismatch hypothesis

Larval growth and survival are hypothesized to depend upon the degree of temporal overlap between the spawning season and plankton production cycles (Fig. 11.2). The fact that the spawning season of the fish may last for several weeks means that some larvae will survive even in years when the spawning season and plankton production cycles are very poorly matched. The result of this mismatch would be expected to be poor recruitment to the stock and a very weak year class. On the other hand, if the hatching of the fish larvae coincided with the maximum supply of

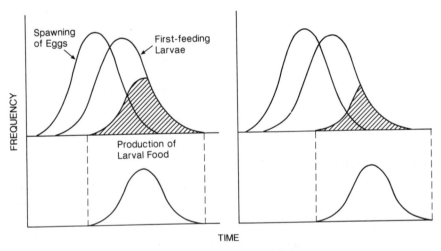

Fig. 11.2 Influence of the timing of the production cycle in relation to the spawning time of a fish species on the potential for survival of the first-feeding fish larvae. Shaded area shows the degree of overlap between the production of larval food and the presence of first-feeding larvae, thereby giving an indication of potential survival under different conditions.

food, then the numbers of fish recruiting to the population would be high. Recruitment will obviously be poorest under conditions when spawning and production cycles are mismatched, and when climatic conditions also result in levels of primary production being low.

Influences of spawning stock size

When the spawning stock is large, the numbers of eggs and larvae produced will be expected to be larger than when there are few mature individuals present. This would apply despite the possible suppressive effects of density-dependent regulation upon adult growth and individual fecundity.

In addition to producing large numbers of eggs and larvae, a large spawning stock would be expected to have a protracted breeding season, lasting for perhaps 2–3 months. Thus, when the spawning stock is large there may be a reasonable degree of temporal overlap between the time of hatching of the larvae and the availablity of prey, even under conditions of a mismatch between the peak times of spawning and plankton production (Fig. 11.3). This would ensure survival of fairly large numbers of larvae, and a reasonable year class strength for recruitment to the population.

In years when the spawning and production cycles were well matched it would be unlikely that food production would be able to support the growth and survival of the very large numbers of larvae hatching from the eggs spawned by a large stock. Under these conditions growth, survival and recruitment would be controlled by density-dependent regulatory factors. Nevertheless, levels of recruitment would still be expected to be good.

Consequently, when there is a large spawning stock, there may only be relatively small differences in year class strengths between years in which the timing of larval hatching is well matched to, and mismatched with, planktonic production cycles (Fig. 11.3). This means that a large spawning stock will tend towards relatively stable levels of recruitment irrespective of the vagaries of short-term changes in environmental conditions.

When spawning stocks are small, a poor match between spawning and plankton production cycles leads to little temporal overlap between the time of hatching of the fish larvae and the time at which food is abundant. A low degree of overlap and, consequently, poor larval survival results in poor recruitment, and a weak year class. In years when the spawning and production cycles are well matched, there will be sufficient food available to support growth and survival of a large proportion of the larvae. The net result of this would be good recruitment and a strong year class. From this it follows that year-to-year variations in the degree of matching of the spawning and production cycles will have a large influence on recruitment to populations that contain few individuals (Fig. 11.3).

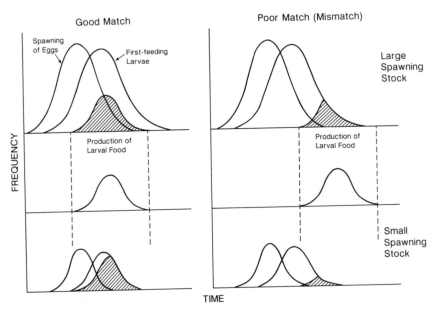

Fig. 11.3 Influence of differences in the timing of the start of the annual production cycle on the potential survival of fish larvae produced by large and small spawning stocks. Shaded areas indicate the potential for survival under the different conditions. Note that when the spawning stock is large, there may be a reasonable chance that quite large numbers of larvae survive even under conditions of a poor match between spawning time and the production cycle. When the spawning stock is small, there are potentially large differences in the chances of larvae surviving between good and poor match conditions. The relative differences in the chances of larval survival are much larger when the spawning stock is small than when it is large, i.e. a large spawning stock will tend to give stable recruitment whereas recruitment from a small spawning stock will tend to be highly variable.

Year class strength and population size

When the spawning stock is small, a strong year class, resulting from a good match between the breeding season and plankton production cycles, may represent a relatively large proportion of the total numbers of fish within the population. Thus, a strong year class would tend to lead to a marked increase in population size, whereas poor recruitment would have little effect on overall numbers. Consequently, fluctuations in the size of small populations are very much under the influence of year-to-year variations in year class strength. The variations in year class strength, in turn, tend to result from short-term, stochastic environmental changes.

On the other hand, when the spawning stock is large, and several year classes are well represented in the population, the numbers of fish recruiting in any given year class will only represent a relatively small proportion of the population as a whole. Thus, the recruiting year class will have little influence on overall population size, irrespective of whether the specific year class is a strong or a weak one. Consequently, large populations tend to be buffered from the large fluctuations that occur in smaller populations as the result of the recruitment of particularly strong or weak year classes. In other words, populations with large spawning stocks tend to be stable, whereas short-term changes in environmental conditions that affect year class strength and recruitment lead to considerably greater fluctuations in the size of smaller populations of fish.

11.4 LIFE HISTORY PATTERNS

In terms of life strategy and survival, a major decision arises in deciding when it is the correct time to stop directing all available resources towards somatic growth and to begin investing resources in reproduction. In other words, the question faced by all individuals and species is: when is the best time to start to reproduce?

In the simplest case, where all resources (e.g. food, space etc.) are in ready supply, a species or breeding group that increases rapidly in numbers to fully exploit available resources will outcompete others. Under conditions of unlimited resources the change in numbers of individuals within the population (N) can be expressed as:

$$\delta N / \delta t = r N \tag{11.1}$$

where r is an expression of the rate of increase and t denotes time.

As the available resources become exploited the possibility for increased population size will be reduced, such that:

$$\delta N / \delta t = r N [(K - N/K)] \tag{11.2}$$

where K is the carrying capacity of the habitat, or the maximum number of organisms (of the species in question) that the particular habitat will support. Thus, as N approaches K, the population growth rate reduces to zero and an equilibrium, or steady state, is reached. Under these conditions, the competitive ability of the species, or breeding group, will be improved more by characteristics that enable it to gain a larger share of the limited resources available, i.e. maximize their own K, than by having the ability to increase rapidly in numbers.

The characteristics leading to success under these different conditions of resource availability will be different. The two extremes have been termed r and K selection. The development of the theory of r and K selection is based on two assumptions about the allocation of a population's resources between competitive and reproductive functions. Firstly, it is assumed that there is a positive relationship between the amount of resources invested in an offspring and the ability of that offspring to survive. The second assumption is that there are only fixed amounts of resources available, resulting in an inverse relationship between the number of offspring produced and the ability of individual offspring to survive.

The r and K selection theory attempts to explain the relationships between life history strategy and the habitat a species occupies. If, for example, the factors causing mortality in an environment are variable, or unpredictable, then the best strategy would be to direct the maximum amount of resources towards reproduction and produce as many offspring as possible (r selection). The contrasting situation is a stable environment in which mortality factors are predictable, and here the optimal strategy would be to produce offspring with good competitive and survival ability (K selection). Since it is assumed that there is a relationship between survival ability of offspring and the resources invested in the individual offspring, this means that K-selected species produce few offspring.

To summarize, an r-selected species is expected to have characteristics which tend to promote productivity, whereas a K-selected species will be characterized by traits giving efficient exploitation of a specific limiting resource. Therefore, specific combinations of population parameters can be identified as being characteristic of an r strategist and the opposite combination would be characteristic of a K strategist.

An r strategist possessing characteristics for high productivity via reproductive activity would be expected to mature early in life, develop rapidly, produce large numbers of offspring for a given body size and have maximum production of offspring at an early age. As a consequence of the allocation of a large proportion of the available resources to reproductive activities, r strategists would also be expected to have small body size, relatively high rates of mortality and a short life span, with the latter either being linked to semelparity, or survival for only a small number of breeding seasons.

In an environment of limited resources and predictable mortality, it will be advantageous to increase allocation of resources to competitive activities provided that the following two criteria are fulfilled. Firstly, reproductive potential must increase with increasing age and size of the fish, and, second, there must be mortality risks associated with the performance of reproductive activities, over and above those experienced by non-reproductive individuals.

Under such circumstances a K strategist would be predicted to delay maturity until late in life, have a relatively slow rate of development, have a low mortality rate, grow to large size, have a long life span and participate in spawning events for several years, i.e. be iteroparous. The K strategist would also be expected to produce relatively few offspring at each spawning event, and may display parental care in order to ensure increased survival of the offspring produced.

Growth models and r–K selection

In fisheries research, data relating to the growth of fish are frequently described using the von Bertalanffy growth model:

$$L_t = L_{max}\{1 - \exp[-k(t - t_0)]\} \qquad (11.3)$$

where L_t is fish length at time t, L_{max} is the asymptotic length, or the theoretical maximum length the fish can reach, and k can be considered as a growth rate function i.e. the rate at which the fish grows to approach L_{max}. Many fish species become sexually mature at a length of approximately 0.7 L_{max}, so k can also be considered to give a description of the developmental rate, i.e. an assessment of the rapidity of the onset of sexual maturation.

The influences of differences in L_{max} and k on growth are shown in the series of curves presented in Fig. 11.4. If a number of species have the same L_{max}, the fish species with the highest k will grow fastest, that is, reach its maximum length at the youngest age (Fig. 11.4(a)). Species with the same k values will, on the other hand, reach their respective L_{max} values at the same age, irrespective of the numerical value of L_{max} (Fig. 11.4(b)). In other words, species with the same k values can be considered to display the same relative rates of development. Examination of the parameters of the von Bertalanffy growth function for a wide range of species suggests that there may be an inverse relationship between L_{max} and k. Further examination of life history data collected for a range of exploited fish species suggests that fish with high k also show a high rate of natural mortality (Table 11.3).

In terms of the parameters most commonly measured in fisheries investigations, an r strategist would have a low age at first maturity, a high value of k and a small L_{max}, a high rate of instantaneous natural mortality, M, and a low maximum age. The K strategist, on the other hand, would be characterized by a high age at first maturity, a low k and a large L_{max}, a low rate of instantaneous natural mortality and a high maximum age.

The r–K selection theory also predicts that fish with a high k and high rates of natural mortality should direct a large proportion of their available resources towards reproduction. A crude estimate of reproductive investment can be made by measuring the gonad index, i.e. the weight of the gonads per unit body weight, just prior to spawning. When gonad index

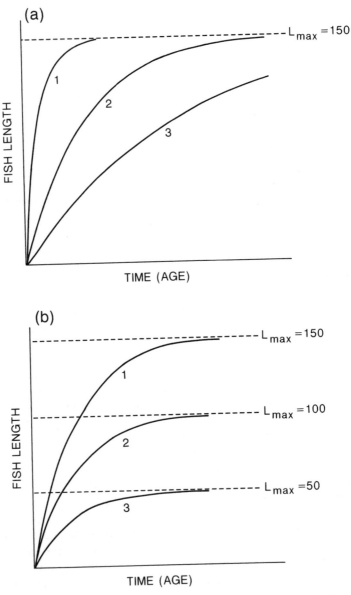

Fig. 11.4 Growth curves for fish illustrating the significance of the different parameters in the von Bertalanffy growth model. (a) Growth curves for fish with different growth rate factors (k) but the same maximum length (L_{max} = 150 cm): curve 1, k = 1.00; curve 2, k = 0.25; curve 3, k = 0.10. (b) Growth curves for fish having the same growth rate factor (k = 0.5) but different maximum lengths (L_{max}): curve 1, L_{max} = 150 cm; curve 2, L_{max} = 100 cm; curve 3, L_{max} = 50 cm.

Table 11.3 The von Bertalanffy growth model parameters (k and L_{max}), and rates of natural mortality (M) for a range of fish species

Species	L_{max} (cm)	k	M
Clupeids/engraulids			
Anchoveta, *Engraulis ringens*	16.9	1.6	1.52
Menhaden, *Brevoortia tyrannus*	20.6	0.74	1.00
Sardine, *Sardina pilchardus*	22.7	0.53	0.67
Herring *Clupea harengus*	40.0	0.38	0.25
Pleuronectids/soleids			
Sole, *Solea solea*	48	0.42	0.25
Plaice, *Pleuronectes platessa*	88	0.13	0.15
Atlantic halibut, *Hippoglossus hippoglossus*	210	0.08	0.16
Gadoids			
Norway pout, *Trisopterus esmarkii*	23	0.59	1.6
Poor-cod, *Trisopterus minutus*	26	0.47	0.9
Haddock, *Melanogrammus aeglefinus*	73	0.28	0.2
Atlantic cod, *Gadus morhua*	120	0.19	0.17

values are plotted against rates of natural mortality, there seems to be a positive correlation. For example, species such as the plaice, American plaice, *Hippoglossoides platessoides*, Atlantic cod and haddock have gonad indices within the range 0.1–0.2 and their rates of natural mortality are below 0.2. On the other hand, species such as the capelin, *Mallotus villosus*, and Pacific cod, *Gadus macrocephalus*, which have gonad indices of 0.25 or over, also have high rates of natural mortality, within the range 0.85–1.3.

Thus, those fish species that are subject to the highest rates of natural mortality have the fastest rates of development (high k), have a relatively small body size (low L_{max}), mature at an early age and invest a high proportion of their available resources in gonadal rather than somatic growth, i.e. they have a large gonad index. These findings are all consistent with the predictions that would be made on the basis of r–K selection theory.

The reproductive and life history strategies of many species appear, however, to fall outside the framework of the *r–K* selection theory. This is, perhaps, because the theoretical approach has been naively focused upon the extremes of what is really a reproductive continuum. In addition the majority of fish species display habitat shifts during the course of the life cycle, and exposure to the factors that have the greatest influence on mortality, growth and developmental rates will differ from one life history stage to another. For example, the eggs, larvae and juveniles will usually be subjected to different levels of predation pressure and will experience different levels of availability of food resources, than larger juveniles and adults.

It is also important to emphasize the limitations of the *r–K* selection concept. The idealized *r* strategist occurs in an ecological vacuum with no density effects and no competition, whereas the idealized *K*-selected species occurs in a completely saturated ecosystem where densities are high and competition for resources is intense. The problems of applying the *r–K* theory to real situations are not trivial but, nevertheless, the theory can give pointers as to how populations of *r*- and *K*-selected species should be exploited and how they will respond to exploitation.

Generally, fisheries based on *r* strategists will be more productive than those based on *K*-selected species. The fish can be exploited at younger ages and at higher levels of fishing mortality. For a given population size, these fisheries should recover more rapidly from overfishing, but, on the debit side, *r*-selected species are likely to be strongly influenced by short-term environmental changes. Thus, fisheries based upon *r*-selected species are likely to be of a boom-and-bust nature, with erratic harvesting levels.

On the other hand, fisheries based on *K*-selected species will be characterized by stable catch levels. This is because the populations comprise many age classes, and there should only be comparatively small natural variations in population size. The *K* strategists are, however, characterized by having relatively low reproductive potential. This means that if overfished, it would require a long time for the stock of a *K*-selected species to re-establish population levels that could again support economically profitable fisheries.

11.5 EXPLOITED FISH POPULATIONS

Removal of fish from exploited populations might, under some circumstances, be so severe as to lead to noticeable reductions in population density. Should levels of fishing be so high as to cause population numbers to fall, then density-dependent regulatory factors should begin to operate in order to return the population to the equilibrium position. For example, increased egg production by the remaining individuals might tend to counteract the effects of removal of a proportion of the population, giving a trend towards re-establishment of the original population size.

If, however, the rate of removal of the fish from the population continued to exceed the rate at which they were replaced, and the population density continued to decline, then the increased resources available to the remaining individuals may influence both rates of growth and reproductive parameters such as age and size of maturity. In other words, the assumption being made here is that the amount of food available to the fish during the course of its life affects growth rate and thereby influences when it will mature.

For example, if the transition to maturity were triggered when an individual reached a certain age, conditions unfavourable to growth would result in a population composed of small mature fish. On the other hand, if the transition were triggered by the attainment of some 'critical size', conditions unfavourable to growth would mean that individuals would reach a relatively high age before they became mature. In practice, it is unlikely that the onset of maturity will be solely dependent upon the attainment of some critical age or critical size, but maturation is more likely to be the result of a compromise between age and size-related factors.

Some support for this has been obtained in laboratory studies carried out to examine the relationships between age and size at maturation in the platyfish, *Xiphophorus maculatus*. Large numbers of platyfish were reared under different conditions of light, temperature and food availability. When the fish grew rapidly there was a minimum age below which they would not mature and there was great variation in the weights of fish maturing at the minimum age. Slower-growing fish matured later, and those fish which grew slowest of all did not mature until they reached a minimum critical weight no matter how long it took. Thus, both the size and the age at which the fish matured could be markedly influenced by rearing environment, with food availability being a major factor.

In the majority of wild fish populations, food availability might be expected to be the overriding factor limiting individual growth rates and determining the timing of the onset of maturity of individuals. In other words most, if not all, individuals would be expected to have the potential to reach maturity at an earlier age if they were given access to increased food resources. A reduction in fish population density would increase the food available to the remaining members of the population, with the consequence of increased rates of individual growth and lower age at maturity. This could not, however, continue indefinitely, and at some stage the fish would reach the point at which they were maturing at the minimum possible age. Any further increase in food resources, due to continued reduction in population density, would not result in a lowered age at maturity but rather would yield an increased size at maturity.

Whilst this is a simplistic representation of the ways in which reductions in population density may interact with food resources to influence growth rates and affect the timing of sexual maturation, it may provide a useful starting point for the understanding of the types of changes observed in wild populations. For example, the European flounder, *Platichthys flesus*, has been heavily fished in the Baltic Sea since about 1900, and in the years between 1900 and 1940 there were changes in growth rates, age and size at maturity. Early in the century, the majority of the females matured at an age of 5 years, with small numbers maturing up to 2 years earlier. In the first quarter of the century growth rates increased, and in the period 1925–1932 there was a reduction of 1 year in the average age at maturity. In later years there was no further decrease in age at first maturity but there was an increase in the size at which the flounder became sexually mature. Thus, the initial reduction in population density due to exploitation may have resulted in a reduction in age at maturity, but continued reductions of population density resulting from fishing may then have led to an increase in size at maturity at a given age.

In summary, exploitation of a previously unfished population may lead to a reduction in population density, with increased food resources being made available to the remaining individuals. The consequences of this might be observed as an increase in individual growth rates and a fall in age at maturity. Should heavy exploitation continue, there may be a further reduction in age at maturity. At some point there will cease to be further decrease in age at maturity, and the fish will start to mature at a young age but with larger body size. Should this be seen, it can be taken as a clear indication that the population has been severely overexploited.

Exploitation and genetic changes

Age and size at maturation are under genetic, as well as environmental, control. For example, the effects of genotype on age and size at maturity have been studied in laboratory experiments with platyfish. Platyfish that were known to be of different genotypes were reared under a range of identical environmental conditions. In spite of the large variations in age and weight at maturity induced by rearing conditions (different temperatures, light conditions and levels of food availability) each genotype had rigidly specified age and size at maturity characteristics that were distinct from those displayed by other genotypes. In other words, fish of some genotypes developed rapidly and matured early at small body size, whereas others reared under the same set of conditions delayed maturation until they were much older and larger. Consequently, there would appear to be the potential for the genetic constitution of a fish population to be changed by selective removal of particular genotypes.

Considerations of life-history theory predict that reduced adult survival will select for earlier maturation and a greater investment of resources into reproduction. On the other hand, reduced survival of juveniles will select for the opposite set of characteristics. Thus, heavy exploitation of adults, leading to high rates of mortality amongst the largest and oldest fish would be expected to give a selection pressure resulting in a population of fish that matured at an earlier age, invested more in reproduction, and produced more and smaller offspring than fish from populations subjected to high juvenile mortality.

The predictions given above have been confirmed in studies of wild populations of guppies, *Poecilia reticulata*, exposed to different levels of predation pressure. Guppies from sites with high rates of predation on adults mature at an earlier age, have higher reproductive effort and have more and smaller offspring per brood than those found at sites where the predators take juvenile, rather than adult, fish. The genetic basis of these observations has been confirmed using both transfer (i.e. moving guppies from one site to another) and laboratory studies conducted over several generations. In summary, the results of these studies provide convincing evidence that differences in age-specific survival may be important in moulding life-history patterns.

In heavily exploited wild populations, intensive fishing over a period of many generations generally leads to a continuous removal of the individuals that are largest and easiest to catch. This leads to decreases in population density which should promote increased growth rates, but such a removal of the oldest and largest fish might also give a selective advantage to early maturing fish of small body size. Ultimately, this could result in genetic changes within the fish population leading to a population of small early maturing individuals, with the concomitant decline in the economic value of the fishery.

It has been suggested that such changes may have occurred in some populations that have been subjected to intensive fishing pressure. For example, the intensification of the commercial Nile tilapia, *Oreochromis niloticus*, fishery in Lake George, Africa during the 1950s resulted in a decrease from 900 to 400 g in the mean size of landed fish. At the same time, the length at maturity declined from 29 to 18 cm. These adverse changes have been attributed to alterations in the genetic composition of the population as a result of the selection pressures imposed by overexploitation. Similar genetic arguments have been invoked to explain the decline of some salmonid fisheries in which there has been a gradual shift towards the migratory fish returning to their home rivers at an earlier age and smaller size than previously.

Similarly, the heavy exploitation of the plaice in the North Sea since the early 1900s may have been an important contributory factor to the changes in the reproductive characteristics observed amongst these fish. Commercial exploitation can be considered to have imposed an additional mortality (fishing mortality), over and above natural mortality, on the oldest and largest fish.

The biology of the North Sea plaice has been studied for almost a century, and substantial changes in rates of growth, age at maturity and fecundity have been recorded. For example, rates of growth of juvenile plaice have increased in recent times, probably reflecting an increase in food availability due to reduced population density. This represents a phenotypic response, rather than being a reflection of genetic changes within the population. On the other hand, there is evidence that both the length and age at sexual maturation have decreased since 1900. Both of these changes would be predicted from life-history theory in a situation where fisheries were exerting a selection pressure due to reduced adult survival.

Thus, factors resulting in alterations in densities of populations, and selective removal of different size or age groups, are expected to lead to changes in the reproductive parameters of fish. Initially there may be changes in age or size at maturity that are reversible in the short term, but continued disturbance over a period of several generations may lead to progressive genetic changes of a more permanent nature.

11.6   CONCLUDING COMMENTS

Natural fish populations tend to fluctuate in size. Should, however, any large perturbations occur, density-dependent mechanisms may operate to return the size of the population to the equilibrium level, about which it oscillates. Thus, despite the fact that fluctuations are to be expected in population sizes, one can nevertheless consider multi-age fish stocks as being relatively stable, with overall numbers fluctuating comparatively little.

Populations that contain many year classes, and several overlapping generations, tend to show damped oscillations around the equilibrium population size, whereas populations that consist of a single, or very few, year classes may show fluctuations in numbers of almost two orders of magnitude.

The size of recruiting year classes tends to be highly variable, and year class strength seems to be determined at an early stage in the life history. Food availability appears to be a major determinant of larval growth, development and survival, and food availability, in turn, is governed by the physical and biological factors at work on the production cycles of the prey organisms.

Variability of recruitment is lower when spawning stocks contain large numbers of individuals than when spawning stocks have been reduced to small numbers of mature individuals. In addition, large populations tend to be buffered from the large changes in numbers that occur in smaller populations as the result of the recruitment of particularly strong or weak year classes. In other words, populations with large spawning stocks tend to be stable, whereas short-term changes in environmental conditions that affect year class strength and recruitment lead to considerably greater fluctuations in the size of smaller populations of fish.

This creates problems for the fisheries biologist since management models tend to be based upon assessments of adult stock size. These management models incorporate estimates of recruitment based upon averages taken over lengthy time periods. In other words, the management models are best suited to the monitoring of long-term changes in fish populations, and they are not designed to predict the possible influences of short-term fluctuations in year class strength. In order to forecast year-to-year fluctuations in recruitment, it will be essential to obtain information about the factors of major importance to larval growth and survival. The ability to predict the timing and magnitude of production cycles would seem to hold the key to the quantitative study of recruitment processes. This is not likely to be successfully achieved until the roles of the various physical, climatic and biological controlling factors become established with a greater degree of certainty.

FURTHER READING

Books

Beverton, R.J.H. and Holt, S.J. (1957, reprinted 1993) *On the Dynamics of Exploited Fish Populations*, Chapman & Hall, London.

Calow, P. (ed)(1987) *Evolutionary Physiological Ecology*, Cambridge Univ. Press, Cambridge.

Cushing, D.H. (1981) *Fisheries Biology: A Study in Population Dynamics*, Univ. Wisconsin Press, Madison.

Cushing, D.H. (1982) *Climate and Fisheries*, Academic Press, London.

Roff, D.A. (1992) *The Evolution of Life Histories: Theory and Analysis*, Chapman & Hall, London.

Stearns, S.C. (1992) *The Evolution of Life Histories*, Oxford Univ. Press, Oxford.

Review articles and papers

Adams, P.B. (1980) Life history patterns in marine fishes and their consequences for fisheries management. *Fish. Bull. US*, **78**, 1–12.

Bailey, K.M. and Houde, E.D. (1989) Predation on eggs and larvae of marine fishes and the recruitment problem. *Adv. mar. Biol.*, **25**, 1–83.

Charnov, E.L. and Berrigan, D. (1991) Evolution of life history parameters in animals with indeterminate growth, particularly fish. *Evol. Ecol.*, **5**, 63–8.

Elgar, M.A. (1990) Evolutionary compromise between a few large and many small eggs: Comparative evidence in teleost fish. *Oikos*, **59**, 283–7.

Fogarty, M.J., Sissenwine, M.P. and Cohen, E.B. (1991) Recruitment variability and the dynamics of exploited marine populations. *Trends Ecol. Evol.*, **6**, 241–6.

Policansky, D. (1983) Size, age and demography of metamorphosis and sexual maturation in fishes. *Amer. Zool.*, **23**, 57–63.

Reznick, D.A., Bryga, H. and Endler, J.A. (1990) Experimentally induced life-history evolution in a natural population. *Nature*, **346**, 357–9.

Rijnsdorp, A.D. (1993) Fisheries as a large-scale experiment on life-history evolution: disentangling phenotypic and genetic effects in changes in maturation and reproduction of North Sea plaice, *Pleuronectes platessa* L. *Oecologia*, **96**, 391–401.

Roff, D.A. (1991) The evolution of life-history variation in fishes, with particular reference to flatfishes. *Neth. J. Sea Res.*, **27**, 197–207.

Shepherd, J.G. and Cushing, D.H. (1990) Regulation in fish populations: myth or mirage? *Phil. Trans. R. Soc.*, **330B**, 151–64.

Stearns, S.C. (1976) Life-history tactics: A review of the ideas. *Q. Rev. Biol.*, **51**, 3–47.

Stearns, S.C. (1989) Trade-offs in life-history evolution. *Funct. Ecol.*, **3**, 259–68.

Chapter twelve

Human impacts on aquatic environments

12.1 INTRODUCTION

Large lakes, the seas and oceans have long been considered to be appropriate dumping grounds for the wastes of human societies, and until recently it was thought that the oceans of the world had an almost infinite capacity for absorbing waste materials. In other words, dilution was the solution to pollution. Thus, the lack of concern in the past about the discharge of highly toxic wastes, such as those containing mercury and DDT, can be attributed to the then-prevailing conception of the oceans as a bottomless receptacle.

The discharge of human sewage and garbage into rivers and coastal waters is still commonly practised throughout the world. The sewage, which adds particulate matter and nutrients to the recipient water body, may or may not have had some treatment before discharge. More insidious than sewage and garbage, which are visible, are the various toxic chemicals that find their way into aquatic systems. These chemicals are often transferred through food chains and exert their effects in places far removed in time and space from the source.

Thus, despite the fact that it has become increasingly apparent that aquatic environments do not have an unlimited capacity to absorb and degrade the wastes discarded by human society, several billion tonnes of waste materials continue to be discharged into aquatic environments each year. The slow but continuous change that has occurred means that the vast majority of the rivers of the world can no longer be considered to be in their natural states with regard to their dissolved and particulate matter loads, and rivers that drain industrialized areas have had their compositions substantially altered by human activities. Similarly, detrimental changes to coastal waters have been observed as a consequence of river, industrial and domestic sewage discharges, and the direct dumping of wastes.

It is the purpose of this chapter to discuss a few of the more significant sources of aquatic pollution and their effects or potential effects on fish and other aquatic organisms.

12.2 ENVIRONMENTAL STRESSORS

When fish, and other organisms, experience environmental disturbances which lie outside the normal range the effects may be dramatic. In the case of a severe disturbance there may be almost instantaneous mortality, and long-term exposure to less severe disturbances can also result in death of certain individuals within the population. Short-term exposure to stressors may result in changes that, whilst not being lethal, impair the ability of the fish to function normally.

The physiological systems of fish can be challenged, or stressed, by a wide range of biological, chemical and physical factors. Sublethal stressors elicit a complex of changes that can, in the long term, potentially lead to impairment of growth, reproductive performance, or disease resistance. Chronic exposure to these sublethal stressors can, therefore, result in reduced reproductive success or decreased survival of individuals, which may have serious consequences for the survival of entire populations.

Exposure of fish to environmental stressors induces a characteristic series of endocrine, and other, responses that are termed *primary*, *secondary* or *tertiary* depending upon the level of biological organization being monitored. Thus, primary stress responses describe changes at the endocrine level, whereas tertiary responses refer to those changes that can be easily seen by observing the animals.

Primary stress response

Adverse stimuli detected by the sense organs lead to the activation of afferent neural pathways that run in the sympathetic nervous system from the hypothalamus to the chromaffin tissue of the head kidney. Direct stimulation of the chromaffin tissue leads to the release of catecholamines. The catecholamines are released rapidly from the chromaffin tissue following the exposure of a fish to a stressor, and plasma levels of catecholamines may be elevated by over an order of magnitude within the course of a few minutes. The liberation of catecholamines into the bloodstream leads to them being widely distributed in the body, where they initiate a series of secondary effects on other organ systems, such as the cardiovascular and respiratory systems.

Other neurons within the hypothalamus run to the adenohypophysis of the pituitary gland. A neuropeptide is transported from the hypothalamus to the pituitary within this system, and release of the neuropeptide stimulates production and secretion of adrenocorticotropic hormone (ACTH). The ACTH is released into the blood and travels via the systemic circulation to the interrenal cells of the head kidney. Within the interrenal, ACTH stimulates the production and release of corticosteroid hormones, particularly cortisol. Since there are close links between the hypothalamus and the release of ACTH and cortisol, acting as a cascade, it has become usual to refer to this series of interactions as being governed by the hypothalamic–pituitary–interrenal axis (HPI axis)(see also Chapter five). A major target organ for corticosteroid action in fish appears to be the liver. Cortisol enters the cells and binds to nuclear receptors. This results in gene activation, with the production of a series of enzymes which have a range of metabolic effects.

Secondary stress responses

The release of catecholamines and cortisol triggers a broad suite of biochemical and physiological changes. The metabolic effects may include hyperglycaemia, hyperlacticaemia, depletion of tissue glycogen reserves, lipolysis and inhibition of protein synthesis.

Osmotic and ionic disturbances may occur as the result of diuresis and the loss of electrolytes from the blood. There may also be changes in haematology, with erythrocytic (red blood cell) changes and a reduction in numbers of white blood cells (leucopenia). The catecholamines have marked influences on the cardiovascular system, leading to changes in patterns of blood flow and increased gill perfusion, and alterations in the oxygen transport capacity of the blood. On the other hand, the corticosteroids are known to stimulate the ion-transporting mechanisms present in the gills and kidney. These effects are of particular importance in the period of recovery from stress, when the fish attempts to maintain oxygen supplies to the tissues, and to regain osmotic and ionic equilibrium.

Tertiary stress responses

The secondary stress responses, which are induced by the rapid endocrine changes, lead swiftly to the rapid mobilization of energy reserves. Thus, the secondary stress responses are believed to have evolved as adaptive mechanisms that enable the animal to meet the increased energy demands imposed by exposure to environmental stressors. Typically these changes persist for only a few hours or days, following acute exposure to the stressor, and do not, therefore, result in any serious deleterious effects to the animal. In contrast, chronic exposure to stressors can induce a number of pathological changes, and can lead to reductions in reproductive success, depression of growth rates, and decreased disease resistance.

It is known, for example, that exposure of fish to many different types of environmental stressors may result in increased susceptibility to a wide range of common pathogens, e.g. viruses, bacteria, fungi and protozoans. This stress-induced increase in susceptibility to disease organisms can be mimicked by the administration of physiological doses of corticosteroids to otherwise unstressed fish.

Similarly, exposure to environmental stressors, such as acid waters and sublethal levels of chemical toxicants, has been shown to reduce the reproductive success of a range of fish species, and this seems, also, to be mediated via the corticosteroids. Both exposure to stressors, and the elevation of plasma cortisol levels due to cortisol implantation, have been shown to lead to suppression of the pituitary–gonadal axis, resulting in a reduction in circulating levels of sex hormones in salmonids (see also Section 9.9). Chronic exposure to stressors may lead to vitellogenesis being impeded, ovarian and egg size being reduced, and the survival of eggs and offspring of stressed fish being reduced in comparison with those of unstressed fish.

Thus, when fish are exposed to environmental stressors a hierarchy of responses is initiated, and if the stress is severe or long-lasting, successively higher levels of biological organization become affected. The cumulative effects of prolonged exposure to environmental stressors at sublethal levels may affect disease resistance, growth and reproductive success of individuals, which in turn may reduce recruitment to successive life stages and, eventually, cause population numbers to decline.

12.3   XENOBIOTICS: TOXICITY AND BIOACCUMULATION

Hundreds and thousands of different chemicals can be found in association with water. Xenobiotics are defined as chemicals that are foreign to biological systems, and many of the xenobiotics that are discharged into water bodies may be toxic to aquatic organisms. The various pollutants and toxic chemicals can enter aquatic environments by several routes. These routes include direct precipitation, surface and subsurface water run-off, sewage discharges and industrial wastewater outfalls. Water bodies can contain naturally occurring substances like carbon dioxide and hydrogen sulphide at levels that make the environment unsuitable for habitation by fish, or the water source may be contaminated by toxic levels of manufactured organic compounds such as oil and petroleum residues, pesticides and polychlorinated biphenyls (PCBs).

Many of the herbicides and pesticides in use today, e.g. cholinesterase inhibitors such as organophosphates and carbamates, are short-lived – that is, they are biologically active for only a few days rather than months or years – so effects of modern pesticides are most likely to be observed as acute poisonings or fish kills. On the other hand, herbicides containing arsenic were once commonly in use, and in the past many fungicides were based upon mercury as an active ingredient. Arsenic and mercury retain their toxicity, and water seepage from soils and sediments contaminated with these toxicants can represent a threat to fish inhabiting recipient water bodies. Similarly, many of the chlorinated and hydrocarbon pesticides, such as DDT, are persistent. These compounds will usually enter the food chain at a low trophic level, and will tend to be accumulated, i.e. increase in concentration, in animals at the highest trophic levels within the food chain.

Heavy metals, such as mercury, lead, cadmium and zinc, can enter water bodies in industrial wastewater, but other potential sources of contamination may not be as immediately apparent. For example, even though abandoned mining tips and industrial dumping sites may be situated some distance away from open water they are potential sources of heavy metal pollution. Surface run-off and groundwater seepage carrying heavy metal residues can flow into rivers and lakes from such sites, causing contamination.

There are several characteristics of water bodies that will have a significant effect on the toxicity of a given pesticide, herbicide or heavy metal. Temperature, water pH, alkalinity (carbonate concentration) and hardness (concentration of divalent cations) can, for example, all influence the solubility or toxicity of a chemical. The toxicity of most heavy metals is reduced as pH increases, because at high pH the metals bind to form hydroxide and carbonate complexes. Thus, there are reductions in levels of the toxic metallic ion as pH increases. In most water bodies hardness, alkalinity and pH increase, or decrease, together so the effects of these factors on the toxicity of metals tend to be closely interwoven. A number of metals may bind to organic matter, leading to reduced toxicity, so the toxicity of heavy metals may be modified by the levels of organic matter present in the water body.

Bioaccumulation and detoxification

When fish are exposed to potential toxicants the chemicals may both enter the body over, and cause damage to, the gill membranes. Toxicants are frequently observed to result in disturbances to iono- and osmoregulation, these disturbances arising, primarily, due to damage to the gills. Gill damage may, in turn, affect ion and respiratory gas exchange, acid–base balance and the excretion of waste products.

Falls in plasma ion concentrations have often been observed in freshwater fish following exposure to a variety of toxicants, including heavy metals, pesticides and chlorinated hydrocarbons. There are several factors that contribute to the drop in the plasma levels of the various ions, but increased gill permeability seems to be of prime importance. An increase in the permeability of the gill lamellae to water and ions would lead to an increased efflux of ions from freshwater fish, and there would also be a concomitant enhancement of the osmotic uptake of water. An increase in water influx would, in turn, lead to increased urine production and thereby further exacerbate the loss of ions. In addition, the inhibition of active ion uptake by the cells of the gill epithelium that accompanies exposure to a range of toxicants contributes further to the negative ion balance.

In addition to entering the fish through the highly permeable gill membranes, toxicants may also be consumed along with the food and be absorbed from the gastrointestinal tract. These chemicals may then be deposited and stored in various tissues of the body, such that tissue concentrations continue to rise with prolonged exposure to the chemical. This *bioaccumulation* is not only dependent upon the rate of uptake of the chemical, but is also influenced by the mechanisms of tissue storage. A major determinant of the bioaccumulation of a toxicant under a given set of exposure conditions is the rate at which the chemical is metabolized, detoxified and cleared from the body.

Not all toxic substances are accumulated in the tissues, and distributions may also differ from tissue to tissue. For example, whilst both mercury and cadmium may accumulate in fish tissues, other metals, such as zinc and copper, are usually deposited and stored much less readily. Fish have a number of mechanisms to prevent accumulation of many substances, and following exposure to potential toxicants a number of biochemical pathways are induced in order to detoxify and excrete the foreign compounds.

Transformation of potentially toxic compounds occurs in all of the body tissues, but the highest levels of the enzymes that metabolize toxicants are found in the liver. The transformation reactions usually involve an initial oxidation, reduction or hydrolysis of the toxicant, followed by a reaction in which the residues are transformed into water-soluble metabolites that can be readily excreted from the body.

One of the major enzyme systems involved in the transformation and detoxification of organic compounds, such as hydrocarbons and pesticides, is the cytochrome P-450 system. This system is not only involved in detoxification reactions, but is also implicated in the biotransformation and metabolism of endogenous compounds such as steroid hormones. The biotransformation of organic xenobiotics occurs in two stages. The first stage involves a metabolic transformation in which an oxygen atom is introduced into the lipophilic organic substrate by the cytochrome P-450 enzyme system. In the second stage, the water solubility of the oxidized xenobiotic is increased via conjugation to glucuronic acid, sulphate or glutathione. The conjugated products are generally both less toxic and more easily excreted than the untransformed compounds.

Detoxification of inorganic compounds, such as heavy metals, occurs via a different set of biochemical processes. Exposure of fish to heavy metals such as cadmium and copper leads to the synthesis of metallothioneins in liver, kidney and gill tissues. Metallothioneins are low-molecular-weight proteins that can bind to metals. The binding properties of the proteins vary, but the various metallothioneins generally have the lowest affinity for zinc and a higher affinity for cadmium and copper. Because exposure to heavy metals induces the synthesis of metallothioneins, these proteins are thought to play an important part in protecting the cell from damage by heavy metal toxicants. The proteins may serve a protective function by sequestering the free ions of the heavy metals, thereby preventing the metal ions from binding to, and inhibiting, cellular enzymes and other functionally important proteins.

One of the main mechanisms by which toxicants, both organic xenobiotics and heavy metals, exert their deleterious effects is via lipid peroxidation. Lipid peroxidation is a chemical process that results in the oxidative deterioration of polyunsaturated fatty acids, and thereby leads to destructive changes in the lipids of the cell membranes. Thus, the net result of lipid peroxidation is the destruction of cell membranes and the disruption of membrane-bound enzyme activities.

The lipid peroxidation chain reaction is initiated by the removal of a hydrogen atom from a polyunsaturated fatty acid. This results in the formation of a lipid radical. Lipid radicals react very rapidly with oxygen to form hydroperoxide radicals, and these, in turn, may remove hydrogen atoms from fatty acids in other lipid molecules. This results in the formation of a molecule of hydroperoxide and a new lipid radical, which continues the chain reaction. The hydroperoxide can be degraded to give a variety of products, including malondialdehyde.

Protection of the cell membrane lipids against peroxidation can be provided by two broad classes of antioxidants, the *preventive* and the *chain-breaking* antioxidants. Preventive antioxidants reduce the rate at which lipid radicals are formed, i.e. they act by hindering the initiation of the lipid peroxidation chain reaction. The most important preventive antioxidants are enzymes (e.g. catalase, glutathione peroxidase) that are capable of reducing hydroperoxide radicals. The chain-breaking antioxidants, on the other hand, owe their antioxidant activity to their ability to bind with and trap hydroperoxide radicals. Ascorbic acid (vitamin C) is an important water-soluble chain-breaking antioxidant, and the most important lipid-soluble antioxidants are the tocopherols (vitamin E and related compounds) (Fig. 12.1).

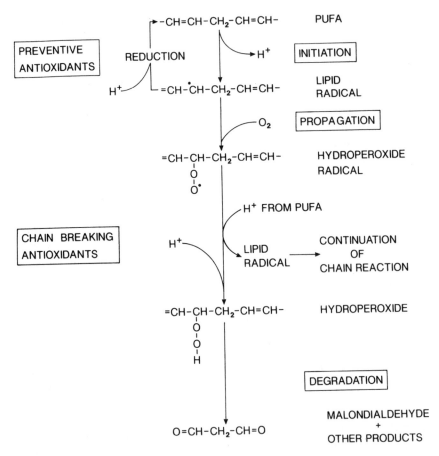

Fig. 12.1 Destruction of polyunsaturated fatty acids (PUFAs) by lipid peroxidation. Note that preventive antioxidants and chain-breaking antioxidants have different points of intervention in the lipid peroxidation pathway.

Toxicity tests

Whether or not exposure to any given chemical proves to be lethal will be dependent upon the concentration of the chemical in the water, the characteristics of the water with regard to pH, hardness, alkalinity and temperature, and the length of time the fish is exposed to the potential toxicant. Consequently, it is not possible to give single values for the 'lethal limits' of fish species. Thus, when lethal limits are reported it is essential that information also be given about the measurement conditions and exposure period. For example, a fish held in 'soft' water may tolerate exposure to a given concentration of a chemical for only a few minutes or hours, whereas the same concentration may not prove to be lethal to fish held in 'hard' water even with longer-term exposure.

The methods used for determination of toxicity (lethal limits) involve the investigation of the effects of combinations of concentrations of the chemical under test (i.e. different dosages of the chemical) and exposure time on groups of fish. In toxicity testing it is most common that the studies involve the examination of mortality of animals exposed to different concentrations of the chemical for a given time period. The results are plotted graphically and the concentration giving 50% mortality is estimated. These types of study give values of median lethal doses (LD_{50})(alternatively called median lethal concentration, LC_{50}) for the chosen exposure time. Experiments carried out to determine the LD_{50} over discrete time periods (usually not more than 96 h) are called acute toxicity tests. Since there is a dosage–exposure time interaction in such studies it is important that information about both the LD_{50} and exposure time be given. Thus, results will usually be presented as 24 h LD_{50}, 48 h LD_{50}, 96 h LD_{50} and so on. The practical value of these results may be limited, both because they apply only to a specific set of conditions and because direct comparisons between different studies may not be possible.

The fact that there is a dosage–exposure time interaction results in a progressive reduction in LD_{50} with longer exposure times. At very long exposure times the LD_{50} will approach a constant value. This is illustrated in Fig. 12.2, which shows that a plot of LD_{50} values against exposure time will give a relationship with a distinct break point. Thus, the LD_{50} decreases as exposure time increases and plateaus at the break point. This minimal LD_{50} represents the *incipient lethal level* and is an estimate of the dose which, theoretically, 50% of the population could tolerate for an indefinitely long period of time. The incipient lethal level is, therefore, estimated by the determination of dosage mortality after a long exposure time. In toxicological studies, for example, it has been common practice to determine incipient lethal levels based upon the results of tests employing exposure times of 96 h.

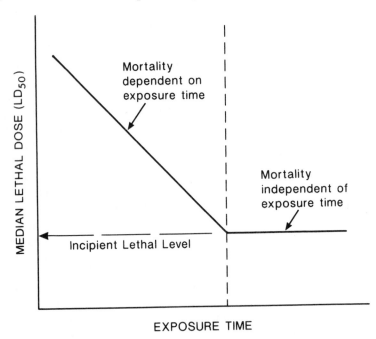

Fig. 12.2 The effect of increasing the length of the exposure time upon the median lethal dosage (LD_{50}) of a toxicant as established from dosage–mortality curves. Note that the LD_{50} decreases as exposure time increases until it finally reaches a plateau. The minimum LD_{50} represented by the plateau indicates the 'incipient lethal level'.

Long-term, or chronic, exposure to low concentrations of toxicants might be lethal to some individuals within the population, or could cause pathological changes. Thus, 'safe' or 'permissible' levels are often considered to be some fraction (e.g. 10%, 1% or 0.1%) of the measured LD_{50} or incipient lethal level. Since most workers only conduct acute toxicity tests, and report LD_{50} values, decisions about the setting of 'permissible' levels of given toxicants are quite often arbitrary, rather than being based upon the results of extensive long-term study.

12.4   EUTROPHICATION

The dictionary definition of eutrophy is healthy nutrition, but the eutrophication of water bodies has come to have negative connotations due to the often excessive promotion of growth of algae and macrophytes that results from nutrient inputs. In order to grow properly, plants require certain levels of available nutrients, and while a variety of such chemicals is required, the ones that typically become limiting are nitrogen and phosphorus. Thus, in the intensive production of crop plants, farmers will almost invariably provide additional sources of nitrogen and phosphorus in the form of artificial fertilizers.

Nitrogen is taken up by plants primarily in the form of nitrate ions (NO_3^-), and phosphorus is usually added to the land in the form of orthophosphate (HPO_4^{2-} and $H_2PO_4^-$). The excess inorganic fertilizer can be washed out of the soil by rainwater and will eventually reach ponds, streams, rivers and other water bodies in surface or subsurface run-off. Phosphorus is usually present in only minute concentrations in natural waters ($1-100$ μg l^{-1}), and it tends to be the first limiting plant nutrient in fresh waters. Nitrogen, in the form of nitrate, nitrite, ammonia and as a constituent of organic compounds, is generally present in higher concentrations ($100-750$ μg l^{-1}), but nitrogen may be the first limiting nutrient in the marine environment.

Various types of primary producers can be found associated with water bodies, including rooted and floating aquatic macrophytes, benthic algae, phytoplankton, floating filamentous algae and autotrophic bacteria. The carbon:nitrogen:phosphorus ratio required by many of the primary producers will be close to 100:16:1, which indicates the potential of even small amounts of phosphorus input to influence primary productivity in the presence of sufficient concentrations of the other two elements. Since phosphorus is usually present at very low concentrations, the phosphorus that is released from decaying plant and animal material will generally be rapidly sequestered by the primary producers and be removed from solution. Eutrophication resulting from an influx of nutrients in the run-off from agricultural land will usually lead to an increase in plant growth, often via the promotion of phytoplankton blooms. Much of the phosphorus will be stored in the plant tissues, but the collapse of a phytoplankton bloom will lead to the release of large quantities of plant nutrients into the water body. Under some circumstances, this can lead to explosive growths of undesirable filamentous algae or macrophytes, which in turn can result in the water body becoming a less suitable environment for fish and other aquatic organisms.

Oxygen depletion

The promotion of plant growth that results from eutrophication can lead to marked fluctuations in levels of dissolved oxygen during the course of a diel cycle. Thus, oxygen tension may rise to over 200 mmHg during the day when oxygen is being released by the plants as a result of photosynthesis. Photosynthesis is the process by which plants convert carbon dioxide into organic matter in the presence of light. Most photosynthetic organisms absorb light of wavelengths within the range 350 to 700 nm, and plant species have a range of pigments that are capable of absorbing light of different wavelengths. Photosynthesis is, however, dependent upon the presence of chlorophyll pigments, at least three forms of which occur in green plants. The process of photosynthesis, leading to the production of organic matter from carbon dioxide and water, with the concomitant release of oxygen, can be summarized by the following reaction:

$$CO_2 + H_2O \rightarrow CH_2O + O_2$$

During the course of the night, when photosynthesis ceases, and oxygen is being consumed by the respiration of both the plants and animals, oxygen tension may fall dramatically to 20–30 mmHg. Some fish species are incapable of surviving in hypoxic waters with such low oxygen tensions and will succumb. Others will attempt to seek out refugia in which oxygen tensions are higher, and some fish will adopt *aquatic surface respiration*, in which they move to the water surface and use the thin oxygenated layer at the air–water interface. Bimodally breathing fish will increase their frequency of aerial respiration as levels of dissolved oxygen decline and fish of some species are able to survive in hypoxic environments due to their ability to respire anaerobically for quite prolonged periods.

The eutrophication of ponds and lakes can lead to the growth of floating plants being promoted to such an extent that they cover the entire water surface. Plant growth of this type can have a number of negative consequences for other aquatic organisms. Firstly, the growth of floating plants will restrict the amount of light penetrating to greater depths in the water column. This will, in turn, hinder the photosynthesis of phytoplankton and benthic algae, possibly resulting in die-backs of these plants. The decomposition of rotting vegetation consumes considerable quantities of oxygen, so aquatic animals that inhabit eutrophic water bodies containing large amounts of decaying plant material may be frequently subjected to hypoxic or anoxic conditions. Secondly, explosive growth of floating vegetation can result in marked decreases in both the surface of contact between the water and the atmosphere, and the level of wind-blown mixing of the surface water layer. Both of these effects of increased surface cover by floating plants will generally have negative effects upon levels of dissolved oxygen.

Organic and other nitrogenous wastes

A significant problem associated with the rearing of livestock is the disposal of liquid and solid wastes, and similar problems arise in connection with the disposal of the sewage produced by humans. These organic wastes can be deposited in sewage lagoons where degradation occurs, but it is not infrequent for untreated or partially degraded sewage to be discharged directly into rivers, large lakes or the sea. Moderate levels of nutrients released from these wastes can support increased growth of phytoplankton and aquatic macrophytes, but high nutrient inputs may lead to an explosive growth of undesirable species of filamentous algae and fungi. In addition, the discharge of animal wastes and sewage into water bodies may be undesirable because of the risk that fish and other aquatic organisms exploited for human consumption could accumulate potential pathogens in their tissues. Pathogenic bacteria are, for example, frequently isolated from manures and sewage effluents, so concerns for public health appear to be real.

Oestrogenic xenobiotics

Sewage effluents and industrial wastewaters may contain a range of substances that mimic the action of female sex steroids. Steroids and 'steroid-like' substances having oestrogenic activity include pesticides (e.g. DDT, kepone and chlordecone), phytoestrogens and several products that result from the biodegradation of detergents and non-ionic surfactants. Phytoestrogens occur naturally in plants, with legumes, such as soya and alfalfa, containing high amounts of these steroid-like substances. The phytoestrogen content of plant tissues tends to increase following fungal degradation.

During sewage treatment, detergents and surfactants (alkylphenol–polyethoxylates) may be degraded first to short-chain alkylphenol–polyethoxylates and finally to alkylphenols. These oestrogenic degradation products are present in sewage effluents, and certain stretches of Western European and North American rivers contain these substances in concentrations of several $\mu g \, l^{-1}$.

The alkylphenols, in common with chlorinated insecticides (e.g. DDT and kepone), are both lipophilic and persistent. These compounds tend, therefore, to accumulate in lipid depots and may reach concentrations of several $\mu g \, g^{-1}$ tissue. Lipid mobilization would lead to release of the stored oestrogenic xenobiotics, allowing them to exert their metabolic effects. These compounds are known to induce the synthesis of vitellogenin by hepatocytes, so the negative repercussions of biologically-active oestrogenic xenobiotics could be particularly severe for male and immature fish.

Nitrogenous wastes

Nitrogenous wastes are excreted by animals in several forms – ammonia, creatine, creatinine, amino acids, urea and uric acid – and nitrogenous compounds are also released during the decomposition of plant and animal remains. The primary nitrogenous excretory products of aquatic animals are ammonia (NH_3) and ammonium ions (NH_4^+). Ammonia is a small lipophilic molecule that can diffuse easily, and relatively rapidly across biological membranes. Diffusion from the blood of the animal to the water is also aided by the fact that the ammonia concentration in the blood plasma will usually be much higher than in the surrounding water. Ammonium ions, on the other hand, are larger than ammonia molecules and they also carry a positive charge. Rates of diffusion of ammonium ions across biological membranes are slower than those of ammonia, and it seems probable that the excretion of ammonium ions is strongly coupled to the movement of other cations. Ammonium can, for example, serve as a counter ion in the uptake of sodium across the fish gill. In fish species, the diffusion of ammonia may account for about 60% of the nitrogen efflux, 20% or so of the nitrogenous excretion may be in the form of ammonium ions, and the remainder will be excreted as urea and other metabolic end products.

Nitrification and nitrite toxicity

In natural waters, ammonia excreted by aquatic organisms may be oxidized, first to nitrite and then to nitrate. Urea may be hydrolysed to ammonia and carbon dioxide, with the resulting ammonia entering the nitrification process. The nitrification process is that by which ammonia is oxidized to nitrate by aerobic bacteria. Both heterotrophic and chemoautotrophic bacteria may play a role in the nitrification process, but *Nitrosomonas* and *Nitrobacter* bacteria are undoubtedly the most important. Aerobic bacteria in the genus *Nitrosomonas* are responsible for converting the ammonia to nitrite (NO_2^-), while *Nitrobacter* bacteria are responsible for the conversion of nitrite to nitrate (NO_3^-). The growth of *Nitrobacter* is inhibited in the presence of ammonia, so under conditions where there are high concentrations of ammonia in the water the nitrification process will not proceed to completion. Under such circumstances nitrite would tend to accumulate, and could eventually reach toxic levels.

Industrial, and other human, activities can give rise to increases in nitrite concentrations in aquatic systems. For example, sewage effluents may contain large amounts of nitrite, and nitrite concentrations of 50 mg l^{-1} or more have been recorded in waters receiving effluents from factories producing dyes, metals and celluloid.

Nitrite is taken up across the gills of fish and combines with haemoglobin in the blood to produce methaemoglobin. Haemoglobin that has been converted to methaemoglobin is unable to bind with, and transport, oxygen. Thus, an increase in methaemoglobin, and a concomitant reduction in haemoglobin, usually leads to a decrease in the oxygen-carrying capacity of the blood. In fish with methaemoglobinaemia the blood will be chocolate brown, rather than red. Nitrite appears to enter the blood across the gills using the same transport systems as other monovalent anions, e.g. chloride and bicarbonate, so that the toxicity of nitrite to fish may be reduced in the presence of adequate concentrations of chloride in the surrounding water.

12.5 FRESHWATER ACIDIFICATION

The safe range of environmental pH for the survival of fish is approximately 5–9, but productivity is maximized in waters of pH 6.5–8.5. The pH of sea water is generally in the range 7.8–8.2, and the large buffering ability of sea water means that the effects of acid inputs on pH are negligible. Mixtures of weak acids and their salts are called buffers. Several buffer systems exist in water, including those associated with phosphate, borate and carbonate. Buffer systems resist pH changes because the ionization equilibrium between the weak acid and weak base changes in a manner that allows them to bind hydrogen (H^+) and hydroxide (OH^-) ions that are added to the system.

In the oceans, lakes and rivers, carbonate–bicarbonate represents the primary buffering system. Following an input of acid, hydrogen ions will combine with bicarbonate to form carbonic acid, a weak acid, which dissociates incompletely to give carbon dioxide and water:

$$H^+ + HCO_3^- \rightarrow H_2CO_3 \rightarrow H_2O + CO_2$$

Bicarbonate ion can dissociate to produce a hydrogen ion and a carbonate ion (CO_3^{2-}), and conversely, carbonate can combine with a hydrogen ion to produce bicarbonate. Thus, the keys to the system are the carbonate and bicarbonate ions that are present in the water. Acid inputs result both in the combination of bicarbonate ions with hydrogen ions, and in the conversion of carbonate to bicarbonate. If all the carbonate present in solution is converted to bicarbonate, calcium carbonate ($CaCO_3$) will dissociate to provide additional carbonate ions.

Under some circumstances the rate of calcium carbonate dissociation may not be sufficiently rapid to meet the challenge imposed by acid input. Alternatively, the available pool of calcium carbonate may be insufficient to meet demand. In such instances, the pH of the water body may fall dramatically as hydrogen ions are added to the system.

Thus, the buffering capacity of fresh waters is dependent upon the carbonate–bicarbonate system, and buffering capacity is, therefore, related to the carbonate content of the water. Since freshwater bodies vary markedly in carbonate and ionic contents, buffering capacities are also quite variable. Consequently, the pH of fresh waters covers a wider range than that of oceanic waters, and the effects of acid inputs can be much more marked in rivers and lakes than in the sea. Acidification will be most dramatic in water bodies that lie over, or receive run-off from, bedrock that is poor in carbonate and other soluble minerals that can contribute to buffering capacity.

Quite dramatic falls in pH of lakes and rivers can result from the periodic discharge of acidic industrial wastes. Such discharges can lead to acute respiratory distress, and disturbance of iono- and osmoregulatory performance in fish species. Respiratory distress may arise as a result of reduced oxygen diffusion over the gill membranes caused by an increase in the thickness of the mucous layer covering the secondary lamellae. An increase in secretion of mucus is a typical response to acute exposure to acid conditions, and following longer-term exposure to low pH there may be a hyperplasia of mucous cells of the gill epithelia. Acute exposure to acidified water also results in an inhibition of hydrogen ion excretion and a reduction in ion uptake. Reductions in the net uptake of ions in acidified water can result in a gradual reduction in plasma ion levels, and the inhibition of hydrogen ion excretion may give rise to acidosis if the fish are in water of low calcium content. The reduction of the ionic content of the plasma may arise as a result of a reduced influx, as with calcium, or both a reduced influx and an increased efflux of the affected ions, as appears to be the case with sodium. Should acute exposure to acidified water result in the death of the fish the most likely cause is osmotic and ionic failure.

Whilst industrial wastewater outfalls may lead to changes in the pH of rivers and lakes, there is also considerable concern about the long-term chronic acidification of water bodies that has occurred as the result of increased industrial atmospheric emissions. Industries (e.g. electric power plants, smelters and steel mills) that use large quantities of fossil fuels (coal and oil) will produce waste emissions that contain large amounts of sulphuric oxides, and the exhaust emissions from road vehicles contain quantities of nitric oxides. Both of these types of gaseous emission will enter the atmosphere and can later fall as acid precipitation.

'Acid rain' will fall in areas located downwind of large industrial and urban complexes, and it is this acid precipitation with its content of sulphuric and nitric acids that causes the most severe acidification of surface waters. The acid rain water may enter streams and lakes directly, as surface run-off, or after percolation through the soil and subsoil.

Acid rain that falls directly into rivers and lakes will impose a direct challenge to the carbonate–bicarbonate buffer system, and acid precipitation that percolates through soil will result in leaching of base cations (calcium, magnesium, sodium and potassium) and dissolution of calcium carbonate present in the bedrock. Thus, in the long term there will be a gradual depletion of base cations from the soil, and the erosion of bedrock will increase. As the base ions become depleted other ions, such as aluminium, are leached out of the soil in greater quantity.

Aluminium is extremely toxic to fish at a water pH of 5, so the entry of aluminium into water bodies already affected by acid precipitation may lead to mortality. Exposure of fish to aluminium in acid water (pH 5–5.3) can cause severe disruptions of the ion-exchange mechanisms, resulting in reduced net uptake of ions such as calcium and sodium by freshwater fish species. Thus, mortality may arise due to iono- and osmoregulatory failure.

Aluminium toxicity to fish and other aquatic organisms is much reduced as pH is increased. Consequently, attempts have been made to counteract the effects of acid rain by the 'liming' of water bodies in order to increase the buffering capacity and thereby increase and stabilize pH.

12.6 SUSPENDED SOLIDS

Suspended solids are defined as small pieces of particulate matter that occur in the water column. This particulate matter usually consists of mixtures of sediment particles, e.g. fine sand, silt and clay, and particulate organic matter, e.g. detritus composed of plant and animal remains, bacteria and fungi. The higher the concentration of suspended solids in the water column the more turbid the water becomes, and the shallower will be the depth of penetration of light. Thus, if turbidity becomes very high as a result of the presence of large quantities of suspended solids in the water column, primary productivity may be reduced because the depth of light penetration is insufficient to sustain effective photosynthesis.

Sediment particles composed largely of silt and clay may become suspended in the water column as a result of current flows or wind mixing, or such particles may be washed into the system from the land following periods of heavy rainfall. In some cases, larger particles of sand or gravel may also become suspended in the water, but these particles will soon be deposited unless there is continuous turbulent mixing of the water column. Thus, varying levels of suspended solids will occur naturally in oceans, rivers and lakes, the exact quantities observed at any given time being related to the prevailing weather conditions, currents and turbulence in the water column.

Industrial activities, such as mining and earthworks, lumber and sawmills, smelters and coking plants, and sewage works produce wastewater discharges and slurries that may contain high levels of suspended solids. The increased water turbidity that results from such discharges can lead to reduced productivity due to retardation of growth of phytoplankton, benthic algae and aquatic macrophytes.

The presence of suspended solids in the water can also irritate, or cause mechanical damage to, the gills of fish and other aquatic animals. There may, for example, be mechanical clogging of the gills, leading to increased 'coughing' by the fish as it attempts to clear the obstruction. Alternatively, irritation of the gill membranes can lead to increased mucus production with a concomitant reduction in oxygen diffusion. The net result of clogging and increased mucus production may be respiratory distress.

If large quantities of suspended particulate matter are introduced into a river or lake, subsequent settling of the material may lead to the burial of the eggs and larvae of those fish species that lay demersal eggs. It is not uncommon, for example, for salmonid redds, and the nests of centrarchid species, to become buried under the sediment transported into a water body in the aftermath of periods of heavy rainfall. Similarly, the sedimentation of suspended solids introduced into rivers and lakes in industrial wastewater discharges can lead to nest burial and the suffocation of developing eggs and larvae.

The settling out of particulate matter will lead to the accumulation of sediment with time, and such accumulation can result in substrate modification. For example, the large pores in a gravel substrate can become clogged with fine sand and silt, thereby changing the substrate's characteristics and rendering it unsuitable for redd construction by spawning salmonids.

12.7   TEMPERATURE

Many polar species of fish frequently occur in waters at temperatures close to freezing, and survive temperatures that would kill tropical species. Thus, polar species are active in ice-cold water and some may be killed when the temperature exceeds 10–15°C, whereas fish of some tropical species may die when water temperature falls below about 15°C. It is, therefore, clear that there are differences in the temperatures tolerated by fish from different geographical areas, but the thermal tolerances of a given fish, or species, are not rigidly fixed.

The limits of temperature tolerance are modified by the previous thermal history of the fish. Thus, fish held for an extended period under cold conditions are more cold tolerant and more heat sensitive than fish held in warm water. This means that the temperature tolerance limits of fish will often be found to vary on a seasonal basis: in winter a fish will tolerate, and may even be active, in waters of a temperature that would prove fatal to the same fish in summer.

Whilst acclimatization of fish to different environmental conditions leads to some modifications in temperature tolerance limits, fish can generally be categorized as belonging to stenothermal cold-water, stenothermal warm-water or eurythermal species. Eurythermal species are tolerant of a wide range of temperatures, whereas stenothermal species tolerate a much more restricted temperature range. Salmonids, for example, are stenothermal cold-water species having upper lethal temperatures of about 25°C, whereas the goldfish, *Carassius auratus*, is extremely eurythermal and can tolerate temperatures from below 5°C to in excess of 35°C. As a generalization it seems likely that freshwater fish species may have wider temperature tolerances than marine fish. This seems to be the case because many freshwater fish inhabit shallow ponds, lakes and streams in which there may be quite large temperature fluctuations both on a seasonal basis, and in the shorter-term.

Fish, in common with other organisms, respond to a rapid elevation in temperature by synthesizing a group of proteins known as *heat-shock proteins*. Synthesis of the heat-shock proteins is induced following exposure to temperatures 5–15°C above the normal environmental temperature, and the synthesis of these proteins increases the thermal tolerance of the animal to subsequent exposure to elevated temperatures. Thus, the heat-shock proteins are thought to protect the cells against the pathological effects of heat, although their modes of action are not known with any degree of certainty. Some of the heat-shock proteins may regulate transport functions within the cell, whereas others may play a structural role.

Whilst heat-shock proteins are synthesized following acute exposure of the animal to a temperature increase, there is increasing evidence that these proteins may also be synthesized in response to the exposure of the animal to a range of stressors (e.g. heavy metals and other xenobiotics, radiation). Thus, the induction of heat-shock protein synthesis may represent a general cellular response to exposure to adverse environmental conditions.

Major sources of 'thermal pollution' include industrial wastewater outfalls and the 'cooling water' discharged by power plants. Electricity is generated in fossil-fuel and nuclear power plants by boiling water to produce steam that is then used to turn the generators. The steam is then passed through condensers where it is cooled and converted back to water. The water is then returned to the boilers for reuse. Water is circulated around the condensers to cool the steam, and the temperature of this cooling water becomes elevated by several degrees as the steam is cooled.

Many power plants use reservoirs, rivers or estuaries as sources of cooling water, and the plants will usually be constructed to have intake canals from which they draw their cooling water, and discharge canals into which the heated condenser water is pumped after use. In fossil-fuel plants, the condenser water usually comes directly from, and is discharged into, a reservoir, large lake or river. In nuclear power plants, the boiler water and the water used to condense the steam are both held within a containment vessel to prevent the escape of radioactivity. A third water jacket, with water from an outside source is used to cool the containment vessel, and it is this water that is discharged into the environment.

Thus, the warming of recipient water bodies that accompanies the discharge of heated water from power plants, and other large industrial complexes, may lead to them becoming unsuitable habitats for stenothermal cold-water species, such as salmonids. The greatest problems may, however, not arise because of the warming of the water body *per se*. The greatest threats to aquatic animal life may be posed by the large fluctuations in temperature that can result as a consequence of the periodic shutdown of generators, or plants, for service and repair. In addition, it is sometimes necessary for power plant operators to flush chlorine or other toxic chemicals through the intake pipes and condensers in order to eliminate fouling organisms. In some cases the levels of toxicants that enter the recipient water body in the discharge water can have detrimental effects upon fish and other aquatic organisms.

As the cooling water passes through the condensers of a power plant, it is not only heated but is also placed under pressure. Gases that are dissolved in the water can, therefore, become supersaturated. All the dissolved gases can become supersaturated, but it is nitrogen that is of particular concern. Fish that are exposed to supersaturated water can develop 'gas bubble disease' as the nitrogen comes out of solution in the tissues. Such bubbles, or emboli, can lodge in the blood vessels, can develop in the gills or under the skin, and may under some circumstances lead to mortality.

12.8 CONCLUDING COMMENTS

Fish display a wide range of behavioural, physiological and biochemical adaptations to meet the challenges imposed by the changes that can occur within aquatic environments. These adaptations may range from avoidance reactions, in which the fish swim away from the source of the environmental stressor, to the initiation of complex suites of physiological changes requiring the integration and coordination of input signals from a variety of receptor systems.

In recent years there has been increasing concern about the potential impacts of human activities on aquatic environments. This has led to an increased focus on monitoring aquatic habitats, and occasionally remedial measures have been taken in an attempt to reverse the detrimental changes that have been observed. In most cases, however, activities have been restricted to the carrying out of toxicity tests, with some attempt then being made to establish guidelines for *water quality criteria*. These criteria are usually taken to refer to the highest concentrations of specific substances that are not expected to cause an appreciable negative effect on the aquatic environment.

To what extent the tolerance limits determined by toxicity testing have practical application under natural conditions is open to question because fish will rarely be faced by the challenge of a single stressor in isolation. For example, chemical effluents will seldom contain a single toxicant, but will be 'chemical cocktails' of a mixture of potentially harmful substances. Whilst each substance may be present in the mixture at a sublethal concentration, it is possible that interactions between the different components of the mixture may produce deleterious effects resulting in mortality. Similarly, industrial effluents pumped into rivers and other water bodies may differ in temperature from that of the receiving body, with the result that such effluents could impose the simultaneous challenges of 'acute' temperature change, reduced oxygen availability and increased concentrations of chemical toxicants.

Multiple-toxicant, or stressor, experiments are rarely performed, but it is clear that such interactions do exist. A major problem with multiple-stressor studies is that the numbers of possible interactions of this type are myriad. Thus, whilst the multiple-factor approach deserves more attention and exploration it may be extremely difficult to decide which of the possible interactions are likely to produce the most dramatic effects and, therefore, warrant closest attention.

FURTHER READING

Books

Adams, S.M. (ed)(1990) *Biological Indicators of Stress in Fish* (*Am. Fish. Soc. Symp.* 8), American Fisheries Society, Bethesda, MD.

Forbes, V.E. and Forbes, T.L. (1994) *Ecotoxicology in Theory and Practice*, Chapman & Hall, London.

Jobling, M. (1994) *Fish Bioenergetics*, Chapman & Hall, London.

Mason, B.J. (1992) *Acid Rain: Its Causes and its Effects on Inland Waters*, Clarendon Press, Oxford.

Rand, G.M. and Petrocelli, S.R. (eds)(1985) *Fundamentals of Aquatic Toxicology*, Hemisphere, Washington, DC.

Rankin, J.C. and Jensen, F.B. (eds)(1993) *Fish Ecophysiology*, Chapman & Hall, London.

Review articles and papers

Beitinger, T.L. (1990) Behavioral reactions for the assessment of stress in fishes. *J. Great Lakes Res.*, **16**, 495–528.

Beitinger, T.L. and McCauley, R.W. (1990) Whole-animal physiological processes for the assessment of stress in fishes. *J. Great Lakes Res.*, **16**, 542–75.

Buhler, D.R. and Williams, D.E. (1988) The role of biotransformation in the toxicity of chemicals. *Aquat. Toxicol.*, **11**, 19–28.

Coutant, C.C. (1987) Thermal preference: when does an asset become a liability? *Env. Biol. Fishes*, **18**, 161–72.

Eddy, F.B. and Williams, E.M. (1987) Nitrite and freshwater fish. *Chem. Ecol.*, **3**, 1–38.

Hodson, P.V. (1988) The effect of metal metabolism on uptake, disposition and toxicity in fish. *Aquat. Toxicol.*, **11**, 3–18.

Lewis, W.M., jun. and Morris, D.P. (1986) Toxicity of nitrite to fish: A review. *Trans. Am. Fish. Soc.*, **115**, 183–95.

Pelissero, C. and Sumpter, J.P. (1992) Steroids and 'steroid-like' substances in fish diets. *Aquaculture*, **107**, 283–301.

Pickering, A.D. (1993) Endocrine-induced pathology in stressed salmonid fish. *Fish. Res.*, **17**, 35–50.

Wendelaar Bonga, S.E. and Lock, R.A.C. (1992) Toxicants and osmoregulation in fish. *Neth. J. Zool.*, **42**, 478–93.

Author index

Species index

COMMON NAMES

Subject index